HISTORY IN MATHEMATICS EDUCATION

New ICMI Study Series

VOLUME 6

Published under the auspices of the International Commission on Mathematical Instruction under the general editorship of

Hyman Bass, President   Bernard R. Hodgson, Secretary

History in Mathematics Education

The ICMI Study

Edited by

JOHN FAUVEL

The Open University,
United Kingdom

and

JAN VAN MAANEN

University of Groningen,
The Netherlands

KLUWER ACADEMIC PUBLISHERS
DORDRECHT / BOSTON / LONDON

£55.99

A C.I.P. Catalogue record for this book is available from the Library of Congress.

ISBN 0-7923-6399-X

Published by Kluwer Academic Publishers,
P.O. Box 17, 3300 AA Dordrecht, The Netherlands.

Sold and distributed in North, Central and South America
by Kluwer Academic Publishers,
101 Philip Drive, Norwell, MA 02061, U.S.A.

In all other countries, sold and distributed
by Kluwer Academic Publishers,
P.O. Box 322, 3300 AH Dordrecht, The Netherlands.

Printed on acid-free paper

Printed in the Netherlands.

Contents

2. Philosophical, multicultural and interdisciplinary issues

3. Integrating history: research perspectives

4. History of Mathematics for Trainee Teachers

Introduction

John Fauvel & Jan van Maanen

When the English mathematician Henry Briggs learned in 1616 of the invention of logarithms by John Napier, he determined to travel the four hundred miles north to Edinburgh to meet the discoverer and talk to him in person. The meeting of Briggs and Napier is one of the great tales in the history of mathematics. According to William Lily, who had it from Napier's friend John Marr, it happened when Napier had given up hope of seeing his long-awaited southern guest:

It happened one day as John Marr and Lord Napier were speaking of Mr. Briggs "Ah John", saith Marchiston, "Mr. Briggs will not come." At the very instant one knocks at the gate. John Marr hastened down, and it proved Mr Briggs, to his great contentment. He brings Mr. Briggs to my Lord's chamber, where almost one quarter of an hour was spent each beholding the other with admiration, before one spoke: at last Mr. Briggs began: "My lord, I have undertaken this long Journey purposely to see your Person, and to know by what Engine of Wit or Ingenuity you came first to think of this most excellent Help unto Astronomy, *viz.*, the *Logarithms*; but, my Lord, being by you found out, I wonder nobody else found it out before, when now known it is so easy." He was nobly entertained by Lord Napier, and every summer after this, during Lord Napier's being alive, this venerable man, Mr. Briggs, went to Scotland to visit him.

The many layers of significance of this story make it an invaluable resource for mathematics teachers at all levels. For younger pupils, the idea of two grown men sitting looking at each other in silence for fifteen minutes on first meeting is sufficiently strange to provoke mirth and a vivid sense of how important mathematical ideas were to them. Pupils need no knowledge of logarithms to recognise from this that mathematics is something which has been invented by people at particular stages of history, not something which has always been there. Questions arise for young pupils too about the practicalities of life in old times, about travelling long distances as well as how before the days of photographs and television people generally had no accurate idea of what each other looked like unless they met in the flesh. Some may notice that in ancient times people were sometimes called by their name ('Napier') and sometimes by where they lived ('Marchiston'), as Napier lived in a castle called Merchiston Castle. It is in elementary and middle school, too, that teachers can introduce pupils to another of Napier's inventions, his 'rods' or 'bones' for speeding up multiplication. These lay bare the structure of multiplying in the decimal place-value numeral system (Hindu-Arabic numbers, as we call them) in a way which deepens student understanding and memorisation of the process.

Older pupils who are beginning to learn about logarithms are reinforced in understanding their importance, through reflecting on the lengths to which Briggs went in wanting to meet and admire their discoverer. Or were logarithms invented, not discovered? Teachers can explain how arduous calculations could be before logarithms, and tell pupils of Kepler's remark that thanks to Napier the astronomer's life-span had been doubled. This invention is a microcosm of the activity of mathematicians down the ages: the *point* of mathematics is to make things happen more easily and to save people trouble. (This revelation will be quite surprising to some pupils!—or at least to their parents with unhappy memories of their school mathematics lessons.) The possible benefits of the story work on a number of levels. Once students know, for example, how happy the astronomer was when multiplication of two ten-digits numbers reduced to a simple addition, they will never have a problem in remembering which is the correct rule: $\log ab = \log a + \log b$, not $\log(a+b) = \log a \times \log b$.

Senior students will begin to recognise just how significant logarithms are: that a device for easing the activity of calculating turns out to be one of the most influential and far-reaching of ideas in all of mathematics, a function of immense power and reach which pulls together ideas from different areas of mathematics. This illuminates another general truth about the amazing power of mathematics, the way different parts of it reinforce each other. Here, it is little short of miraculous how ideas from ancient Greece (curves from slicing cones, called conic sections), from early seventeenth century Scotland, and from later in the seventeenth century (a general method for finding the areas bounded by curves) all come together to generate a complex of mathematics of great power, and the student who is trained to understand and share in these ideas is immensely empowered as a result.

Trainee teachers reflecting on the story can absorb all these resonances and also notice what the story of Briggs's meeting with Napier tells us about the psychology of learning mathematics: it is every pupil's experience that once some difficult idea has been learned it seems so natural that you cannot understand why you did not understand it before! The concept of an "Engine of Wit or Ingenuity" is a very deep one. The apparent tension in this phrase between mechanical and psychological images is characteristic of the seventeenth century, prefiguring perhaps the 'mechanical philosophy' promoted by René Descartes and others a few decades later.

There are lessons for those designing mathematics education syllabuses too. The curriculum designer will appreciate that an apparently straightforward observation made by several mathematicians from Archimedes onwards, that multiplying numbers can correspond to adding powers of another number, or more simply that geometrical and arithmetical series can run in parallel, took many centuries to be recognised as a key perception to build upon for calculational purposes. The curricular implications may be (put in a rather general way) that what seems simple after the event can pose difficulties for students until they are prepared for new ways of looking at things.

This one short tale from four centuries ago can in this way be seen to lay the grounding for a number of valuable interactions between teacher and student in the mathematics classroom over several school years. A teacher able to support, encourage and lead students in this way through their school career is a better teacher: better prepared, better resourced, more empowered. History, we might say, is an Engine of Mathematical Wit. This story, and the pedagogical reflections which it generates, are to this extent a microcosm of what we hope the present book will achieve.

The background to this study

Does history of mathematics have a role in mathematics education? This book has been made by people who believe that the answer is positive, that the history of mathematics can play a valuable role in mathematical teaching and learning. It is the report of a study instigated by the International Commission on Mathematical Instruction (ICMI). We describe later how the study was carried out, but first sketch the problem setting of the study, the general background of concerns from several quarters which have led to a flourishing of work in this area in recent decades.

Mathematicians, historians and educators in many countries have long thought about whether mathematics education can be improved through incorporating the history of mathematics in some way. This arises from the recognition that mathematics education does not always meet its aims for all pupils, and that so long as some students emerge from their education with less understanding of mathematics than might be useful for them, or indeed with an actual fear or phobia about mathematics, then it is worth exploring possible avenues for improving the process. Nor have they only thought about the possibility of using history; many teachers in classrooms across the world have tried out various pedagogic possibilities. It soon emerges that there is a wide range of views and experiences of how history of mathematics can help. Some educators believe that mathematics is intrinsically historical: so learning the subject must involve its history, just as studying art involves learning about art history. Others see a number of ways in which history can aid the teacher's, and thus the learner's, task, from the apparently banal (such as giving more information about the names students may meet—which, by the way, are often wrong attributions in any case, as in the cases of Pascal's triangle and L'Hôpital's rule, not to speak of Pythagoras' theorem) to a deeper way of teaching mathematics in a historical vein.

It is not only teachers who are concerned with perceived failings in school and college mathematics. Parents, employers and politicians all vie repeatedly in urging attention to the system's ability to deliver enough students passing mathematics examinations. Whatever the truth behind such fears and concerns, resolving them is evidently a political matter, and thus adoption of the contribution offered by this Study, to improve mathematics education through the provision and use of historical resources, is a political choice to be made or influenced at any or all of the several layers of decision-making in complex modern societies.

The ICMI Study

ICMI, the International Commission on Mathematical Instruction, was established in 1908 at the International Congress of Mathematicians held in Rome, its first chair being Felix Klein. After an interruption of activity between the two World Wars, it was reconstituted in 1952 as a commission of the International Mathematical Union

(IMU). The IMU itself was formed at the 1920 International Congress of Mathematicians, held in Strasbourg. The history of these international bodies is thus closely linked with twentieth century internationalisation of mathematical activity, in particular with the efforts of mathematicians to re-energise international co-operation after major wars, as part of the healing and reconciliation process and in a spirit of optimism about building a better future for everyone. In 1972, at the second International Congress on Mathematical Education in Exeter, UK, the idea was developed of an International Study Group on the Relations between History and Pedagogy of Mathematics, which was formally affiliated to ICMI at the 1976 International Congress (ICME-3) at Karlsruhe, Germany. HPM has continued ever since to explore and advise on these relations through the activities of its members, who are mathematics educators, teachers and historians across the world., who are mathematics educators, teachers and historians across the world.

Since the mid 1980s ICMI has engaged in promoting a series of studies on essential topics and key issues in mathematics education, to provide an up-to-date presentation and analysis of the state of the art in that area. The tenth ICMI Study, whose report is presented in the present volume, was conceived in the early 1990s in order to tease out the different aspects of the relations between history and pedagogy of mathematics, in recognition of how the endeavours of how the *HPM* Study Group had encouraged and reflected a climate of greater international interest in the value of history of mathematics for mathematics educators, teachers and learners. Concerns throughout the international mathematics education community began to focus on such issues as the many different ways in which history of mathematics might be useful, on scientific studies of its effectiveness as a classroom resource, and on the political process of spreading awareness of these benefits through curriculum objectives and design. It was judged that an ICMI Study would be a good way of bringing discussions of these issues together and broadcasting the results, with benefits, it is to be hoped, to mathematics instruction world-wide.

ICMI Studies typically fall into three parts: a widely distributed *Discussion Document* to identify the key issues and themes of the study; a *Study Conference* where the issues are discussed in greater depth; and a *Study Volume* bringing together the work of the Study so as to make a permanent contribution to the field. The current study has followed this pattern.

The *Discussion Document* was drawn up by the two people invited by ICMI to co-chair the Study, John Fauvel (Open University, UK; HPM chair 1992-1996) and Jan van Maanen (University of Groningen, Netherlands; HPM chair 1996-2000), with the assistance of the leading scholars who formed the International Programme Committee: Abraham Arcavi (Israel), Evelyne Barbin (France), Jean-Luc Dorier (France), Florence Fasanelli (US, HPM Chair 1998-1992), Alejandro Garciadiego (Mexico), Ewa Lakoma (Poland), Mogens Niss (Denmark) and Man-Keung Siu (Hong Kong). The Discussion Document was widely published, in for example the *ICMI Bulletin* **42** (June 1997), 9-16, and was translated into several other languages including French, Greek and Italian. From the responses and from other contacts, some eighty scholars were invited to a Study Conference in the spring of 1998, an invitation which in the event between sixty and seventy were able to accept.

The *Study Conference* took place in the south of France, at the splendid country retreat of the French Mathematical Society, CIRM Luminy (near Marseille), from 20 to 25 April 1998. Local organisation was in the hands of Jean-Luc Dorier (University of Grenoble). The scholars attending were from a variety of backgrounds: mathematics educators, teachers, mathematicians, historians of mathematics, educational administrators and others. This rich mix of skills and experiences enabled many fruitful dialogues and contributions to the developing study.

The means by which the Study was advanced, through the mechanism of the Conference, is worth description and comment. Most participants in the Conference had submitted papers, either freshly written or recent position papers, for the others to read and discuss, and several studies were made available by scholars not able to attend the meeting. These, together with whatever personal qualities and experiences each participant was bringing to the Conference, formed the basis for the work. Apart from a number of plenary and special sessions, the bulk of the Conference's work was done through eleven working groups, corresponding, in the event, to the eleven chapters of the Study Volume. Each participant belonged to two groups, one meeting in the mornings and one in the afternoons. Each group was led by a convenor, responsible for co-ordinating the group's activities and playing a major part in the editorial activity leading to the eventual chapters of the book. Each group's work continued for several months after the Conference, with almost everyone participating fully in writing, critical reading, bibliographical and other editorial activities.

This way of group working for a sustained period towards the production of a book chapter was a fresh experience to many participants, since the pattern of individual responsibility for separate papers is a more common feature of such meetings and book productions. In this instance the participants proved remarkably adept at using the new structures to come up with valuable contributions to the development of the field, all the more valuable for their being the results of consensual discussions and hard-written contributions, which have been edited and designed into the present Study Book.

Authorship of contributions

As just explained, this ICMI Study adopted a style of collective group work in which international teams worked together on the various issues, each led by a convenor, whose reports form the basis of the chapters in this book. We have experienced this as a very useful and productive way of working for the teachers, educators and researchers involved, who were able to share insights, experiences and ideas, and develop strategies together for future progress in the field. It follows from the working style that it is not quite as straightforward as usual to attribute responsibility and authorship to particular sections of text. As will be seen, each chapter is credited to a team, listed in alphabetical order, headed by the name of the chapter co-ordinator. Within the chapters, sometimes names may appear as responsible for subsections and sometimes not. In the construction of the book some sections retained individual responsibility (while commented on and modified by the help of the rest of the group), and others were by the end of the process a genuinely group or sub-group collaboration (while initially drafted by an individual, as is almost always the case).

An intercontinental discussion at the ICMI Study conference in Luminy, France: Vicky Ponza (Argentina), Daina Taimina (Latvia), Florence Fasanelli (USA), Chris Weeks

Among the considerations here are that readers often find it easier and more welcoming to consider a particular text as written by a person rather than a collective; and that a named author is able to use the word "I" in a text, which is a user-friendly form of address, where appropriate and natural, rather than the forced third-person or first-person-plural style of scientific texts. Another consideration is, of course, that individuals should receive credit for their contributions, particularly in the institutional imperatives of today. But the overall message to readers is that this book represents an act of collective scholarship all of whose contributors shared in its production.

The purpose of the ICMI Study

This book has several functions, namely to
(i)   survey and assess the present state of the whole field;
(ii)   provide a resource for teachers and researchers, and for those involved with curriculum development;
(iii) indicate lines of future research activity;
(iv)   give guidance and information to policy-makers about issues relating to the use of history in pedagogy.
These functions are variously carried out through the eleven chapters which follow. Each chapter has a very short abstract, which is not only a summary of the ensuing

chapter but can be seen to form a sentence or paragraph in a story which as in some Victorian novel can be seen as "The Argument Of This Book".

The argument of this book

People have studied, learned and used mathematics for over four thousand years. Decisions on what is to be taught in schools, and how, are ultimately political, influenced by a number of factors including the experience of teachers, expectations of parents and employers, and the social context of debates about the curriculum. The ICMI study is posited on the experience of many mathematics teachers across the world that its history makes a difference: that having history of mathematics as a resource for the teacher is beneficial. School mathematics reflects the wider aspect of mathematics as a cultural activity.

From the philosophical point of view, mathematics must be seen as a human activity both done within individual cultures and also standing outside any particular one. From the interdisciplinary point of view, students find their understanding both of mathematics and their other subjects enriched through the history of mathematics. From the cultural point of view, mathematical evolution comes from a sum of many contributions growing from different cultures.

The question of judging the effectiveness of integrating historical resources into mathematics teaching may not be susceptible to the research techniques of the quantitative experimental scientist. It is better handled through qualitative research paradigms such as those developed by anthropologists.

The movement to integrate mathematics history into the training of future teachers, and into the in-service training of current teachers, has been a theme of international concern over much of the last century. Examples of current practice from many countries, for training teachers at all levels, enable us to begin to learn lessons and press ahead both with adopting good practices and also putting continued research effort into assessing the effects.

The use of history of mathematics in the teaching and learning of mathematics requires didactical reflection. A crucial area to explore and analyse is the relation between how students achieve understanding in mathematics and the historical construction of mathematical thinking. The needs of students of diverse educational backgrounds for mathematical learning are increasingly being appreciated. Using historical resources, teachers are better able to support the learning of students in such diverse situations as those returning to education, in under-resourced schools and communities, those with educational challenges, and mathematically gifted students.

An analytical survey of how history of mathematics has been and can be integrated into the mathematics classroom provides a range of models for teachers and mathematics educators to use or adapt. Further specific examples of using historical mathematics in the classroom both support and illustrate these arguments, and indicate the ways in which the teaching of particular subjects may be supported by the integration of historical resources.

The study of original sources is the most ambitious of ways in which history might be integrated into the teaching of mathematics, but also one of the most rewarding for students both at school and at teacher training institutions. The integration of history is not confined to traditional teaching delivery methods, but can often be better achieved through a variety of media which add to the resources available for learner and teacher. A considerable amount of work has been done in recent decades on the subject of this study, which is here summarised, in the form of an annotated bibliography, for works appearing in eight languages of publication.

Acknowledgements

In any enterprise such as this the support, vision and confidence of a number of kind people is invaluable for making the project happen. As well as the contributors who worked so hard and without whom this Study would not have happened, we want here to thank the former chairs of HPM, Ubiratan D'Ambrosio and Florence Fasanelli, for their vision of such a study; the successive secretaries of ICMI, Mogens Niss and Bernard Hodgson, for their continued support and enthusiasm for the project; Jean-Luc Dorier for his faultless and energetic organisation of the Study Conference; the Société Mathématique de France for generously making available the facilities of its splendid conference centre at Luminy for the Study Meeting; Joy Carp and Irene van den Reydt of Kluwer for their flexible and constructive help in ensuring the book production took place so efficiently; Liz Scarna of the Open University for her sterling electronic assistance behind the scenes; and all friends and families of all the contributors for tolerating and encouraging a production which took up more time than they may have expected or welcomed.

Preface to the second edition

The title that John Fauvel and I originally proposed for this book was "What engine of wit". We sketched the context of these words in the Introduction to the first edition. The early appearance of the second edition indicates that this volume has found a welcome audience. We owe this to the insight and collaboration of all contributors, but the foundations were John's. He was ill when the first edition was launched, in a session of the 9th International Congress on Mathematics Education (Tokyo, August 2000), but later that year he was able to fully and proudly enjoy its appearance.

Then, on 12 May 2001, John died. His 'good thinking' lives on in his writings. He was our 'Engine of wit'.

Jan van Maanen, 28 August 2002

Chapter 1

The political context

Florence Fasanelli

with Abraham Arcavi, Otto Bekken, Jaime Carvalho e Silva, Coralie Daniel, Fulvia Furinghetti, Lucia Grugnetti, Bernard Hodgson, Lesley Jones, Jean-Pierre Kahane, Manfred Kronfellner, Ewa Lakoma, Jan van Maanen, Anne Michel-Pajus, Richard Millman, Ryo Nagaoka, Mogens Niss, João Pitombeira de Carvalho, Circe Mary Silva da Silva, Harm Jan Smid, Yannis Thomaidis, Constantinos Tzanakis, Sandra Visokolskis, Dian Zhou Zhang

Abstract: *People have studied, learned and used mathematics for over four thousand years. Decisions on what is to be taught in schools, and how, are ultimately political, influenced by a number of factors including the experience of teachers, expectations of parents and employers, and the social context of debates about the curriculum. The ICMI study is posited on the experience of many mathematics teachers across the world that its history makes a difference: that having history of mathematics as a resource for the teacher is beneficial.*

1.1   Introduction

People have studied, learned and used mathematics for over four thousand years, although it is only relatively recently that mathematics has been taught, in most countries, to a high proportion of the population. With the establishment of universal education, more widespread attention has been focused on just what was taught and why. These decisions are ultimately political, albeit influenced by a number of factors including the experience of teachers, the expectations of parents and employers, and the social context of debates about the content and style of the curriculum.

The present ICMI study is posited on the experience of many mathematics teachers across the world that the history of mathematics makes a difference: that having history of mathematics as a resource for the teacher is beneficial. Increasingly a number of local and national governments, and other bodies

John Fauvel, Jan van Maanen (eds.), *History in mathematics education: the ICMI study*, Dordrecht: Kluwer 2000, pp. 1–38 responsible for curriculum design and expectations, are persuaded by the arguments of these mathematics teachers that it is worthwhile to incorporate history of mathematics within mathematics education.   Detailed consideration of these arguments will be found later in the book. The next section (§1.2) in this opening chapter summarizes, therefore, the experience of a number of countries across the world (sixteen in all) in relation to the political guidelines governing inclusion of history of mathematics in the school mathematics curriculum. A further critical area is what happens in the textbooks written to deliver the curriculum. Section 1.3 is a case study looking in detail at how curriculum and textbooks can absorb a historical dimension, in the case of one particular country, Poland. Broader issues of the ways in which historical information can be integrated into textbooks are looked at later in the book (§7.4.1). In assessing the current role of history in mathematics education a further area of critical importance, of course, besides curricula and textbooks, is what happens in teacher training colleges.   This is explored only briefly in this chapter as it is discussed in some detail in a later chapter (§4.2).

Section 1.4 presents a policy statement around the introduction of a greater historical dimension in the mathematics curriculum, with some ideas for promoting it further. It is important to bear in mind that all of the ideas discussed in the rest of this book depend for their practical implementation on the development of a political consensus in the many countries and educational systems across the world. While this ICMI Study volume is not a text in practical political science, one of the aims of the Study is to inform and guide policy-makers about the incorporation of history in pedagogy, a task in which all readers, as concerned citizens as well as wearing a range of other hats, may choose to become involved. The final section of this opening chapter presents some quotations, illustrating how these matters have been thought of by mathematicians, advisers and other opinion formers over the past two centuries, to support the arguments for using the history of mathematics while learning and teaching mathematics at all levels.

1.2   What part does history of mathematics currently occupy in national curricula?

1.2.1   Argentina

In the Educación General Básica, the curriculum laying down what is required for all pupils up to the age of 14, the Argentinian Ministry of Culture and Education gives eight foci for what mathematical studies at school are intended to achieve. These include conceptual comprehension, pleasure in doing mathematics, the value of new technology, the internal cohesion of mathematics, the significance and functionality of mathematics at work, the habit of setting and solving problems in a variety of settings, and finally "the value of mathematics in culture and society, in history and the present." Nowhere in the official documents are there found statements about utilising the history of mathematics within the curriculum, although teachers and faculty individually express such an interest and hold annual national meetings to pursue knowledge of history.

1.2.2   Austria

In the Austrian syllabus, the general teaching goals for grades 9-12 state that the students should "know about the change of mathematical concepts in the historical development as well as in their personal development." More specifically, in the 9th grade students should know about the change of the concept of function; in the 10th grade they should know the historical meaning of logarithms and in grade 11 they should learn historical aspects of the calculus. None of this is compulsory, however. In school books for grades 5 to 8, there are historical notes, ranging from a few lines up to several pages, in connection with trigonometry, complex numbers, and limits of a sequence as well as with other topics. Thus some lines are included about historical figures such as Al-Khwarizmi, Archimedes, Cardano, Eratosthenes, Galileo, Omar Khayyam, Pythagoras, and Adam Ries.

1.2.3   Brazil

From 1931 to 1954, Brazil had a mandatory national curriculum for secondary school mathematics, and from 1946 to 1954 a mandatory curriculum for elementary school mathematics. From 1954 onward, the regulations were changed so that each state can establish its own curriculum. Nevertheless, tradition, inertia, and the fact that textbooks define, in practice, the actual curriculum assure a homogeneity among the curricula of the individual states.

In 1997, after wide discussions and consultations, the ministry of education issued 'parameters' for the first four years of schooling. In 1998 similar parameters were established in an analogous way, for grades 5 to 8, and for grades 9 to 11 of secondary school. The parameters are not mandatory, but there have been in the late 1990s a considerable number of requests, from state offices of education, for a national curriculum for Brazil as a whole. The Ministry of Education has chosen not to establish a mandatory national curriculum, but the national parameters have to some extent taken on this role.

In the parameters for grades 1 to 8, there is a strong emphasis on the history of mathematics, and on the fact that mathematics is not just a body of knowledge, but also of processes and practices that were slowly created in response to human needs and curiosity. The parameters also call attention to the fact that mathematics should not be treated separately from other school subjects, nor indeed from broader concerns with the environment, health, etc. Within mathematics, too, teachers are urged to try to foster integration of arithmetic, geometry, and measurements. Four resources are listed for doing mathematics inside the classroom: problem solving, history of mathematics, information technologies, and games. Specifically in relation to the history of mathematics the parameters say:

The history of mathematics, by means of a process of didactic transposition and together with other didactic and methodological resources, can offer an important contribution to the process of mathematics teaching and learning. By revealing mathematics as a human creation, by showing necessities and preoccupations from different cultures in different historical periods, by establishing comparisons between mathematics concepts and processes of past and present, the teacher has the possibility of developing more favorable attitudes and values to the student facing mathematical knowledge. In several situations, having history of mathematics as a resource can clarify mathematical ideas that are being constructed by students, especially to give answers to some questions and, in this way, contribute to the constitution of a critical look over the objects of knowledge.

Thus teachers are told *why* it will be beneficial to use history, but are given little guidance on *how* to do so.

1.2.4   China

Once the Chinese people won their real independence in 1949, the government launched a movement of patriotism, and asked mathematical educators to foster pupils' patriotic thought by means of incorporating more knowledge of Chinese history of mathematics. This led to researches into the ancient history of mathematics. As a consequence, when Chinese historians of mathematics were invited to compile new textbooks, a number of mathematical results could be re-named after their ancient Chinese equivalents or the Chinese authors who discovered them. For example, before 1949 the Gou Gu Theorem was called Pythagoras theorem, the Yang Hui Triangle was Pascal's Triangle, and the Zu Geng Principle was Cavalieri's Principle.

In China, 95 percent of schools adopt the nation-wide unified mathematics textbooks (1996), in which 16 items are concerned with the history of mathematics:

1. Decimals (4-grade). Chinese ancient mathematics.
2. 'Pi' (5-grade): Liu Hui (about 263), and Zu Chong Zhi (429-500).
3. Equations (7-grade): the *Nine chapters of arithmetic*, a Chinese classical work of the first century, explained by Liu Hui.
4. Negative number (7- grade): *Nine chapters of arithmetic*
5. The origin of geometry (7-grade): Egypt, Euclid, Mo Zi.
6. Parallel axiom (7-grade): Euclid. Lobachevsky.
7. Mathematical symbols (7-grade): Multiplication sign (Oughtred 1631), decimal point (Clavius, 1593)
8. A story of Gauss (7-grade)
9. Gou Gu Theorem (8-grade): Zhao Shuang
10. The discovery of irrational number (8- grade): school of Pythagoras.
11. The history of quadratic equation (8-grade). *Nine chapters of arithmetic*, Diophantus, Yang Hui, Viète, Buddhist Yi Xing
12. Pi (8-grade): Liu Hui, Zu Chong Zhi, Ludolph van Ceulen.
13. The area of triangle (8-grade): Qin Jiu Shao (1202-1261), Heron (about 62)
14. Construction with rules and compasses (9-grade): Greek mathematics
15. Zu Gen Principle (c. 500) (11-grade)
16. Binomial coefficients (12-grade): Yang Hui (about 1250)

Eight items in this list are from China, just half of the total number. Few Chinese teachers use the historical material as aids to mathematics teaching in the classroom, however, except for some paragraphs directly concerning education for patriotism.

In normal colleges and universities, where teachers are trained, there is intended to be an optional course on the history of mathematics (45 classroom hours). However, because of the lack of mathematical historians to teach the subject, many universities are unable to offer a course of mathematics history when the students elect to do it. Most knowledge of history of mathematics that Chinese teachers have is from other mathematics courses. Recently, however, the research of history of mathematics has made rapid progress in China, and now more than one hundred historians of mathematics are working in institutions for teacher training.

A lot of historical events lead teachers to formalism, abstraction and absolutism. In China, Marxist philosophy is an important political course. In addition, Marx had an unpublished work, his *Mathematical manuscripts,* in which many mathematical problems were explored.   In particular, Marx talked about the logical basis of Newton's calculus. In the curriculum of the Masters degree in mathematics education, there is a basic course on philosophy and history of mathematics. Therefore many Chinese mathematicians and teachers pay more attention to the logical aspect of mathematics, and explore such topics as Russell's paradox and Cantor's set theory. A number of books explain the historical context and details of three mathematical crises: the discovery of irrational numbers; infinitesimal calculus; and the paradoxes of set theory; as well as the familiar three schools of logicism, formalism, and intuitionism.   For this reason many university students name Kurt Gödel as a mathematical hero.   In contrast to their enthusiasm for formalism, abstraction and absolutism, history of mathematics textbooks in China pay less attention to applied mathematics. Maxwell's equations, for example, are usually ignored in histories which discuss the 19th century.

The mathematics curriculum in China maintains the standards of a formal, rigorous, deductive system. Most mathematics teachers believe that training in logical thinking is the core of mathematics teaching, and that any informal approach will be harmful to pupils. In 1996, the Chinese Education Ministry published the programme of Mathematics Curriculum, in which only one sentence concerned the history of mathematics: "Aid by history of mathematics to foster pupil's patriotism." The *Programme of mathematics teaching and learning* which the Education Ministry published in 1996 pointed out that "By presenting of ancient and modern achievements in China, the pupil's sense of national pride and patriotic thought is aroused."

1.2.5   Denmark

The history of mathematics played a fairly minor role in Danish mathematical curricula at all levels up until the 1970s. Even though the history of mathematics was represented as personal choices of topic in the past (and still is today) at the university level where courses are taught, and quite a few students write Master's theses in the subject, there was no influence of history on the teaching and learning of mathematics at large. At the school level, no historical component at all was given, apart from names attached to theorems. Only in one text book system were anecdotes included as spices to the diet, but not made the object of teaching or learning.

Changes began in 1972, when Roskilde University was established with the purpose of bringing some innovation in tertiary education in Denmark. The history of mathematics was included in the mathematics programme right from the beginning. It was, and remains, an underlying idea in the Roskilde programme that mathematics is a discipline that exists, evolves, and is exercised in time and space; that is, in history and society. Rather than requiring students to take specific history courses, they are required to include historical considerations in their studies. It should be noted that these studies are strongly based on projects.

In the mid 1970s, the academic upper secondary school system (grades 10 to 12) underwent a rapid expansion that almost caused a crises as far as mathematics was concerned. Instead of continuing to address a rather limited elite of 10 percent of the youth cohort, about 30 percent were addressed, and this has now expanded to 50 percent. It was clear that such a great portion of the population could not be expected to swallow the very theoretical diet previously taught. During 1979, mathematics educators conducted a series of meetings and in-service courses for teachers throughout the country where they presented some ideas of what could be done to cater for the much broader audience now enrolling in advanced secondary education. Mogens Niss suggested the 'historical aspect of mathematics', among other aspects, as a part of defining the curriculum in *dimensions* of mathematics rather than the *topics* of a traditional syllabus.

A lot of experimentation was begun by schools, and by individual teachers throughout the country, including ways to include the historical aspect of mathematics in its teaching and learning. When, later in the 1980s, the Parliament decided to establish a general reform of the structure of the curriculum of the upper secondary school, the Ministry of Education began to promote a slightly modified form of the Standard Experimental Curriculum. This is still in force (1999), with a few modifications in the organizational structure. The current curriculum document includes these statements:

Students have to acquire knowledge of elements of the history of mathematics and of mathematics in cultural and societal contexts. [. . .] Some of the main [mathematical] strands are to put in perspective by considering elements of the history of the topics dealt with, and— to a lesser extent—aspects of the epoch, culture or society in which those topics were developed.

The fact that the historical aspect of mathematics was made a compulsory component of the upper secondary curriculum had implications for the university studies in mathematics that (at Masters level) prepare upper secondary school teachers. It simply became a requirement for the employment of a university graduate in mathematics in an upper secondary school that his or her university studies have included elements of the history of mathematics. So all Danish universities were forced to introduce such elements in their programmes (which some of them did only reluctantly).

In the primary and lower secondary school levels no historical elements are included in the curriculum either officially or unofficially, and this too has consequences for the education of teachers for those levels (which takes place in independent teacher training colleges). It is certainly not the case that the historical aspect of mathematics is given prominent position in mathematics education at all levels in Denmark.

The reason why elements of the history of mathematics were introduced, as a non-negligible component of upper secondary mathematics education, and later of university mathematics programmes, was not that individuals or associations did a lot of clever and efficient canvassing and lobbying to influence the authorities. Instead, a combination of historical conditions and circumstances paved the way for the changes which are described here. Two factors seem to have been essential: firstly, the task and role of the upper secondary school in general, and of mathematics in particular, became subject to drastic changes that called for reform. Nobody believed that things could have continued unaltered. Secondly, mathematics educators who had thought about new principles for the design of mathematics curricula and gained experiences from innovative teaching and learning, were available with ideas that might be explored as possible means to solve some of the problems encountered.

1.2.6   France

The main level of the French syllabus in which history of mathematics is involved is the tertiary level. France has a centralized education system that officially prescribes the various courses of instruction that students follow. In this system, some 50,000 of the best students undergo their first two years of tertiary education, at the end of which they sit competitive examinations. These gain them entry to engineering schools, for the great majority, or to the *écoles normales superieures* in order to become researchers and teachers at either the tertiary or secondary level.

In mathematics, the same teacher teaches one class, of about 45 students, for some sixteen to twenty hours a week. The basic class either works as a whole group or is divided into subgroups according to the activity. For two to five hours a week, the students work on exercises in groups of a half or a quarter of the class, with or without the computer. One hour a fortnight they work in groups of three for oral questions.

This system underwent an important reform in 1995, affecting both its structure and its programmes of instruction. This reform aims to reduce the importance of mathematics with regard to other disciplines, to bring more coherence with physics and engineering science into their learning and to develop a spirit of initiative in the students. The mathematics syllabus is in two parts with accompanying comments. The first sets out general aims while the other deals with the topics of linear algebra, calculus, and geometry. The students must know how to use both calculators and programs which perform symbolic manipulations. There are differences in content, according to the particular course of study, but the educational objectives are the same. Mathematical education must "simultaneously develop intuition, imagination, reasoning and rigour."

The history of mathematics was not neglected in this reform:

It is important that the cultural content of mathematics should not be simply sacrificed to its technical aspects. In particular, historical texts and references allow the analysis of the interaction between mathematical problems and the construction of concepts, and brings to the fore the central role played by scientific questioning in the theoretical development of mathematics. Moreover, they show that the sciences, and mathematics in particular, are in perpetual evolution and that dogmatism is not advisable.

Another innovation through these 1995 reforms was the introduction of project work. The history of mathematics is also mentioned in relation to this: "The study of a subject brings an increasing depth of theoretical understanding together with experimental aspects and applications as well as the application of computing methods. It may include an historical dimension." In the first year, students choose their project freely. In their second year they must fix their area and subject in a very wide framework. At their final assessment each student presents a page long summary and speaks for 20 minutes before two examiners. This project work allows teachers the freedom to introduce the history of mathematics.

1.2.7   Greece

The educational policy adopted in Greece on the relation between the history of mathematics and the teaching of mathematics takes place through several institutions. The Institute of Pedagogy, an official institution of the Ministry of Education, has responsibility for planning curricula and producing textbooks for primary and secondary education. Unlike in some other countries, each subject taught in Greek schools has only one official textbook, so their content is of particular importance. In secondary education, mathematics textbooks written during the period 1987-1993 are still in use in 1999. Almost every chapter in these books ends up with a historical note, printed on a different colour of paper, which is strictly separated from the rest of the mathematical content of the chapter. These notes cover in total 104 pages out of the 2500 pages of the official mathematics textbooks used in the six grades of secondary education.

   According to the 'Guidelines for Teaching Mathematics' edited by the Institute of Pedagogy, the official aim of supplying textbooks with historical notes is to stimulate students' interest and love for mathematics. There are also brief recommendations to the teachers for using these notes in the classroom and for encouraging discussion on them. But this tends not to happen. Although teachers consider some historical notes very interesting, the fact that they are presented in isolation from the rest of the text makes them appear to be useless and having nothing to do with the real problem of teaching and learning mathematics. The teachers themselves have little historical experience or confidence, since there is at present an almost total absence of history of mathematics in either pre-service or in-service experiences of teachers. (But see §4.3.1.2 for discussion of an exception to this general experience, in a primary pre-service context.) This negative practical response from teachers to the historical notes in current textbooks has caused some concern to the Institute of Pedagogy's officials, who are considering a proposal for changing the position of historical material in textbooks. In 1999 a new mathematics textbook for science-oriented students of age 17 was published, with historical material incorporated into the various chapters as historic introductions which will necessarily be taught in the classrooms.

1.2.8   Israel

There is no recommendation in official documents about the teaching of history of mathematics in Israeli schools. However, several initiatives have been undertaken by universities and academic institutions to develop materials suitable for both classroom use and teacher education courses, both pre-service and in-service. These materials are used in teacher college courses and in-service programs and some are being slowly incorporated by teachers into their classrooms.

1.2.9   Italy

In Italy the association of history with mathematics teaching has a long tradition. This tradition is linked to the past influence of important scholars in the field of history of mathematics and epistemology, who were concerned both with problems of mathematical instruction and of teacher education. Evidence of this historical tradition in teaching is seen in the publication in Italy around 1900 of a mathematical journal for students in which the history of mathematics was one of the basic topics treated.

This Italian orientation towards a historical perspective in teaching is also present in the new Italian mathematical programmes, which are very centralized and have a national examination. For students aged 14-16 the official programme states that "At the end of the first two years of upper secondary school the student has to be able to put into a historical perspective some significant moments of the evolution of mathematical thinking." This has been in place since 1923. In 1985 there was an experimental new programme which included the following statement: "The results of research in the historical/epistemological field offer the best inspiration for stimulating students to create conjectures, hypotheses, problems on which the teacher may develop his/her teaching." For the students aged 11-14 the concept since 1979 has been that: "The teacher has to orient the student towards a reflection on the historical dimension of sciences." In 1990 a compulsory two-year post-graduate qualification for those aspiring to teach was set up, requiring didactics, history, and epistemology in the curriculum. There is support of the revival of interest in historical and epistemological themes from the Ministry of Universities and the National Research Council.

In teacher education, history of mathematics is not compulsory, although in the post-laureate courses where prospective teachers specialize, courses in the history and epistemology of mathematics are on offer. All teachers, though, have degrees in mathematics or physics, not pedagogy, and in the university, courses in history of mathematics are often offered. These are usually very technical since they are mostly aimed at forming researchers in the history of mathematics.

In practice there are few books written in or translated into Italian on the history of mathematics. The result is that the teachers' historical culture may be confined to what is written as optional notes in the students' textbooks. Nevertheless, a considerable amount of information has been gathered about how and why teachers use history in their classrooms in Italy: to use paradoxes for eradicating students' false beliefs on mathematical concepts; to discuss critical concepts of mathematics starting from history such as variable; to investigate the students' beliefs about the historical development of mathematics; to use original sources in geometry. It is apparent that some teachers have clearly focused on certain mathematical objectives of their teaching which can be conveniently pursued through history.

1.2.10 Japan

School curricula in Japan from elementary schools up to senior high schools are strictly controlled by the government, through the Ministry of Education and Culture. The curriculum is reformed periodically, at least every ten years. In the last curricular reform, in the mid 1990s, there was a major change in the basic view of what the compulsory curriculum in mathematics should contain.

Mathematical topics were divided into two categories: the basic compulsory core subjects, and optional subjects which could be freely chosen. This opened paths of free choice in the Japanese national curriculum for the first time, although the options were limited to several predetermined subjects. But it had another and much deeper pedagogical effect, to move beyond the traditional way of piling up new mathematical knowledge upon old. Now, if some knowledge is indispensable for solving a problem, students are to gain that knowledge at the stage that they themselves recognize the need for it. For example, if they come across a maximum-minimum problem of a function of degree more than four, they will realize they need to factorize a polynomial of degree more than three. They will then be well motivated, or such is the theory, to study the basic factor theorem or remainder theorem of polynomials. For this reason the basic discussion of formulae, including manipulations of complex fractional formulae and fundamental exercises related to the theory of polynomials, is moved out of the core curriculum.

Other topics were introduced in the core course: elementary probability theory and discussion of progressions. Another core topic was finite discrete mathematics, introduced because of its growing importance in the digital information age. But the most basic part of high school mathematics, numbers and formulae, was transferred into the optional category.

So the new categorization between core and optional subjects regards the importance of traditional subjects in a quite different way. It is true that the traditional systematic way of teaching, impelled to an extent by logical efficiency, did not meet the needs of all learners, and true too that school teachers will be inclined to teach in much the same way as before. Nevertheless, the changed balance gives the impression of being rather too daring, of throwing away fundamental knowledge from the core curriculum. It does not appear to be a curriculum development framed in a historically alert way. If the reformers had had a greater knowledge of the revolutionary significance of algebraic symbolism discovered and developed in the 17th century, and the immense influence it had in the whole of mathematics, they might have been a little more cautious about seeming to abandon the historical legacy of mathematics for fashionable topics of the moment.

It appears, then, as though not only does history of mathematics play very little role in the explicit mathematics curriculum in Japan, it did not play as important a role in the construction of that curriculum as, it could be argued, is desirable. It is argued later in this book that historical awareness has an important role to play in the construction of curricula even where there is no explicit historical content in the curriculum itself.

The first generation of students educated under the new curriculum has already entered university. There is a widespread feeling among university teachers that these students, even those majoring in the exact sciences, have less understanding of basic skills and concepts (such as equation, function), while being able to carry out routine actions such as finding derivatives and primitive functions. Such concerned teachers see the new curriculum as a factor contributing to the apparent weakening in students' mathematical understanding. It remains to be seen whether the next round of curricular reforms will absorb the lessons of this book and bring the advantages of a knowledge of history to the construction of the Japanese curriculum.

1.2.11 Netherlands

At present history of mathematics has no structural position in mathematics education in Dutch secondary schools. Yet, there was a vivid tradition in the history of mathematics during the twentieth century. Several textbooks for secondary schools had chapters on the history of the subject matter. The mathematics teacher and historian of mathematics Eduard Jan Dijksterhuis promoted the history of mathematics as a subject in the training of mathematics teachers, and at some places history was even an optional part of the upper secondary level curriculum. But in present mathematics curricula, history of mathematics has no longer a place as a subject in its own right.

On the other hand, historical notes are found in some mathematics textbooks, in the form of biographical information or historical introductions to topics. Most of the time history is perceived as an extra which can be left out (and probably is left out by many teachers), much as an illustration with a caption but no textual reference. In most texts references to history are rare or totally absent. A survey done in 1992 on how four school texts treat Pythagoras' theorem revealed that in two of them (*Exact Wiskunde* and *Wiskundelijn: Wiskunde* is the Dutch for 'mathematics') there was no historical reference for this topic; one (*Moderne Wiskunde*) added the solitary remark "Pythagoras was a Greek philosopher (about 500 B.C.)"; and one (*Sigma*) remarked of the theorem "It is thought that a Greek mathematician, Pythagoras (580-496 B.C.), was the first to prove it." While the text-writer clearly had a little historical information, references of this kind seem somewhat perfunctory.

In the curricula for primary education and lower secondary (general) education nothing is said about history of mathematics. One can safely say that it plays no official role whatsoever. For upper secondary education (16-18 years of age, followed by about 35% of the population), a new programme ('Wiskunde B') has been in preparation. The curriculum proposals at different levels all have one remark in common: the student should be able to "identify (historical) situations where mathematics played an important role" (the parentheses are in the original text). For the more difficult topics further historical integration is suggested: it is asked that the student should have some knowledge about the historical development of the calculus and of the historical roots of geometry.

It is not clear that how much influence these well-meaning remarks will have, at least in the short term: they are too general, and there is no guarantee that any historical understanding acquired by students will be reflected in the contents of the state examinations. The state examination form only half of the final examination, however; the other half is done by the schools themselves. In these school examinations a new dimension has been prescribed: students should write papers, collect materials, study relevant literature, give presentations, and so on, for mathematics as well as for other subjects. Mathematics teachers are not used to this, and are rather worried about it. History of mathematics can offer rich materials for these kind of activities, and so well-trained and resourced teachers have good classroom opportunities. But there remain no certainties for the use of history in mathematics education.

Besides these curricular opportunities there are further positive signs for the future role of history in mathematics education. Notably, research into the area is going on, with some PhD students at Dutch universities. In summary: history of mathematics used to have a stronger position in Dutch mathematics education than it has currently, but things are looking up.

1.2.12 New Zealand

The New Zealand Mathematics Curriculum is designed around six specified strands: Number, Measurement, Mathematical Processes, Statistics, Geometry and Algebra. None of the strands explicitly emphasises or encourages the inclusion of history of mathematics. However, a real encouragement to include history as a part of the process of enabling students' mathematical nurturing is to be found in the

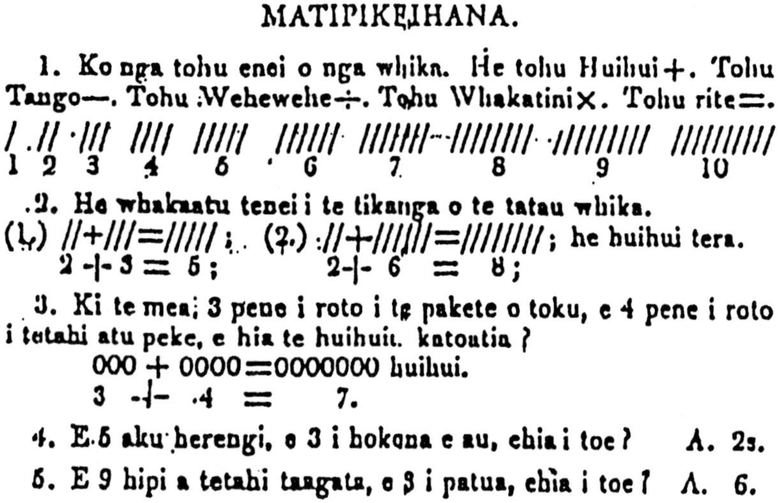

MATIPIKEJHANA.

1. Ko nga tohu enei o nga whikn. He tohu Huihui+. 'Tohu Tango—. Tohu Wehewehe÷. Tohu Whakatinix. Tohu rite=.

/ .// ·/// //// //// ////// ////// - ////// · ////// //////////
1 2   3   4   5   · 6   7.   8   9   10

.2. He whakaatu tenei i te tikanga o te tatau whika.

(1.) //+///=//// ;. (2.) .//+/////=///////; he huihui tera.
  2 -|- 3 = 6 ;   2-|- 6 = 8 ;

3. Ki te mea; 3 pene i roto i te pakete o toku, e 4 pene i roto i tetahi atu peke, e hia te huihui. katoatia ?
         000 + 0000=0000000 huihui.
          3 -|- .4 =   7.

4. E.5 aku herengi, e 3 i hokona e au, ehia i toe ?   A. 2s.

5. E 9 hipi a tetahi tangata, e 3 i patua, ehia i toe ? A. 6.

Figure 1.1: From the first printed Maori arithmetic, by Henare Taratoa, 1858

Development Band of the Curriculum. This band is an official addition to the basic curriculum, designed for use in enriching the mathematical experiences of gifted and talented students. Ministry of Education publications written in support of the Development Band recognise and encourage an understanding of the changes in the ideas and practice of mathematics that have occurred over time .

Some individual teachers of mathematics in New Zealand are known for their enthusiasm in regard to utilising history of mathematics as a part of their normal method of teaching. Papers and workshops offered by them at conferences, for example at the biennial conferences of the New Zealand Association of Mathematics Teachers, are regarded increasingly, by other conference participants, as interesting and innovative. One of the better known New Zealand mathematics publications, the *Mathematical Digest*, regularly carries historical information and suggestions for ways of including history in teaching or thinking about the mathematics covered in each of the Curriculum strands.

1 Me mōhio te ākonga ki te tuhi i te roanga atu o tētahi *tauira raupapa*, ka   *sequential pattern*
   whakaahua ture ai;

   • he torotoro, he hanga, he whakaahua, he tuhi i te roanga atu o tētahi
     tauira raupapa tau, *mokowā* rānei. (Mā konei ka mōhio te ākonga ki te   *space*
     āhua o te tauira mutunga-kore); Hei whakatauira:
     TAUIRA RAUPAPA TAU: 2, 1, 3, 2, 1, 3, 2, 1, 3, ...

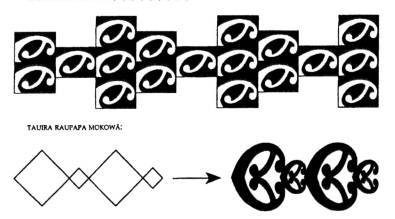

   TAUIRA RAUPAPA MOKOWĀ:

   • he hanga, he tauira tau tāruarua ki te tātaitai.

Figure 1.2: a 'tauira raupapa' (sequential pattern), as included in the New Zealand Mathematics Curriculum

Local and international culture-focused studies have begun to influence the ways in which those using the Mathematics Curriculum reflect historical differences. The incentive to adopt the changes in attitude that this requires has been intensified in New Zealand by the legal requirement that all official statements reflect bi-culturally sensitive attitudes. The term 'bi-cultural' is used in New Zealand to denote the recognition of the life and culture of the first settlers of the land, Maori, in relation to more recent settlers.

The present Mathematics Curriculum, implemented in 1994, appeared in a Maori version (see figure 1.2) as well as in an English one, but this simply highlighted the fact that translation is not enough to make sense of a culture's historical mathematical expression. Policy makers and teachers alike are presently searching for more appropriate ways of acknowledging the history and practice of Maori mathematical needs and methods, so that they, as well as the mathematics of other cultures, will have a similar status to that presently enjoyed by the mathematics emanating from the history of the mathematical needs and methods of Europeans.

1.2.13 Norway

National curricula have guided Norwegian school work in mathematics since 1827. A component reflecting culture and history has been made more explicit in the most recent Norwegian curricular reforms, those of 1994-98, through the efforts of Norwegian researchers and mathematics educators inspired by the Danish thinking described above (§ 1.2.3). In the new 1994 curriculum for grade 11 (16-17 year olds), for instance, the common goals (those describing attitudes, skills and perspectives which should penetrate the whole course for all students) include the following:

Goal 8: Mathematics as cultural heritage. Pupils should gain insights into the history of mathematics and know some of its importance for our social and cultural life. Pupils should know some main themes from the history of mathematics, the roots of mathematics in different cultures, some typical tools in the mathematics of these cultures, the importance of mathematics for the techno-scientific culture, and examples of the interplay between mathematics and art.

It is interesting to note that the preliminary published version of this goal (dated April 1993) attracted some criticism in the media: "Something here is fundamentally wrong. The main point of mathematics is, and should be, to solve mathematical problems." Other arguments advanced from this perspective were that such a generalised goal was difficult to assess, and that teachers were not trained to teach it. On the other hand, the criticism was also an opportunity for counter-arguments to be brought forward: that history reflects an important part of the national heritage, that it helps explain how mathematics is the basis for other subjects, that pupils can better understand features of the conceptual development of mathematics, and that it inspires pupils and helps to humanise the subject in their eyes.

The new elementary school curricula for grades 1-10 (pupils aged 6-16), brought forward in 1997, are also attentive to the inclusion of historical material. The sixth common goal for mathematics is "that pupils develop insight in the history of mathematics, and in the role of mathematics in culture and science." Some specific examples are given in the detailed spelling-out of the curriculum. In grades 8-10, for example, pupils "should have some knowledge about the main features of number systems used by different cultures", and "should experience aesthetic aspects of geometry through practical examples in architecture, art and handicraft, and see this in a cultural and historical connection."

The publication of textbooks for the Norwegian curriculum requires national approval and the history of mathematics is not yet fully integrated into the five currently-approved texts. Although there is some lip-service to historical and cultural issues, for example comparing the Hindu-Arabic numeral system with those of the Babylonians, Romans and Egyptians, the analysis is rather shallow and pedagogical opportunities are lost through not discussing comparative structural advantages and disadvantages of the systems. The presentation is like storytelling and does not explain or discuss problematic issues.   Another theme treated notionally but inadequately is the solution of equations: although pupils become aware that solutions for different kinds of equation were developed in the past, there is no discussion of the transition from verbal to symbolic solutions or the role of symbolism in facilitating the later expression of these solutions. The textbook writers seem to underestimate the magnitude which the step from arithmetic to symbolic algebra represents for each student, and not to understand the help which historical parallels can present here for both teachers and pupils. These are but two examples of missed opportunities in the treatments seen in the textbooks. It is clear that the historical aspect needs to be developed further to really become an integrated area of inspiration for teachers and pupils.

At the teacher education level, reforms began to be implemented in 1998 in which the support for prospective teachers in relation to the historical/cultural dimension of the curriculum was to be strengthened. Mathematics 1, for example (a course obligatory for all student teachers, representing one half of their studies for one year) contains the statement:

Mathematics has its history in all cultures and societies. It shows its development from ancient geometry to fractals, from astronomical calculations of the Mayas to Newton's and Einstein's mathematical models of the universe, and through the development of number systems and ways of reckoning. Historically, mathematics has developed in interchange with problems from other sciences and subject areas of society. The field also develops on its own premisses and by posing its own problems. Students should (e.g.)

- know the historical development of numerals, number systems and geometry
- be able to describe ethnomathematics as expressed in the daily life of some peoples
- be able to give examples of how mathematics influences Norwegian society and culture
- identify and explain mathematics in music, drama, art, architecture and handicraft.

1.2.14   Poland

In Poland, the school year 1999/2000 brought important changes to the system of general education for pupils from the ages of 7 to 19. Hitherto there was a '8 + 4' system: eight grades of elementary education (pupils 7-15 years old) followed by four grades of secondary education (students 15-19 years old). From 1999, this was progressively replaced by a new '6 + 3 + 3' system: six grades of elementary school (pupils 7-13 years old), followed by three grades of gymnasium (pupils 13-16 years old), and finally three grades of lycee (secondary school, for students from 16 to 19 years old). The new system will be fully in place by 2004..

This reform was not restricted only to administrative changes, but essential changes of school curricula were introduced. The new *Curriculum Basis for General Education* was elaborated for every educational stage. This document, from the Ministry of National Education, specifies the basic knowledge and basic skills which are needed at a given educational level. This document serves as a point of departure for constructing the *curriculum* for school subjects and for elaborating *standards of learning outcomes* at the end of each educational level. A curriculum can be prepared by educators, subject experts or teachers and must be accepted by the Ministry of National Education. A teacher can choose, from many such proposals already prepared, whichever which seems the most suitable for their pupils. Curriculum proposals are usually accompanied by suggestions about the textbooks and various didactical materials, which help teachers to work with pupils more effectively.

In the case of mathematical education, this way of preparing and organising work with pupils has been developed over several years, especially at elementary school level—that is, creating mathematics curricula based on the common core and preparing school-books adequate to them. In secondary school there is little diversity; this educational level is rather more traditional, mainly because of the urgency of passing examinations for the secondary-school certificate and then university entrance examinations.

The main idea of the new educational changes was to place general education within a framework of what are called 'key competencies', such as: planning, organising, evaluating one's own learning, effective intercommunication in various situations, working in a group, problem solving ability, efficient using of information technology. In relation to mathematical education, the *Curriculum Basis for General Education* includes for every educational level a list of essential mathematical skills to develop and mathematical notions to form, those which are deemed necessary from the point of view of general education and of developing the key competencies. In this context there is no entry (key word) in this document connected with history of mathematics. The important role of history of mathematics in mathematics education will happen on the level of the realisation of the educational aims.

There are some ten curriculum proposals for mathematics in the new elementary school and gymnasium, prepared on the new basis, most of which include a fairly perfunctory attention to history. The textbooks proposed to support these curricula similarly tend to include at best a few biographical notes and rather basic historical information. One curriculum proposal and accompanying textbook series, however, called *Mathematics 2001*, is an exception in including rather more history and with a more considered integration of historical and mathematical learning materials.

1.2.15 United Kingdom

Traditionally, schools and teachers in the UK had autonomy about what to teach and how to teach it, even though the examination system and the text books dictate the syllabus in secondary schools to some extent. There are several examination boards each of which is competing for customers and increasingly the government have encouraged the notion of education as a market place. Schools are funded on numbers of students and parents are encouraged to shop around for the best school. Examination results from each school must be published. It is therefore in the school's financial interests to gain the best possible exam results for their children. To ensure the best results teachers must decide which exam board to use. Many of those working in schools, however, would say that the home background of the children is what makes a big difference to the results; it is certainly true that the areas with high-priced housing seem to be very much over-represented in the schools with the best exam results. Even under a state system of education there are ways to buy better education for your children.

There is no requirement for history of mathematics to be included in the syllabus. The word 'history' does not appear in the National Curriculum for

---

Reflecting on Chapter 8

What you should know

- how the work of Descartes compares with that of Pierre Fermat
- evidence of Descartes's approach to negative and imaginary solutions of equations
- how Descartes constructed a normal to a given curve
- the relationship between algebra and geometry in *La Géométrie*.

Preparing for your next review

- Bring your answer to Activity 8.1 to the review together with half a page in your own words describing the revolutionary features of *La Géométrie*.
- Answer the following check questions.

1   Prepare a presentation explaining how the work of Descartes was influenced by van Schooten, Viète and Fermat.

2   Use Descartes's method to find the equation of the normal to $y^2 = 4x$ at $(1,2)$.

Figure 1.3: From Nuffield's 'History of mathematics' option (1994) the final question of Chapter 8

mathematics. Text books are not vetted by any official body, and vary in the extent to which they include history of mathematics, but there is very little history in any set of text books. (An exception to this statement is a option on the history of mathematics which ran within the Nuffield 'A' level scheme for some years during the 1990s—but it stands out by its exceptionality.)

As in many other countries, mathematics graduates are not keen to enter teaching where the morale is low, as are pay and working conditions, whereas in other fields of employment mathematics graduates are found to be attractive and paid accordingly. This gives the good mathematics student more of a choice than that in some other disciplines. On the other hand, many departments of education in the universities are obliged to accept students with qualifications which are not the best, with either low grades in their mathematics degree or graduates in other disciplines (but who have studied some mathematics). The latter are required to take a two-year post-graduate degree, whereas mathematics majors take a one-year course. To become a teacher of mathematics for ages 11-18 a student would be expected to be a graduate with a post-graduate teaching certificate. Once a teacher is qualified there is the possibility of teaching outside the discipline.

The 1998 National Curriculum for teacher training makes no mention of the history of mathematics but is essentially concerned with the content of the curriculum and teaching styles. It is concerned with students' mathematical attainment and requires the institution to carry out a subject audit for all those on the course and to ensure that by the end of the course all students reach the required standard. For students aged 3 to 11 teachers must have completed a subject study in mathematics and be able to demonstrate that they can reach the required standards in mathematical understanding. Again, there is no mention of the history of mathematics in the National Curriculum for Teacher Training.

1.2.16 United States of America

In the USA, there is a great deal of variability since most educational decisions are taken at a state level: according to a 1996 publication by the American Association for the Advancement of Science,

State departments of education or state education agencies are continuously grappling with how to create an equitable education system that includes flexible policies and practices which take into account the needs of each student

In school reform in mathematics, the state education agencies use the National Council of Teachers of Mathematics Standards; hold planning seminars for school superintendents and district administrators, school principals and administrators, lead teachers, and community and business leaders; strengthening-service; and write and disseminate curricula and guides for use in the planning an implementation of curricula by school districts. 99 percent of high school graduates had studied mathematics in high school between 1982 and 1992. However, only 68.4 percent studied algebra; 48 percent geometry, 37 percent algebra II; 12 percent trigonometry; and 4 percent calculus.

Making changes in curriculum varies from state to state. In Florida, for example, at the top administrative level the State Board of Education has to approve the State Standards or Framework. The plan in place since 1996 is to have trainers come in to work in each district to align the local curriculum with the new standards. All decisions on usage are local. Florida teachers are not required to have majored in mathematics to teach the subject. Other states, however, such as Michigan, have certification requirements which ensure that new teachers are well prepared. It is the colleges and universities which makes higher impositions on graduating students.

The position of the various societies with regard to history of mathematics in teacher education varies. The position of the National Council of Teachers of Mathematics (NCTM, with 120,000 members) is that

Students should have numerous and varied experiences related to the cultural, historical, and scientific evolution of mathematics so they can appreciate the role of mathematics in the development of our contemporary society and explore these relationships among mathematics and the disciplines it serves. . . . It is the intent of this goal—learning to value mathematics— to focus attention of the need for student awareness of the interaction between mathematics and the historical situations from which it has developed and the impact that interaction has on our culture and our lives.

This view is made explicit in regard to calculus:

As students explore the topics proposed in this standard, it is important that they develop an awareness of, and appreciation for, the historical origins and the cultural contributions of calculus.

The NCTM standards have made a major impact nationally and locally. The revised standards will be published in 2000. The NCTM has long supported the contribution history can make to mathematics education, notably in its celebrated and influential thirty-first yearbook, the 542 page *Historical topics for the mathematics classroom* (NCTM 1969), in which a number of leading historians and mathematics educators came together to provide an overview and reference resource for mathematics students and teachers.

The Mathematical Association of America (MAA, with 18,000 members) recommended in 1991 that teacher education classes include the history of mathematics, and it too has a long and distinguished record in promoting and publishing books on the history of mathematics in relation to mathematics education. (The second president of the MAA, indeed, was the historian and educator Florian Cajori.) Not all organisations have been so supportive, however; the Mathematical Sciences Education Board of the National Research Council did not mention either history or culture in their 1990 recommendations for reshaping school mathematics.

1.3   History of mathematics in curricula and schoolbooks: a case study of Poland

Ewa Lakoma

The framework for current developments in Polish school education is described above (§1.2.14), where it was noted that some ten curriculum proposals for mathematics in the new elementary school and gymnasium have been put forward. Generally these include only such quasi-historical entries as the Roman notation of numbers, or knowing and applying the Pythagorean Theorem. Most textbooks written according to these proposals merely include some biographical notes on the most famous mathematicians—Pythagoras, Euclid, Plato, Descartes—and information on the most famous historical facts concerning school mathematics: who introduced the rectangular coordinates system, what is the Pythagorean triple, what are Platonic solids, and so on (see Nowecki 1996-9; Pawlak 1999). In the Polish mathematics curriculum there are two theorems which traditionally have to be considered at school level: Pythagoras' theorem (direct and opposite) and the theorem of Thales (direct and opposite). (The theorem of Thales is essentially Euclid's *Elements* vi.2, that a line parallel to one side of a triangle cuts the other two sides proportionally.) But the labels 'Pythagoras' and 'Thales' in themselves do not constitute history.

From among the variety of proposals for the new mathematics curriculum it is useful to look at a project called *Mathematics 2001*. Its curriculum, textbooks, and other didactical materials include relatively more mathematical history than other proposals (Dabrowski 1999a, 1999b). This project uses the history of mathematics

John Fauvel, Jan van Maanen (eds.), *History in mathematics education: the ICMI study*, Dordrecht: Kluwer 2000, pp. 19-29

as an origin for didactical situations, which can be interesting for pupils, and as a source of original simple reasonings, which can be readily understood and also turn out to be helpful for today's pupils. The history of mathematics serves as a source of information on various ways of mathematical thinking and arguments. Besides a list of topics to learn, the curriculum *Mathematics 2001* presents a list of the student learning outcomes expected at the given educational level. Moreover, it includes a list of examples of didactical situations and concrete tasks to solve by pupils. Among these didactical situations and outcomes we can find relatively many elements of the history of mathematics. I present some examples below, first in the curriculum and then in the schoolbooks for that curriculum, to illustrate the range of topics which can draw upon history. General issues about the range of ways in which historical material can be incorporated into textbooks are looked at in a later chapter (§7.4.1).

1.3.1   History of mathematics in mathematics curricula

– Elementary school (4th grade):

Topic 411 *Numbers and their properties*: the authors propose, among examples of didactical situations and tasks to solve, that pupils could compare various systems of writing numbers, for example a system with the base 5 and the Aztec numeral system.

Topic 413 *Algorithms of arithmetic operations*: consider ways of calculating numbers using Chinese abacus.

Topic 414 *Properties of numbers, properties of divisibility*: use graphical representations of numbers (rows of stones, rectangular shapes or notched rectangles) and manipulate them in order to justify discovered properties of numbers.

Topic 415 *Numbers and their properties*: use tangrams, distinguishing their parts and describing their size, and adapt them to introduce the concept of fraction.

Topic 454 *Measure*: tangrams are used in order to introduce methods of measuring an area.

– Elementary school (5th grade):

Topic 552 *Measure, area of triangle, quadrangle*: build figures of various shapes using the same pieces of tangrams, and express their observations concerning an area.

– Elementary school (6th grade):

Topic 613 *Properties of numbers, properties of divisibility*: consider Pythagorean triples, find generators, and search for relations with triangles (proportional triples versus triangles with similar shape). Also pupils can be asked to explore magic squares, and describe ways of their transformations.

Topic 642 *Geometric transformations, isometries*: build isometric figures, assembling pieces of tangrams.

Topic 664 *Algebra, discovering & formulating regularities*: use the algorithm known as the sieve of Eratosthenes to find all prime numbers less than 100. Gymnasium (1st grade)

Uczniowie mogliby:
- analizować różne dowody twierdzenia Pitagorasa i ustalać, które z nich są bardziej przekonujące np.

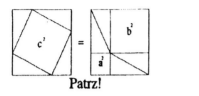

Patrz!   **Licz!**

- szukać w literaturze różnych dowodów twierdzenia o kącie środkowym i wpisanym oraz porównywać je;
- poszukiwać luk i usterek w różnych rozumowaniach;

Figure 1.4: "Behold!" and "Calculate!"

Topic 106 *Theorem, assuming, thesis, proof*: analyse and discuss various proofs of the Pythagorean theorem, e.g. they can discuss the justifications involved in *Look at it!* (*Patrz!*) and in *Calculate it!* (*Licz!*)—see figure 1.4.

– Gymnasium (2nd grade):

Topic 232 *Geometric figures and their properties, trigonometric proportions*: find, in some materials on history of mathematics, information about how Thales estimated from the sea coast a distance between ships on the sea.

Topic 251 *Measure, π number, circumference, area of a circle*: find out information on π, i.e. on how the circumference of a circle was measured in the past. Pupils can be also asked to find in literature how Erathostenes computed a radius of the Earth, and to compare his results with data known at present.

Topic 241 *Homothety and similarity of figures*: consider the theorem of Thales and justify it by means of similar triangles.

1.3.2   History of mathematics in mathematics school-books

In all books we can find some biographical notes on the most famous mathematicians, and notes concerning the origins of various mathematical notations, eg of the sign for equality, of the sign for squaring (ie power 2), or of the square root sign. Moreover, many of them include brief notes on the historical development of mathematical activities or mathematical ideas in a range of cultures, such as Egyptian mathematics, Chinese mathematics, Hindu mathematics, Greek mathematics, the school of Pythagoras, Euclid and the *Elements*, and the origins of algebra.

The series of school-books *Mathematics 2001* (Lakoma 1996, 1997a, 1998; Zawadowski 1999) includes, in comparison to other textbooks (eg Novecki 1996-9; Pawlak 1999) a relatively large component of history of mathematics. First we look at examples from the *Mathematics 2001* series, and then will illustrate some examples from other textbooks.

– Elementary school textbook (4th grade):

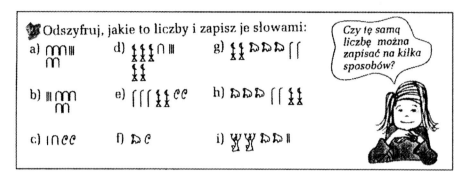

Figure 1.5: From the module 'How did Egyptians calculate?'

Module 2: "How did Egyptians calculate?" (see figure 1.5). Topic *Numbers and their properties*. Pupils get acquainted with Egyptian symbols of numbers notation, and are asked to discover ways of writing numbers by means of these symbols. They try to answer: what was the Egyptian system of writing numbers? What is a system of writing numbers today?

Module 11: "Calculating sticks" (see figure 1.6). Topic *Numbers and their properties*: the algorithm of multiplication by 'calculating sticks' is presented here as elaborated by John Napier in the 17th century (often called 'Napier's rods' or 'Napier's bones' to make multiplying numbers an easier process. Pupils are asked to discover how this works, and to analyse and understand the algorithm of multiplication. Then they consider contemporary algorithms for multiplication and choose the most suitable for them.

– Elementary school textbook (5th grade):

Module 2: "How did Hindu multiply numbers?". Topic *Numbers and their properties*. The subject serves as a point of departure to develop a skill of multiplying numbers. Pupils are asked to discover the *Hindu algorithm*, to analyse it and to find pros and cons of this way of multiplying.

Module 4: "Number sieve". Topic *Properties of numbers, properties of divisibility*. Pupils are asked to find numbers which are divisible by 2, 3 and so on. In this way they get know the method of finding prime numbers known as 'Erathostenes' sieve'.

Module 15: "Pros of multiplication table". Topic: *Numbers: adding and subtracting fractions*. Pupils become acquainted with Egyptian fractions and are asked to present some fractions as sums of Egyptian fractions.

Module 19: "What are matches for?". Topic *Geometric figures: equilateral triangle, isosceles triangle*. Pupils read about some discoveries of Thales: The two angles at the base of an isosceles triangle are equal; two intersecting straight lines form two pairs of equal angles; the diameter of a circle divides it into two equal parts. In this context pupils are asked to discover and analyse some further properties of triangles.

– Elementary school textbook (6th grade):

Module 27: "Time for a puzzle". Topic *Discovering regularities*. Pupils are asked to discover and analyse regularities of mathematical activities, using various

Dlaczego ten sposób mnożenia jest dobry?
Przyjrzyjmy się raz jeszcze.

237 × 4

12 dziesiątek
to 120,
a 8 setek
to 800!

28
120
+ 800
948

A jak mnożymy dziś?
Można tak:

$$237 \times 4 =$$

$$= 200 \times 4 + 30 \times 4 + 7 \times 4 =$$

$$= 800 + 120 + 28 = 948$$

lub tak:

lub tak:

Figure 1.6: Evaluation of methods for multiplication based on a comparison with Napier's rods

graphical representations of numbers, they also discuss Pythagorean triples (as was suggested in the curriculum).

In modules 29 and 30, pupils have the opportunity to do simple examples of classic geometrical constructions by means of compasses and a ruler.

— Gymnasium schoolbook (1st grade):

Reader 1: "What is a theorem?" Topic *Theorem, assumption, thesis, proof.* Pupils read a text with information about the earliest theorems, mainly from Greece. *Theorema* means 'that which is seen', so the first proofs served as tools, leading to catching a sight, meeting with an illumination. As an example Greek 'pebble arithmetic' is shown, for justifying properties of numbers.

Module 7: "Secants and tangents". Topic *Geometric figures*: pupils are asked to make a poster presenting the mathematical fact, discovered by Thale*s,* that *a triangle*

1. Napis na nagrobku Diofantosa z Aleksandrii:

Pod tym nagrobkiem spoczywa Diofant — a dzięki przedziwnej sztuce zmarłego wiek jego zdradzi Ci ten głaz: $x =$

Chłopcem przez szóstą część życia pozostać Bóg mu pozwolił, $\dfrac{x}{6}$

lica pokwitły mu zaś, kiedy dwunasta znów część życia mu minęła, $+\dfrac{x}{12}$

a znowu gdy przebył część siódmą, młodą małżonkę w dom dobry wprowadził mu Bóg, $+\dfrac{x}{7}$

która, gdy pięć lat minęło, małego mu powiła synka. $+5$

Ale okrutny los chciał, że kiedy syn ledwie wiek Ojca w połowie osiągnął, ponury zabrał go Hades. $+\dfrac{x}{2}$

Kojąc ogromny swój ból, szukał Diofant wśród liczb jeszcze przez cztery lata pociechy, aż rozstał się z życiem. $+4$

Diofantos żył i pracował w Aleksandrii w drugiej połowie III wieku n. e. Jego prace przyczyniły się do rozwoju algebry.

Taki nagrobny napis nazywa się epitafium. Zapisz to epitafium — zagadkę w postaci równania. Uprość je. Sprawdź, że Diofantos żył 84 lata.

Figure 1.7: The ancient puzzle of the life span of Diophantus is an excellent opportunity for students to solve linear word problems.

whose vertices are situated on a circle, with one of its sides a diameter of the circle, is right-angled.

Module 18: "From a problem to an equation" (see figure 1.7). Topic *Algebraic language*. Analysing problems of *Diophantos*, and of *Bhaskara* is a point of departure for gaining a very important skill, to express a mathematical word problem in algebraic symbols. Presenting old ways of thinking helps pupils to analyse particular steps of the process of this translation. Pupils already know examples of simple equations. Now they develop their skills to read mathematical texts and express a problem in symbols, leading to an algebraic equation.

Module 22: "Let's cut a square!" Topic *Pythagoras' theorem*. Pupils are asked to build a puzzle. Thanks to it they are able to discover Pythagoras' theorem. This important theorem is presented, and analysed. In module 23 pupils are also asked to consider some situations coming from every day life in which it is useful to apply this theorem (e.g. parking a car). They also try to discover a method of finding segments whose length is a square root. How to get a square root of 10, of 17 as quickly as possible?

Reader 7: "The Pythagorean legend". Topic *Properties of numbers, discovering regularities*. This is a story about mathematical ideas associated with the Pythagoreans. The algorithm of alternative subtracting, to find common divisor of two numbers, is presented. When applied to a side and a diagonal of the regular pentagon, this algorithm does not stop, leading to the conclusion that the side and diagonal of a regular pentagon are incommensurable: their lengths are numbers which are irrational.   This story lets pupils become acquainted with important problems of ancient mathematics. The aim of the reader is to let pupils know a fascinating adventure in the history of mathematics and to present ways of thinking which can be interested and understandable for a pupil at this educational level.

– Elementary school textbook (8th grade, old structure; Lakoma 1997 b):
   Module 40: "Trousers of Thales". Topic *Theorem of Thales*. Pupils are asked to analyse a sequence of figures, which illustrate succeeding steps of mathematical reasoning leading to the proof of Thales' theorem.
   Module 7: "Squaring of a circle" Topic *Measure, area of a circle*. Pupils become acquainted with an information on a classic problem of squaring of the circle. This information is a point of departure for estimating and discovering a method of calculating an area of a circle.
   In this book we can also find examples of classic geometrical constructions by means only of ruler and compasses. Pupils have also opportunity to analyse Platonic solids.

Other examples of school-books including history of mathematics

– Secondary school textbook (4th grade, old structure; Walat 1990):
   Although this book is out of print, being replaced by other series of textbooks, it is worth presenting the main ideas and to show examples of historical elements included in it. It was addressed to students preferring humanistic subjects, like languages, history, philosophy, psychology, fine arts etc. Although this book was written according to the curriculum of the old secondary school, which did not included elements of mathematics history, it was decided by the authors to present one third of mathematical material in a form of historical investigations. In this part of the book students could find a lot of old mathematical texts, written in original language or translated into Polish. Students usually were asked to read a text, to analyse it and to follow the argument, or to apply it in some situations. They were often asked to compare old mathematical methods with these methods in use today. The historical texts chosen for the students to have opportunity to read were fundamental to the historical development of mathematics. Students could read fragments of the following works:
   Euclid, *Elements*, book i, some of books ii and iii, written in old Polish (a translation from 1817).
   Cardano, passages concerning the development of algebra. Students were asked to interpret his algebraic description and to translate it into today's language of algebraic symbols.
   Rene Descartes, passages from *La géométrie*. Students were asked to analyse the rule of signs, which gives information about the number and the position of the roots of a polynomial equation, and to apply it in some simple cases. They also had the opportunity to read the first book of *La géométrie* in French or in parallel Polish translation. Students could follow the method of mathematical reasoning proposed

by Descartes. The authors briefly explained the main ideas of this method just above the original text. Students were asked to understand Descartes' way of solving quadratic equations using ruler and compasses.

Nicolaus Copernicus, extracts from *De revolutionibus orbium coelestium* presented in Latin (figure 1.8) with parallel Polish translation (figure 1.9). Students got to know the theorem of Ptolemy: the product of diagonals of a quadrangle inscribed in a circle is equal to the sum of products of the opposite sides. Then they were asked to work out why Pythagoras' theorem can be deduced from the theorem of Ptolemy.

Students could also find a lot of old mathematical methods from the traditional

Theorema fecundum.

SI quadrilaterum circulo infcriptum fuerit, rectangulum fub diagonijs compraehenfum, aequale eft eis, quae fub lateribus oppofitis cotinentur. Efto enim quadrilaterum infcriptum circulo A B C D, aio, quod fub A C & D B diagonijs continetur, aequale eft eis quae fub A B, C D, & fub A D, B C. Faciamus enim angulum A B E, aequale ei qui fub C D. Erit ergo totus A B D angulus, toti B B C aequalis, affumpto B B D, utricq communi. Anguli quocq fub A C B, & B D A fibi inuice funt aequales in eodem circuli fegmento, & idcirco bina triangula fimilia B C B, B D A, habebunt latera proportionalia, ut B C ad B D, fic B C ad A D, & quod fub A C & B D aequale eft ei, quod fub B C & A D. Sed & triangula A B B & C B D fimilia funt, eo quod anguli qui fub A B B, & C B D facti funt aequales, & qui fub B A C, & B D C eandem circuli circumferentiam tufcipientes funt aequales. Fit rurfum A B ad B D, ficut A B ad C D, & quod fub A B & C D eaquale ei, quod fub A B & B D. Sed ia declaratu eft, quod fub A D, B c tantu effe, quantu fub B D, & B C Coniunctim igitur quod fub B D & A C aequale eft eis, quae fub A D, B C, & fub A B, C D. Quod oftendiffe fuerit oportunum.

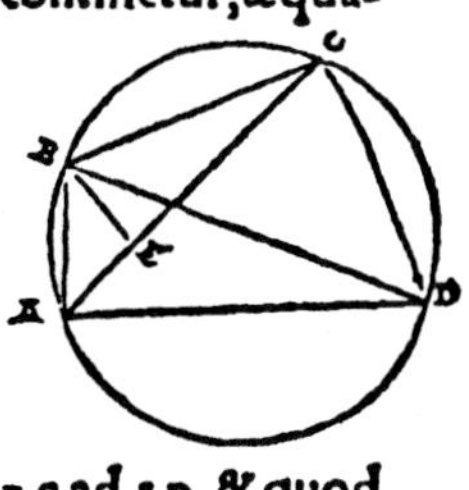

Figure 1.8: Copernicus proves Ptolemy's theorem, from De revolutionibus *(1543)*

canon of mathematical knowledge, for example the 'galley method' of dividing numbers Students were asked to apply this method of dividing numbers and to compare it with a method that they use today.

- Secondary school textbook (3th grade, old structure; Walat 1988):

The earlier third-grade textbook in the same style also contains many references to history of mathematics. Studying properties of numbers leads to using Gauss' method or graphic representations of numbers for calculating sums of many components. Students can also consider polyhedrons; they were encouraged to analyse some examples like the three-dimensional stellated polyhedra and the Platonic solids described by Luca Pacioli in *De divina proportione,* and also are able

TWIERDZENIE DRUGIE

Jeżeli czworobok wpisany zostanie w koło, to
prostokąt rozpostarty na przekątnych, równy
jest tym (sumie tych), które rozpostarte są
na przeciwległych bokach. Niech będzie zatem
czworobok wpisany w koło ABCD, twierdzę, że
ten (prostokąt) który na AC i DB jest
rozpostarty, równy jest tym (sumie tych),
które są na AB, CD, & na AD, BC. Utwórzmy
zatem kąt ABE, równy temu, który
(rozpościera) CBD. Będzie więc cały kąt ABD,
całemu kątowi EBC równy, przyjmując, że EBD,
jest dla obu wspólny. Kąty rozpostarte na
ACB, & BDA sobie nawzajem są równe (bo
chwytają) ten sam segment koła (.) Oba
trójkąty podobne BCE, BDA, mają boki
proporcjonalne, jak BC do BD, tak EC do AD,
& (stąd prostokąt) rozpostarty na EC & BD
równy jest temu który na BC & AD. Lecz i
trójkąty ABE & CBD są (tak utworzone, że
są) równe, & te (kąty) na BAC, & BDC ten sam
(kawałek) obwodu koła chwytające są równe.
Zatem jak uprzednio AB do BD tak jak AE do
CD, i ten (prostokąt) na AB & CD równy jest
temu na AE & BD. Lecz już zostało
powiedziane, że ten na AD, BC taki jest
jak na BD, & EC (.) Razem zatem ten który
(rozpostarty jest) na BD & AC równy jest tym
które (sumie tych które rozpostarte są) na
AD, BC & na AB, CD. Co pokazać trzeba
było.

Kopernik przedstawia tu dowód *twierdzenia Ptolemeusza* o tym, że dla czworokątów wpisanych w koło iloczyn przekątnych równa się sumie iloczynów przeciwległych boków. Odcinki Kopernik oznacza zawsze dwoma dużymi literami, a zamiast zwrotu „iloczyn przekątnych" używa zwrotu „prostokąt zbudowany na obu przekątnych". Przyjrzyj się tekstowi Kopernika i porównaj jego styl ze stylem Kartezjusza.

11.2. W jaki sposób z twierdzenia Ptolemeusza wynika twierdzenie Pitagorasa?

Figure 1.9: Polish translation of the passage in figure 1.8, with the question relating Ptolemy's theorem to Pythagoras

to discover Euler's formula. They also read brief information on the thirteenth book of Euclid's *Elements*, in which the construction of the Platonic solids is presented. Analysing logarithms give an opportunity to present methods of calculating them, due to Jost Bürgi, John Napier and Henry Briggs. A short computer program was used here, to generate various values of the logarithmic function.

1.3.3   Final remarks

Textbooks used now at secondary school level do not include elements of the history of mathematics. Their structure is much more rigid, authors preferring the form of a formal lecture try to explain mathematics as simply as possible. This brief review has allowed us to see that there are two different points of view on the place of mathematics history in curricula and school-book, according to whether historical elements are encouraged or discouraged. What is the reason for such a polarisation of standpoints? Careful analysis of those didactical approaches which use history of mathematics, and those which do not, allows us to hypothesise that an attitude towards history of mathematics in mathematics education very much depends on a general viewpoint towards mathematics learning and mathematics teaching.

There are, broadly speaking, two contradictory cognitive styles in mathematics education. One is seen in the work of Euclid, the other in Descartes. The style of Euclid was presented in *Elements*, perhaps the world's oldest textbook. It is a systematic presentation of mathematics: definitions, axioms, theorems, proofs, theorems, proofs, . . . Formulating mathematical theory in such a dogmatic frame became the canon of knowledge for many centuries. This rigid style, albeit replaced for educational aims by equivalent texts written in a form more suitable for students, has its votaries even today. In the work of Rene Descartes, by contrast, there is no such style. Descartes presented mathematics as a fascinating description of his adventures in connection with solving mathematical problems (see for example Fauvel 1988). It is symptomatic that he presented his fundamental work *La géométrie* as an example of activity, supplemented to his *Discours de la Methode* He showed the reader ways to solve a given problem, then he posed several others connected with it and sketched their solutions in such way that a reader had opportunity to solve them individually. Descartes introduced convenient algebraic notation, which spread out very quickly and is also used today.

Euclid also introduced terms and concepts which we still use, in what is called 'school geometry'. However the difference between the styles is fundamental: in Euclid's works we find a logically built structure of knowledge, whereas Descartes provided us with essays on natural ways of mathematical reasoning which lets us construct the world of mathematics. Both these styles of presenting mathematics correspond with different cognitive styles characteristic of people's learning. We can call these styles the *dogmatic* style and the *discursive* (nearly: discours-ive) style. Schools, in order to be effective, have to adapt themselves to the cognitive abilities of the students, and have to take note of the pattern of their cognitive development. When this fundamental demand is taken into account in mathematics education, the discursive style is generally preferable.

History of mathematics can play a very useful role in mathematics education, but the way in which it is used very much depends which style of education we prefer. Mathematics history in education can be presented as a set of curious details, which can arouse students' interest in mathematics. In this context it can be used in both styles of education. History of mathematics can also create a context for introducing mathematical concepts, in ways which encourage students to think. Historical solutions let students to continue simple ways of thinking and to develop them individually. The different points of view which are possible to present in historical contexts give students the opportunity to develop the art of discussing, to justify their own opinions, to present their own reasoning to other people. Historical cases encourage students to repeat individual attempts to solve problems. All these activities are very useful for forming mathematical concepts and developing mathematical thinking. Thus, history of mathematics seems to be especially useful when we prefer a discursive style of education.

We can risk posing this hypothesis: the more attention we pay to regard pupil's cognitive development, the more useful becomes history of mathematics in creating and realising didactical proposals.

References for 1.3

Dabrowski M., Piskorski P., Zawadowski W. 1999a. *Mathematics 2001: Curriculum for elementary school, grade 4-6* (in Polish), Warsaw: WSiP

Dabrowski M., Piskorski P., Zawadowski W., 1999b. *Mathematics 2001: Curriculum for gymnasium, grade 1-3* (in Polish), Warsaw: WSiP

Fauvel, John 1988. 'Cartesian and Euclidean rhetoric', *For the learning of mathematics* **8**, 25-29

Lakoma E., Zawadowski W., e.a., 1996. *Mathematics 2001, textbook for the 4th grade of elementary school* (in Polish), Warsaw: WSiP

Lakoma E., Zawadowski W., e.a., 1997a. *Mathematics 2001, textbook for the 5th grade of elementary school* (in Polish), Warsaw: WSiP

Lakoma E., Zawadowski W., e.a., 1998. *Mathematics 2001, textbook for the 6th grade of elementary school* (in Polish), Warsaw: WSiP

Lakoma E., Zawadowski W., e.a., 1997b. *Mathematics 2001, textbook for the 8th grade of elementary school* (old system) (in Polish), Warsaw: WSiP

Nowecki B., e.a., (collective work), 1996-1999. *Blue Mathematics, series of textbooks for elementary school and gymnasium* (in Polish), Bielsko-Biala: Kleks

Pawlak Z., e.a., (collective work), 1999. *Mathematics step by step, series of textbooks for elementary school and gymnasium* (in Polish), Lodz: Res Polona

Walat A., Zawadowski W., 1988. *Mathematics, textbook for the 3rd grade of secondary school* (in Polish), Warsaw: WSiP

Walat A., Zawadowski W., 1990. *Mathematics, textbook for the 4th grade of secondary school* (in Polish), Warsaw: WSiP

Zawadowski W., e.a., (collective work), 1999. *Mathematics 2001, textbook for 1st grade of gymnasium* (in Polish), Warsaw: WSiP

1.4 Policy and politics in the advocacy of a historical component

Science is a political issue because scientific research and achievements are of great importance for all people in the world. Mathematical education is a political issue because it is one of the essential channels for science to be approached by all people. The history of mathematics shows how fundamental is the link between mathematics, mathematical education, and the general conceptions of people at each time and each place. It shows, too, how the content of any mathematics curriculum represents a choice, essentially a political choice, that has been made; and by the same token is apt to change when other political influences are in place. This section states the position of the group of the ICMI Study who were charged with reaching a strategic view on the political context, outlining some strategies that can be employed to ensure that history is incorporated into teaching.

Should a historical dimension be incorporated in the official mathematics curriculum? And if so, how can it be made to happen? As the present study clearly shows, there is wide variation in just what "a historical dimension" can be taken to mean. The many different ways in which history of mathematics can enter the educational process are discussed in depth in other chapters of this book, the possibilities ranging from anecdotal support in on-going teaching (§7.3.1, §7.4.1), to using original sources in the classroom (chapter 9); from influencing the structure of a curriculum (§8.5.2) to exploring history of mathematics for trainee teachers (chapter 4), for example by gaining further insights into the development of students' understanding (chapter 5). But precisely because there are so many possibilities, there is scope for confusion and muddle in the advice tendered to educational policy formers, as well as in their reactions to well-meant solicitation.

It is important therefore to be alert to any possible problems which a historical dimension could be thought to generate, and indeed to examine policy proposals carefully from the perspective of a devil's advocate. In this way counter-arguments and misunderstandings can be anticipated (as was seen in the example of Norway, §1.2.13 above).

Such awareness of arguments countering a role for history, as well as knowledge of the benefits and potential of a historical dimension, has to be exercised in the context of the wide range of opinion-formers and policy advisers in many countries nowadays. A number of different groups are concerned with, and have greater or lesser influence over, decisions about what is taught in schools: classroom teachers, head teachers, school authorities, educational theorists and researchers, parents, local politicians, national politicians, publishers, journalists and other influences on public opinion. These groups do not all have a common interest in solving educational problems in the same way. (For example, a classroom teacher who is convinced of the value of gaining a fuller historical awareness through in-service training may find a different reaction from the head teacher who is having to make budgetary decisions as well as pedagogical ones.)

Several difficulties might be anticipated at the classroom level, let alone at other levels of opinion formation. To incorporate history within a mathematics curriculum might be thought to consume more time, involve more effort, distract pupils from the task of exploring and gaining confidence in their handling of mathematics itself; it might indeed be thought a somewhat alien intrusion, introducing a quite different world from that of mathematical inquiry. Furthermore, it might be something beyond the competence of the teacher, who may indeed have chosen to pursue mathematics precisely because it didn't involve reading about history, or writing joined-up sentences. Teachers without some historical training may feel nervous and ill-prepared, and worry about how to access historical sources as well as their competence in handling them. There may be concerns about the danger of replacing a flexible strategy, in which teachers can use history as a resource and teaching aid as appropriate, with a rigid curricular imperative in which the history exam becomes as terrifying to contemplate as the maths test. It could be that introducing a new pedagogical principle in an unimaginative or bureaucratic way may do more harm than good. (Such concerns are raised in the work of Jean-Pierre LeGoff, discussed in § 3.2 below.) To seek to impose the wider use of history may precisely miss the point, that it is the enthusiasm of the teacher which effects the most successful teaching, even in the most knowledgeable teacher, not the way of delivering the subject matter. There are a variety of such concerns which may need to be resolved as we consider just what results we want to accomplish: namely, the successful delivery of a better mathematics education for all students everywhere.

That said (and these matters will be returned to periodically in the course of this volume), it will be helpful to draw attention here to some possibilities for the political development of the ideas put forward in this book. This section speaks to the range of opinion formers mentioned above, and outlines ways in which they might be approached and with what message. The experience of different countries is very different both in political structures and in the length of chain of communication between, say, a classroom teacher and the ministry of education, so this section is not a prescriptive framework but a reminder of possibilities of influencing those who control what is in the curriculum and are in a position to influence any role of history of mathematics within the experiences of pupils and students of mathematics.

1.4.1   Political authorities (at all levels)

Bureaucracies work both through personal contacts and position papers, so a useful strategy would be to prepare a summary of the Study Book to be presented to each authority (for example, by the ICMI representative in each country). This executive summary should be translated into a number of languages and published in mathematics and education journals. It is as well to make sure that authorities are aware of the three following important issues: national contributions will always be recognised; the study of history of mathematics will attract students to study the exact sciences; and to reassure that the study of history of mathematics is not a substitute for the study of mathematics but a resource within its better delivery. Throughout the political process a robust assertive tone, sensitive to counter-arguments and determined to anticipate and rebut them, will gain respect and influence.

1.4.2   Teacher associations

It is vital to call attention to the importance of imbuing professionals with knowledge of history of mathematics, and more especially with sensitivity towards the arguments of this Study, involved in the national committees. In several countries, special sessions on history of mathematics and its use in the classroom are scheduled at the national and regional meetings. A related task is to encourage historically oriented articles, and discussions of the classroom use of historical material, in journals. Meetings can profitably be held with officers of teachers' associations to summarize and explain the contents of the Study Book. The importance should not be underestimated of including in reward systems for teachers due recognition of those who work to develop their skills in a historical direction; the argument may be strongly put that such teachers are better teachers, better informed and better motivated, as a result of this kind of in-service training.

1.4.3   Professional mathematics associations

There is growing enthusiasm among professional mathematicians for history of mathematics, through a greater awareness of its important roles (though this has long been perceived by leading mathematicians: see the quotations by Lagrange, Abel, De Morgan and others, §1.5 below). It is notable that some major international meetings, such as the International Congress of Mathematicians, have invited historians to make a major plenary address, and it is reasonable to work for such a presence at other meetings too. At a number of regional, national and international meetings, special sessions on history of mathematics are increasingly popular features, and in journals historically oriented articles are increasingly encouraged and seen.

Studies on the history of contemporary mathematics should be encouraged and also—this is particularly important—simpler versions aimed at students; the success of recent accounts of, for example, chaos theory and Andrew Wiles' solution of Fermat's Last Theorem shows that where there is a will, simplified accounts of recent history can be achieved and resonate with students, teachers and the general public.

1.4.4   Tertiary teachers

Encourage the designation of grants to study history of mathematics and its integration in teaching; for example by the production of study units. The promotion of seminars on the history of mathematics can reinforce the efforts of faculty members to widen the number of colleagues friendly towards history. The reward system for both college faculty and school teachers needs to be revised so that tenure and promotion criteria are seriously considered for those whose research activity includes research on the use of history of mathematics in mathematics classes, and the disseminate of these findings to mathematics teachers and faculty.

A key emphasis at tertiary level is teacher education, since teachers will implement the curriculum with ease and pleasure only if they are familiar with it and fully understand it. History of mathematics should be included in teacher education programmes, so that teachers have the flexibility of using it as and if they choose to do so: it is then knowledge over which they have ownership. These issues are discussed further in chapter 4.

1.4.5   Parents

Parents can be among the most worried of opinion formers, both through their own bad experiences with mathematics in the past and through concerns about modern education generally.   By the same token, any strategy which leads to noticeably better results or greater enthusiasm among pupils should gain strong parental support.   The promotion of mathematics awareness with an historical aspect, such as through texts in national newspapers about aspects of the history of mathematics, is helpful, as are other ways of popularizing mathematics and its history though a range of media (books, plays, newspapers, films, TV programmes).   Teachers have found that mathematics awareness evenings, weeks, or other events of a semi-social nature are an excellent way of boosting parental interest in their childrens' mathematical studies, and that a historical dimension to the content of these events produces remarkable growth in confidence and support.

1.4.6   Textbook authors

Some authors may need encouragement and help both in accuracy and relevance of historical references and in thinking through the pedagogical challenges of incorporating historical material.   This task, on a country by country basis, is occupying the attention of several leading mathematics educationists with historical interests.

1.5   Quotations on the use of history of mathematics in mathematics teaching and learning

There has been interest over several centuries in the relations between the history of mathematics and the teaching and learning of mathematics, as will become apparent from various discussions in the course of this book.   Many leading mathematicians and teachers down the ages have expressed, in different ways and for different reasons, ideas about the relationships between mathematics and its history.   We conclude this chapter with a number of quotations illustrating this theme, for reference and interest as well as to strengthen the case of those arguing today for a stronger incorporation of history in the educational process.   Each quotation could be analyzed in detail for its presuppositions, pedagogical attitudes, social context and contemporary value—which might provide useful discussion material for groups in, say, initial teacher training—but here we leave such analysis as an exercise for readers and their students.

Portugal 1772

From the Statutes of the Portuguese University of 1772, written by Jos Monteiro Da Rocha, in the section on the first-year curriculum.

1 In order that the lessons of the mathematics course be done in good order, and with profit from the students: the reader of geometry, to whom belong the disciplines of the first year, before entering the lessons proper to his chair, will read the general prolegomena to the mathematical sciences.

2 In them he will make a brief introduction to the study of these sciences: showing the object, division, and its general prospect: explaining the method that they use; the usefulness, and excellence of it: and making a summary of the main accomplishments of its history through its most remarkable times. They are: from the origin of mathematics, until the century of Thales and Pythagoras; from this until the foundation of the Alexandrian School; from this until the Christian Era; from this until the destruction of the Greek Empire; from this until Descartes; and from Descartes until the present time.

3 This summary shall be proportionate to the capacity of the students, so that it predisposes them, and encourages them to enter the study with pleasure. Because of it the reader will not enter in the detailed description of the discoveries that were made in the said sciences in different times and places; because it cannot be understood, unless the same sciences have been studied; and then they will not need the voice of the master, to be instructed in history. He will recommend nevertheless very much to his disciples, that according to their progression in the mathematics course that they should be instructed particularly in it: showing them, that the first thing that must be done by somebody that wants to study in the progress of mathematics, is to instruct himself in the discoveries made before him; in order not to lose time in discovering for the second time the same things; nor in working in the tasks and undertakings already carried out.

Métodos para o traçado de cónicas

Método do jardineiro

O chamado *método do jardineiro* para a construção de elipses é bem conhecido, e apresentamos apenas a reprodução de uma gravura do matemático holandês van Schooten, publicada no seu livro de 1646, *De organica conicarum sectionem in plano descriptione tratactus* (fig. 7). Na figura 8 mostramos uma figura obtida a partir de um modelo do mesmo método feito no *Sketchpad*.[17]

Aconselhamos o leitor a tentar reproduzir no *Sketchpad* os métodos análogos para construir uma hipérbole (diferença constante em vez de soma constante) e uma parábola (igualdade de distâncias ao foco e à directriz).

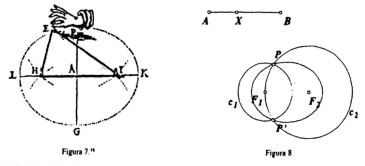

Figura 7.[18]   Figura 8

Figure 1.10: Portugal still encourages teachers to incorporate history in mathematics education. Here an age-old method for drawing ellipses is illustrated by images from a 17th century text and from the dynamic geometry software package Sketchpad, in the teacher training text by Eduardo Veloso, Geometria: temas actuais, *1998*

And in the description of the second year algebra curriculum we find:

3 To facilitate better his entrance in it, and assure the fruit of the lessons: the professor will begin with the respective prolegomena: giving a detailed idea of its purpose, and the means it

applies to obtain its goal: showing its origin, and progresses: and making a summary of the history of the said algebra through its most notable times.

4 In particular he will show the reason that the ancients, although they knew the fundamental rules of analysis and were endowed of such great skill, have not extracted from it the amazing advantages that the moderns have discovered; lacking the instrument of analysis that is algebra.

France 1790s

Joseph Louis Lagrange (1736-1813), the leading pure mathematician in France after the French Revolution, was co-opted to teach mathematics to trainee school-teachers at the Ecole Normale.   This passage occurs in his lectures to them on logarithms (J. L Lagrange, *Lectures on elementary mathematics*, Open Court 1901, 22):

Since the calculation of logarithms is now a thing of the past, except in isolated instances, it may be thought that the details into which we have entered are devoid of value. We may, however, justly be curious to know the trying and tortuous paths which the great inventors have trodden, the different steps which they have taken to attain their goal, and the extent to which we are indebted to these veritable benefactors of the human race. Such knowledge, moreover, is not a matter of idle curiosity. It can afford us guidance in similar inquiries and sheds an increased light on the subjects with which we are employed.

Norway 1820s

Niels Henrik Abel (1802-1829), Norway's greatest mathematician who died tragically young, wrote this in the margin of one of his notebooks:

It appears to me that if one wants to make progress in mathematics one should study the masters.

England 1865

Augustus De Morgan (1806-1871): from his inaugural address as first president of the London Mathematical Society, 16th January 1865.

I say that no art or science is a liberal art or a liberal science unless it be studied in connection with the mind of man in past times. It is astonishing how strangely mathematicians talk of the Mathematics, because they do not know the history of their subject. By asserting what they conceive to be facts they distort its history in this manner. There is in the idea of every one some particular sequence of propositions, which he has in his own mind, and he imagines that that sequence exists in history; that his own order is the historical order in which the propositions have successively been evolved. The mathematician needs to know what the course of invention has been in the different branches of Mathematics; he wants to see Newton bringing out and evolving the Binomial Theorem by suggestion of the higher theorem which Wallis had already given. If he be to have his own researches guided in the way which will best lead him to success, he must have seen the curious ways in which the lower proposition has constantly been evolved from the higher.

Italy 1871

Eugenio Beltrami (1835-1899), professor of rational mechanics at the University of Bologna (Giornale di matematiche **11** *(1873), 153)*

Students should learn to study at an early stage the great works of the great masters instead of making their minds sterile through the everlasting exercises of college, which are of no use whatever, except to produce a new Arcadia where indolence is veiled under the form of useless activity.

England 1890

From the presidential address of J W L Glaisher (1848-1928) to Section A of the British Association for Advancement of Science, 1890

In any treatise or higher text-book it is always desirable that references to the original memoirs should be given, and, if possible, short historic notices also. I am sure that no subject loses more than mathematics by any attempt to dissociate it from its history.

USA 1896

Florian Cajori (1859-1930; *A history of elementary mathematics with hints on methods of teaching,* New York: Macmillan, 1896, v):

The education of the child must accord both in mode and arrangement with the education of mankind as considered historically; or, in other words, the genesis of knowledge in the individual must follow the same course as the genesis of knowledge in the race. To M. Comte we believe society owes the enunciation of this doctrine—a doctrine which we may accept without committing ourselves to his theory of the genesis of knowledge, either in its causes or its order." [Herbert Spencer, *Education: intellectual, moral and physical,* New York, 1894, p.122] If this principle, which was also held by Pestalozzi and Froebel, be correct, then it would seem as if the knowledge of the history of a science must be an effectual aid in teaching that science. Be this doctrine true or false, certainly the experience of many instructors establishes the importance of mathematical history in teaching.

Germany 1897

Hermann Schubert (1848-1911; Mathematical essays and recreations, Chicago 1898, 32):

The majority of mathematical truths now possessed by us presuppose the intellectual toil of many centuries. A mathematician, therefore, who wishes to acquire a thorough understanding of modern research in this department, must think over again in quickened tempo the mathematical labours of several centuries.

UK 1919

From a Mathematical Association Committee report, 1919:

The Historical aspect of Mathematics has never yet found its fitting place in teaching of the schools. [. . .] Every boy ["(Throughout the report the word BOY is to be taken as referring to pupils of either sex)"] ought to know something of the more human and personal side of the subject he studies. [. . .] The history of mathematics will give us some help in framing our school syllabus. [. . .] Recommendation: That portraits of the great mathematicians should be hung in the mathematical classroom, and that reference to their lives and investigations should be frequently made by the teacher in his lessons, some explanation being given of the effect of mathematical discoveries on the progress of civilization.

Scotland 1929

Dame Kathleen Ollerenshaw (*b. 1912*; 'Living mathematics', *IMA Bulletin* 25 (1989), 50-56; 51):

One memorable experience at St Leonards was the enlightened gift from my housemistress of H.W. Turnbull's lovely small book *The great mathematicians.* Every youngster showing an interest in a particular branch of learning (or other worthy activity) should be given the appropriate book telling of the giants in their field who paved the way. If there was any one moment in my life when I knew that I must specialise in mathematics, with no conceivable alternative, it was when I first read this beautifully written history.

USA 1930

Vera Sanford (*A short history of mathematics*, Boston: Houghton Mifflin, v):

Upwards of a century ago, Augustus De Morgan presented a brief for the study of textbooks in arithmetic in these terms: "A most sufficient recommendation of the study of old works to the teacher, is shewing that the difficulties which it is now (I speak to the *teacher* not the *rule-driller*) his business to make smooth to the youngest learners, are precisely those which formerly stood in the way of the greatest minds, and sometimes effectually stopped their progress." It was not necessary to limit this to teachers nor to the study of textbooks in a particular branch of mathematics, for the struggle of mankind to formulate mathematical concepts, to evolve a useful symbolism, and to solve quantitative questions arising from his environment are of interest to teachers, students, and bystanders as well.

Soviet Union 1931

Mark Yakovlevich Vygodskii (*Foundations of infinitesimal calculus*, Moscow-Leningrad 1931, 5):

I wrote this book because of my deep conviction that none of the existing textbooks puts the key ideas of infinitesimals before beginners with the necessary sharpness and clarity. . . . This is the reason behind the depressing fact that the apparatus of analysis remains a dead apparatus in the hands of the students.

The viewpoint underlying the present textbook is that the learner must be *introduced* to the study of analysis by getting acquainted with its fundamental notions at the stage at which they arise directly from practical needs. Their rigorization and cleansing must be a later issue, initially of secondary importance.

In other word, I am attempting to replace the formal-logical scheme by the historical scheme, or rather, by the historico-logical scheme.

Of course, this does not mean that I propose to take the reader through all the twists and turns of the historical development and reconstruct the chronological order of the evolution of the ideas of analysis. The historical material in my book is not its subject matter but the basis for the exposition.

UK 1958

From a UK Ministry of Education report, 1958:

The teacher who knows little of the history of Mathematics is apt to teach techniques in isolation, unrelated either to the problems and ideas which generated them or to the further developments which grew out of them. [. . .] A knowledge of the arguments and dissensions between great mathematicians might induce healthy skepticism and discussion in the classroom and lead to a firmer grasp of principles. [. . .] One of the most valuable assets which the teacher can acquire from a knowledge of the history of his subject is an appreciation of the influence of current traditions. [. . .] It is important to convey to the pupils the knowledge that much of what is taught today as a finished product was the result of centuries of groping or of spirited controversy. [. . .] Mathematics can be properly taught only against a background of its own history.

Chapter 2

Philosophical, multicultural and interdisciplinary issues

Lucia Grugnetti and Leo Rogers

with Jaime Carvalho e Silva, Coralie Daniel, Daniel Coray, Miguel de Guzmán, Hélène Gispert, Abdulcarimo Ismael, Lesley Jones, Marta Menghini, George Philippou, Luis Radford, Ernesto Rottoli, Daina Taimina, Wendy Troy, Carlos Vasco

Abstract: *School mathematics reflects the wider aspect of mathematics as a cultural activity. From the philosophical point of view, mathematics must be seen as a human activity both done within individual cultures and also standing outside any particular one. From the interdisciplinary point of view, students find their understanding both of mathematics and their other subjects enriched through the history of mathematics. From the cultural point of view, mathematical evolution comes from a sum of many contributions growing from different cultures.*

2.1   Introduction

In the 1980s, mathematics educators and didacticians in many countries felt the need to give a more reliable foundation to educational research through philosophical reflection on the processes involved. What philosophy is suitable for this purpose?

Philosophy must explain mathematical thought not only at the level of research, but also as far as teaching is concerned. It must also explain the development of mathematics in the past: philosophy needs history. But what history is suitable? There is a *history of documents* and a *history of ideas*. The latter needs the former, but didactics and epistemology need the latter. This means that we must avoid the identification of philosophy of mathematics with mathematical logic. Our philosophy must guide and explain educational choices; it must help in a better planning of teaching. It must be open to new reflections. In this sense it could be considered as being almost equivalent to epistemology (Speranza and Grugnetti 1996).

John Fauvel, Jan van Maanen (eds.), *History in mathematics education: the ICMI study*, Dordrecht: Kluwer 2000, pp. 39-62

A cultural perspective on mathematics makes us attend to mathematical histories and to what they tell us about who developed mathematical ideas in different societies (Bishop 1995). Multicultural aspects and interdisciplinary issues become therefore part of epistemological reflections about mathematics education; the relationships between philosophical, multicultural and interdisciplinary issues are very strong. Moreover, the history of mathematics as the history of ideas is strictly linked to (or better, is part of) the history of human beings. In this view we have to analyse the cultural, political, social, economic contexts in which ideas arose

2.2   Philosophical issues

2.2.1   Historical investigation, evidence and interpretation

Differing views on the nature of historical enquiry

The widely held view of mathematics as a pure subject uninfluenced by outside forces is slowly changing, and this is reflected in the changes in the approach to more general historical study. If we agree that history is that branch of knowledge which caters for society's needs to understand particular aspects of the human past, then we express our needs by demanding answers to a range of who? what? when? how? and why? questions. However, as soon as we start to investigate, we find that these questions are not at all easy to answer. Traditionally, history is viewed as a study of carefully delimited aspects of the past employing systematic research in all available sources. The approach can be from a social, political or economic point of view, and necessarily employs a general philosophy (for example, structuralism, Marxism, etc.) in its interpretation. More recently, 'post-modern' history is seen as a set of processes and power relations linking the past to the present, where the interpretations of events and facts are critically interrogated, the underlying assumptions are revealed, the status of texts are called into question, and where groups of people and their conditions are defined and redefined by those in power.

In a similar manner, there have been changes in the way history of mathematics is undertaken. 'Internalist' history of mathematics is recognised by its tendency to see mathematics as a subject isolated from 'external' influences and as a progression of ideas which are improving and becoming more abstract and general with time. In the internalist, sometimes called 'whiggish', account the events of the past are seen as instances of steps towards the present more perfect structures. This kind of history tends to interpret the past in terms of modern concepts. More recently, researchers have tried to take a more holistic view, with mathematics seen as a component of the contemporary culture; the historian's task is then to discover the influences, conditions and motivations (social, economic and political as well as scientific and mathematical) under which problems arose. Admitting these points of view necessarily leads to much reinterpretation of the received wisdom of earlier writers. In the past, most research in the history of mathematics has been carried out by those with mathematical training. In consequence, the interpretation and its writing up has not only utilised technical language, but has tended to employ a narrative which maintains the genre of the hypothetical-deductive style employed in mathematics itself. In this way it has given the impression of an authoritative account of the events in question, where often the historical subject can be criticised for making errors, pursuing a fruitless avenue of enquiry, or not seeing a solution which later seemed obvious.

Historians, on the other hand, realise that there are many different sorts of questions about the past, giving rise to many different sorts of history. The events, structures and processes of the past are known only through the relics and traces of the past, which are themselves politically and conceptually loaded and imperfect. There are difficulties of understanding archaic languages, contemporary technical terms, and the special 'codes' within the available sources, so that any interpretation is cautious and aware that many concepts may carry with them a collection of unsubstantiated assumptions. In perceiving relationships between different events and conditions the historian may have to consider theories derived, for example, from economics, psychology, sociology or anthropology. Furthermore, the account is constrained by conventions of language, genre, mode, argument, and a number of other cultural and social contextual conventions. In this perception theory, sources, and style interact in an iterative way.

Facts and events

The notion of a 'fact' is ambiguous, since it includes the sense of both *event* (meaning whether or not the event took place), and *a statement about an event* (where the concern is with the truth or falsity of an occurrence or statement). In this sense, facts are constructed in the documents which refer to the occurrence of the events, not only by interested parties (contemporary or more recent) commenting on the events or the documents, but also by historians giving what they believe is a true account of what really happened in the past. Therefore it is the facts that are subject to revision and further interpretation, and they can even be dismissed given sufficient reasons.

This view allows us to account for the fact that historiographical consensus about any event is very difficult to achieve. It is always open to revision from another perspective. We not only change our ideas of what the facts of a given matter are, but our notions of what a fact might be, how facts are constructed, and what criteria should be used to assess the adequacy of a given collection of facts in relation to the events which they claim to support. The relation between facts and events is always open to negotiation and re-conceptualisation not because events change with time, but because we change our ways of conceptualising them.

This argument leads to a position of historical relativism in which the truth and authoritativeness of a given account of the past must be assessed in relation to the cultural context and social conditions prevailing at the time, and with respect to the perspective of the current interpretation. As noted above, the interpreter's viewpoint is also involved. The problem is that this position appears to deny a secure and timeless epistemological foundation for history, which still causes concern to some historians. However, it must be recognised that a particular historical investigation in its final written form does not represent a totally authoritative statement or a

secure piece of knowledge. It is a contribution to knowledge which is immediately open to scrutiny, analysis and criticism by fellow historians.

The philosophical problematique

Following this line of argument, the problem of interpretation is even more difficult in the history of mathematics since much of what we now choose to identify as mathematics has been perceived differently by people in the past, so that the history of mathematics is different for different periods and cultures. For example, the interpretation of Mesopotamian tablets by Høyrup, Damerow and others (Høyrup 1996) is radically different from that of Van der Waerden, whose 'Babylonian algebra' appears as the standard interpretation in many histories of mathematics (Van der Waerden 1961). For events even further in the past, Seidenberg has followed the anthropologists' maxim that 'the present use of the tool does not necessarily indicate its origin', and provided evidence that the origins of some techniques that we regard as elementary mathematics can be found in ancient ritual practices (Seidenberg 1962). According to this account, these techniques were then much later adapted to other purposes such as counting, geometrical measurement, elementary mechanics and astronomy. Looking for the beginnings of mathematical ideas may well lead us into new areas quite outside our familiar territory. These examples of conflicting historical interpretation illustrate the need for a careful philosophical analysis of the assumptions and other aspects of the process of doing history of mathematics.

2.2.2　Philosophy of mathematics, old and new

The demise of traditional philosophies

Until relatively recently, scholars who were involved in the philosophy of mathematics could only be found in departments where there was a significant interest in logic and related technical aspects of mathematics. This could be seen as broadly the result of changes in mathematics through the nineteenth century where abstraction and structural aspects dominated the interest in the development of mathematics in many fields. From these beginnings, we can see *Platonism* revisited; the idea of an ideal perfect structure, 'out there' for us to discover seemed very appealing.

Later, various forms of *Formalism* inspired by Hilbert and his co-workers provided many interesting technical challenges within the branches of mathematics. Rigour ruled, and while of course, rigour is important, for some, like Russell, it became the central concern. So much so, that *Logicism* attempted to reduce the whole of mathematics to a cold, clear and unassailable core of technical manipulation without any necessary meaning. We know (through the work of Gödel and others) that this programme was not entirely successful, and its failure began to cast doubts on some of the other areas where Formalism influenced the way that mathematics was conceived (see also Hofstadter 1979).

An approach from an entirely different direction came from one of the principles of Kantian philosophy, and provided the interesting and challenging idea of introducing the role of intuition in the creation of mathematics. Starting from a minimal basis, Brouwer and Weyl put forward the principles of *Intuitionism*, and in a sense, re-introduced the role of the individual creative actor into the making of mathematics (Heyting 1956; Weyl 1949). While the pure theory was interesting, we know it was unable to provide a sufficiently coherent foundation for all the mathematical ideas that were current at the time, and the exclusion of so much of the mathematics that had been built up over so many years was too much to bear.

Bringing the human actor back more firmly into the centre of the stage continued with the work of Wittgenstein, who challenged a range of tacitly accepted assumptions about the nature of the enterprise we call mathematics (Wittgenstein 1956). The semiotic significance of acts, the meanings of words in the language and the social contexts in which these meanings are built up was not at first seen as a challenge to Formalism. However, the influence that Wittgenstein had on his contemporaries was enough to sow the seeds of a fundamentally new look at the nature of mathematics and the way in which mathematics was done.

Definitions of mathematics

Mathematics has meant different things to people at different times. Of course, when we use the word mathematics in this context, we can never be sure that those who talked about it in the past gave quite the same meaning to it as we do today. For the Pythagoreans, in whose world all objects were fundamentally numerable, all was number. For Plato, where 'ideal forms' played a significant part in his world view, mathematics was something else again. The idea of a pure and unassailable truth which had some connection with the real world was for long one of the cornerstones of the definitions of mathematics. In the seventeenth century mathematics became the model of God's universe, and was seen as the supreme science of counting and measurement, and this view became reinforced the more humankind discovered how to use the technical power of mathematics to describe the motions of the stars or the tides of the sea.

There was a radical change in point of view in the nineteenth century with the invention of non-Euclidean geometries and the search for the 'basic laws' of algebra. This led to an emphasis on abstract structures, extending and generalising ideas into other domains, and the bringing together of a number of hitherto apparently unrelated areas under some general unifying concepts. Along with these developments came the realisation that mathematical truth was a matter of consistency of arguments; and further, that mathematics did not necessarily have anything to do with the real world. Definitions became more and more abstract and inclusive, but always it seemed that there was still some aspect of the enterprise that got left out. As a final gesture of defeat, some decided to retreat to the position that mathematics is what mathematicians do. Now, even among research mathematicians there is no consensus on what mathematics is (see Thurston, 1994; Atiyah, 1994).

History and culture

Following Wittgenstein, other ideas slowly began to be brought into the discussions of the nature of mathematics, and a number of these key influences appeared from completely outside mathematics. Almost at the same time, and in their own particular ways, Kuhn, Wilder and Lakatos were concerned about the role of science and mathematics as a function of the cultural context in which the ideas grew. Putting people back into the picture as active creators of the theories was a major step in forming new views about the philosophy of mathematics. Although, of course, the definition of a *paradigm* may be as elusive as the definition of mathematics, the way in which Kuhn argued for the persuasive power of discourse and the psychological basis for the adoption of a change in a theory (Kuhn 1962) was a serious challenge to accepted beliefs about the nature of the historical process in science. Wilder, influenced by anthropological theories, was ambiguous in the way in which he talked about culture, and regarded mathematics somehow as an organic whole (Wilder 1950; Wilder 1968). In identifying generalised internal and external influences on the development of mathematics he located mathematics in a milieu of social, economic and cultural stresses which were seen to determine the directions in which the subject developed. Lakatos, again, (Lakatos 1976) putting forward a quasi-empirical view, showed how the *fallibilism* of Popper and the *heuristic* of Polya contributed to the way in which the step by step development of theories could be described as a process of successive refinement of ideas by modifying the original idea to include (or exclude) any new objects or properties of objects.

While there had always been people writing histories of mathematics, up to now the writing had been more internalist in approach, but from the late 1960s these influences began to make themselves felt, and more 'socially based' writing of history slowly began to appear. Likewise, the writing in the philosophy of mathematics began to change. For example, Kitcher's discussion of the nature of mathematical knowledge (Kitcher 1983) largely relies for its evidence on material from the history of mathematics, and the contributors to Gillies' edited collection of arguments for or against the idea of revolutions in mathematics (Gillies 1992) clearly view mathematics as an ongoing process of reconceptualisation influenced by a range of both internal and external events. Some historians have come to feel that there has been a revolution in the historiography of mathematics, because of the way in which contemporary historical interpretation has come to include social contexts.

2.2.3   The ends of the spectrum

Abstraction versus empiricism?

There is a tension in the ways we view the nature of mathematics. On the one hand, mathematics is a body of abstract knowledge which is available to be learned or rediscovered and then improved upon by any individual. On the other, mathematics arises from problems, which are expressed in the needs of people at a particular time. It has been argued that these two general ways of regarding mathematics may be fundamentally incompatible: mathematics as a set of timeless truths and value free facts may be actually inconsistent with mathematics as a cultural product set in social contexts. However, it is our contention that these are not incompatible, on the grounds that a study of the history of mathematics itself shows that the conception of mathematics as timeless derives from a particular cultural context.

Epistemology can be regarded as the investigation of how and under what conditions our knowledge of the world is formed. Since, furthermore, what we regard as 'knowledge' is a distillation and abstraction of the responses to questions and problems (many of which lie in the remote past of our species), then that part of our present knowledge which we call mathematics consists of the concepts and theories which have been built up in the process of answering certain types of problems in rather special ways. Initially, these problems were essentially practical, but as society evolved, many problems and their solutions became progressively abstract and entered an intellectual world which some neo-Platonist philosophers regard as somehow detached from individuals, having an independent existence. Thus we arrive at a belief that mathematics is 'discovered' in some way, implying an *a priori* existence for a host of mathematical concepts, many of which have yet to be discovered. In this sense the idea of an independent world of mathematical ideas is one generated under particular historical circumstances and transmitted to individuals through socio-educational influences. Furthermore, even if we accept the idea that mathematical ideas are carried and transmitted to individuals in some unconscious way through their culture, many studies have shown that mathematical concepts and processes are very different in different societies, and so the belief in the universality and *a priori* existence of mathematical ideas cannot be sustained.

Invention versus discovery?

Considering the nature of mathematics from a background of the history of mathematics changes the way we conceive the epistemological problems of the development of mathematical knowledge in the individual and in society. The historical approach encourages and enables us to regard mathematics not as a static product, with an *a priori* existence, but as an intellectual process; not as a completed structure dissociated from the world, but as an on-going activity of individuals. Recognising this activity is important for the establishment of scientific parameters for any didactical theory since the historical and cultural dimensions considered by Vygotsky have, until recently, largely been omitted from the established body of psycho-pedagogic theory inspired by many of the followers of Piaget. Following this line of argument, we can arrive at a distinction, based on the experience of studying the history of mathematics, which may help us to resolve the false dichotomy of invention *versus* discovery which besets the neo-Platonists. We can suggest here that concepts are invented, modified and extended in the process of answering problems, (which recognises the originality, creativity and social contexts of the activity of the human mind), and that theorems and proofs are discovered in the process of finding solutions to these problems (which recognises the particular modes of thought and logical patterns which we call mathematics).

2.3   Multicultural issues

2.3.1   Introduction

During the twentieth century, significant changes have occurred in our understanding of the contributions that different cultures have made to our history. It is important in mathematics, as in any other discipline, to be sensitive to new issues. Showing how mathematical thinking and applications developed in different cultures, in response to the needs and thinking of different societies, not only enables a wider understanding of the concepts embodied in mathematics but also encourages greater creativity and confidence in using its various branches. A history that shows the diversity, rather than the universality, of mathematical development adds an exciting dimension to the subject. It allows the world and its history to enter the classroom in a way that works against a narrow ethnocentric view, without denying the extent to which developments have often been embedded in cultural contexts. A multicultural approach both requires and enables us to step into a realm of thinking which challenges our valuing of different styles and branches of the activity we recognise as mathematics.

At the beginning of the twenty-first century we have a much greater understanding of the global nature of mathematical endeavour than previously, and this has considerable implications for the ways in which history can be interpreted and incorporated beneficially in mathematics classrooms. One of the most widely accepted contemporary descriptions of what history is about can be summarised in the phrase *the study of change over time*. This study, however, always carries a fundamental philosophical approach, explicit or implicit, which influences the kind of interpretation put upon the historical events. There are many advantages in contemplating the changes that have taken place in mathematical thinking over the centuries. Historical study allows identification of the cultural factors which enabled one idea to be acted upon but another to be forgotten about, sometimes for centuries, and it also allows a study of the practical applications that the ideas were used to support. Furthermore, it encourages an understanding of the ways in which ideas that benefit one group have been used to benefit other groups, and an appreciation of the ways in which various issues have effected changes within or among different cultures. Searching for these aspects can lead to a greater appreciation of the wealth of ideas that are a part of mathematics, and a greater understanding of the contribution to change made by one's own culture as well as a greater awareness of the contributions made by other cultures. This broadening of perspectives can give a new impetus to teachers and students alike to search within their own background and culture as well as within the cultures of others, and thereby come to understand that what is found is a part of a global heritage rather than merely a national or regional one.

Mathematics is not just text; it lives in the minds of people and can, to an extent, be disclosed by interpreting the artefacts they have produced. These artefacts, inscriptions, instruments, books, and technical devices have been developed in particular places for particular reasons and an understanding of these reasons can help students to relate mathematical ideas to something greater than simply their own immediate environment. Through finding primary sources, making conjectures from the evidence of history, researching from secondary sources, using original instruments and methods, thinking of the ways in which need leads to creativity, students can learn to use the tools of other peoples and other cultures. In this way they can expand their own skills in ways that empower them to express their own intuitive feelings and thoughts, enabling them to use and develop their own aesthetic and creative senses.

2.3.2   Multiculturalism inside the history of mathematics

One of the ways of using the history of mathematics to help interlock ideas, illustrate their development and engage the attention of students is through the use of themes that are problem-based rather than personality-based. The following five examples are included to encourage teachers to find different methods from history for doing calculations which have applications that are both modern and appropriate. The first two examples come from early Egyptian mathematics.

(i) Egyptian multiplication example

Did ancient Egyptian mathematicians multiply in the same way as we do, and if not is their method a useful resource for today's classroom? In their method of duplication and mediation the Egyptian would proceed as follows. In multiplying 17 by 13, for example, the scribe had first to decide which of the numbers he was going to multiply by the other. If 17 was chosen he would then proceed by successively multiplying 17 by 2 (i.e. counting to double each result), and stopping before he got to a number which exceeded the multiplier, 13. He then added such results as would correspond to multiplying by 13 (here, 1 + 4 + 8). This method can used for the multiplication of any two integers, since every integer can be expressed as the sum of integral powers of 2. It is highly unlikely that the Egyptians were aware of this general rule in the form that we give it today, though the confidence with which they approached all forms of multiplication by this process suggests that they were aware of a reliable algorithmic process. This ancient method was widely used by Greeks and continued well into the Middle Ages in Europe.

(ii) Early Egyptian division example

For early Egyptians, as for us, the process of division was closely related to the method of multiplication. In the Ahmes Papyrus a division x/y is introduced by the words 'reckon with y so as to obtain x'. So an Egyptian scribe, rather than thinking of 'dividing 696 by 29', would say to himself, 'starting with 29, how many times should I multiply it by itself to get 696'. The procedure he would set up to solve this problem would be similar to a multiplication exercise. The scribe would stop at 16, for the next doubling would take him past the divisor, 29, and taking the sum of the appropriate numbers from the continued doubling of 29 gives the answer.

People have developed skills in mathematics, just as they have in other subjects, which reflect their cultural needs and values. Mathematics ceases to be relevant if it appears to be at odds with what people believe to be useful or true. The next three examples come from islands and island groups in the southern Pacific Ocean and show other aspects of the way a multicultural approach affects the presentation of mathematical ideas and methods (see Begg *et al* 1996).

(iii) Concepts of distance and area

In many Polynesian languages the idea associated with distance is to do with how long it takes to get somewhere rather than with the linear ground distance, so traditional questions about calculating distance may not convey any useful methods. In New Guinea the value of a piece land has more to do with its productivity potential than with its length and breadth, so the traditional concept of area has little relevance. In lands which have such very different terrains than those of the parts of Europe where concepts and unit terms for describing linear distance and length-and-breadth area were developed, it is more useful to see these traditional calculations methods as a part of European mathematics.

(iv) Beliefs, algebra and statistics

In introducing variables or equations one talks about x standing for an unknown. Many traditional beliefs among Pacific Island peoples link unknowns with magic, evil spirits and things to be avoided. At least in the early stages of algebra, students of these islands would find $— + 3 = 7$ much easier than $x + 3 = 7$. Similarly, consider this question: "If half of all children born are boys and the sex of the child is an independent outcome for each birth, what is the probability of the fourth child in a family being a boy if the first three were girls?" Asking this in some cultural groups could well return the answer that the assumptions are not valid, as the sex of a child depends on God and is not random.

(v) Language

It does not make mathematics multicultural simply to translate European mathematics into the language of an indigenous people. For example, in English 'equals' has different meanings in the context of sets and numbers. Some languages use the same word for a number of equivalent relationships (equals, congruent, equivalent, similar), and so translation does not convey the same breadth of mathematical language. Even with non-mathematical words, one language may not have developed subtlety in the same areas as another.

These examples help us to see how many aspects of mathematics and its development can be discussed when a problem-based approach is enhanced by adding a multicultural   dimension to the teaching of the history of mathematics. Discussion around topics such as those mentioned in the above examples help students to see ways in which differences in history, geographic location, culture and beliefs have influenced developments in mathematics. This, in turn, may well help students to understand better the concepts of multiplication, division, measurement, algebra, statistics, reasoning and so on.   In solving similar problems in several different ways, there could well be a kind of synthesis of mathematical procedures and traditions from a number of different countries and cultures.

This approach does not merely give a way of integrating the history of mathematics effectively, it is also inclusive of students with different intuitive dispositions and insights. It is useful in increasing the range of different mathematical skills which students have available for problem solving. It allows students and teachers to think of mathematics as a discipline of continuous reflection and action influenced by thoughtfulness, reasoning, known procedures, intuitiveness, experimentation, and application to practical situations. It demonstrates that mathematics, like all other subjects that students study at school, is not merely a subject in which one learns a series of irrefutable and unchangeable truths. It opens up ways in which the study of mathematics also contributes to the study of the ways in which people everywhere come to know things, and come to see how new knowledge can be used constructively without discarding or belittling old knowledge of a different time or place.

Using a multicultural approach to unravel some of the threads of the past helps us to understand the limitations that were placed on our perspectives by former labelling methods. For example, the term *Arab mathematics* usually refers to a phase of mathematical development which occurred in very different places, from Baghdad at the time of Harun al-Rashid or al-Ma'mun through to the Iberian peninsula, and was in fact the work of many people from different origins and even different religions who happened to be at the courts of the caliphs. The time span of the development of this section of mathematics was long, and there were too many known intercultural influences to talk simply of 'the Arabs' or of 'Islamic mathematics'. What is important to realise about the mathematics of this period is that it is far from being ethnocentric. Appreciating the mathematics that developed makes us realise that some caliphs, somewhere in Mesopotamia, judged the achievements of science and mathematics as being so important that they invited eminent specialists from all over the world to their courts. The work was undertaken by many learned people from many countries and cultures, who engaged in what amounts to a collective effort, all the way from the Middle-East to the West of Spain. It subsequently inspired several scholars in Europe who not only used the mathematical insights, but also completed translations from Greek and Syriac into Arabic and then into Latin, Hebrew and other languages. So it involved very many peoples, very many cultures, several religions, and many languages, and directly influenced the mathematics that developed in Africa, the Gulf States, and central and western Europe.

The emphasis one places on this detail or that is frequently affected by the philosophical climate of the environment in which one finds oneself. During the latter part of the twentieth century, the extension of the ideals of multiculturalism into the practice of all disciplines has changed the climate in classrooms, in mathematics just as much as in other subjects in schools' curriculum. Retaining a contextual balance in teaching the history of mathematics will mean that horizontal connections are grappled with as well as vertical ones. For example, the ways in which powerful men and expanding states employed mathematicians for economic, colonising and military issues should not be overlooked. History can expose diverse motivations, both external and internal, for why and how mathematics has developed in different societies.

An appreciation of the contribution that multiculturalism has made to our thinking and attitudes gives teachers a good background for judging the ways in which they can extend ideas of what can be learned about mathematics beyond the parameters previously set by European culture and societies, and value other ways of looking at things. An acceptance of multiculturalism has meant that history can now be used to convey a message which corresponds to the general attitude of many philosophers, writers and mathematicians over the centuries: that enthusiasm and creativity is fired by a desire to know, to think, to explore and ultimately to prove enough to move on to the next mathematical challenge, rather than by a desire to find a way to elevate one's own country or culture, or one's own gender or race.

2.3.3   Mathematics as a human enterprise

In the teaching of mathematics there are opportunities for introducing aspects of the history of mathematics through stories and examples from different ethnic and social perspectives. However, each such episode should be carefully prepared, presented and respected within its own context.   For example, labels such as *ethnomathematics* or *women's mathematics* are often useful for bringing attention to particular issues, but can often act to politicise these issues and have the danger of preventing particular groups from sharing in the wider community of mathematics. As we use and teach the global evidence of mathematical ideas developing in relation to contemporaneous need and in conjunction with the interdisciplinary interactions of ideas of each society, we can contribute to the freshness that multiculturalism engenders.

With the increase in the availability of human and intellectual resources which is generated by this philosophical shift, students and teachers of the twenty-first century will be able to see how one culture, or one group of people, or one geographical area has influenced another or added to understanding already gained in a different setting. Mathematics is a human enterprise, a voyage into the realm of human thinking and experimentation, and not a constantly upward movement towards perfection.   One looks to history with the idea of restraining the paraphernalia of one's own culture and national identity so that a broader understanding of the ideas of others can be gained. This is not to say that teachers should avoid identifying the mathematics of their own region. It is one of the contributions made by taking a multicultural approach that regional developments no longer need be treated as separate issues. It is important that students can identify and defend the ideas and nuances contained in the mathematics developed in their particular region, and that they recognise the significance of these ideas in terms of both the time and cultural context in which they appeared, and the kinds of problems that they were developed to solve.

It is important to celebrate the diversity that history can show us, and to recognise that in particular times and places, conditions supported the growth of certain groups of scholars and mathematical ideas which made significant contributions to the establishment of our current body of knowledge. It is also necessary to understand that mathematical rigour is relative to the time and place in which the arguments were first conceived, and that denigrating individuals or groups for a lack of rigour shows a clear misunderstanding of the meaning and purpose involved. Different types of thinkers transcend mere cultural lines; even among people of similar backgrounds, there are wide variations in the ways in which people think. Understanding these differences and finding the many examples in the history of mathematics which illustrate these points can also allow the different thinking and learning styles of students to become recognised. Within one culture there can often be found those who empathise with the mathematics of another culture, or with the ways that others have of expressing mathematical ideas. The Alcazar in Seville was built by an Islamic engineer but commissioned by a Christian ruler who preferred the mathematical beauty of the designs of his Moorish predecessors just as much as he recognised the greater efficiency of the Islamic architectural designs for controlling the climatic influences of temperature and humidity.

A multicultural approach does not seek to lead people to the belief that ultimately every culture or group of people have thought of everything that really matters in mathematics. Nor does it seek to persuade that everyone will benefit from, and be able to use, all mathematical approaches. Studying and understanding the methods that other groups of people have developed in response to their needs may well help students to identify the particular characteristics of the method being taught to them, and thus better understand a particular concept, but it does not have the long-term effect of putting everyone in an equal position. It does, however, open up the possibilities of comparisons and the recognition of diversity. It enables us to see that an exchange of ideas can be made from the security of a mutual concern to explore mathematical concepts and to experience the advantages or the beauty that the application of the concept provides in the environment of the people who use it. Throughout the history of mathematics it is not always possible to decide which particular branch or emphasis will persist. There is no clear way to predict what will happen, or to nominate which movement will retain ascendancy, or to judge which skill will be rendered useless as it is replaced by a new discovery.

Multiculturalism then, in the sense that we have tried to convey here, is the identification and celebration of diversity, the respecting and valuing of the work of others, the recognition of different contexts, needs and purposes, and the realisation that each society makes and has made important contributions to the body of knowledge that we call mathematics. Given this view, the inclusion of a multicultural dimension in our teaching of mathematics makes a significant contribution to humanist and democratic traditions in education.

2.4   Interdisciplinary issues

2.4.1   Introduction

Suppose that you attend a lecture in which a well-known mathematician presents the results of their recent research activity. You might be surprised to see that the lecture room is not crowded, that the size of the audience does not reflect the famous name of the scientist, but consists of a small number of research mathematicians and a few graduate students. During the lecture you come to realise that only a handful of colleagues working in the same specialised field are able to understand the proof of the theorem, despite the lecturer's efforts and unquestionable capability.

People connected with this issue are aware that, quite often, editors of mathematics research journals have difficulty in finding knowledgeable reviewers to assess newly submitted papers. The huge advances in the discipline of mathematics and the high degree of sophistication in all of its branches have resulted in narrow specialisations. No mathematician would nowadays be able or even dare to try to learn all mathematics, to be a kind of universal mathematician.

This is one of the characteristics of modern mathematics which has generated a widespread idea that mathematics is a highly difficult and demanding subject. Its abstract nature and specialised symbolism makes it unattainable for most ordinary people. Another popular view of mathematics, by contrast, is that of a utilitarian subject seen only in the context of its applications. This ambiguous inheritance of mathematics, seen as both a mystical, abstract, difficult subject and as a tool for other disciplines, has contributed to the development of negative attitudes towards mathematics. These beliefs are unproductive in the teaching and learning process, adversely influencing the attitudes of students.

The history of mathematics, however, informs us that this kind of specialisation is a recent development and that the situation was quite different in the past. Going back in history, we can easily see that not only were the various branches of mathematics unified and interrelated, but that mathematics, particularly elementary mathematics, was constructed by humans in an effort to answer real life problems. More than that, those problems were not only mathematical but also indistinguishable from other disciplines to such an extent that it is often not clear whether practical or theoretical problems motivated the development of one or the other. Part of this story is by no means unknown to the teacher. There are many examples where teachers refer to applications of mathematics in physics and other school subjects. Yet these references are often made incidentally and in a passive rather than in a systematic and organised sense, whereas the student may be interested in having first hand experience of those mathematical concepts and methods which were motivated and developed hand in hand with other disciplines.

Specialisation in education too is a modern phenomenon, which results in viewing present day school mathematics as completely separate from other subjects of the curriculum, and school administration and time-tabling of classes also often work against efforts to make links between subjects. Typically, history classes deal with politics and economics but only mention technology in passing. Little, if any, attention is paid to the contributions that mathematics at all levels has made to general economic development, while the political motivations for developments in mathematics go unnoticed. Clearly every subject has its own history, but all their histories are linked within the contexts in which they originated and were used and studied.

The history of mathematics can act not only as the factor linking mathematical topics, to the fuller understanding of both, but also between mathematics and other disciplines and as part of history itself. We consider here (1) how the history of mathematics links with the study of history; (2) how it links topics within mathematics; and (3) how it links mathematics with other disciplines.

2.4.2   History of mathematics and the study of history

With regard to the place of history in the curriculum, one might argue that mathematics plays no role within history. However, there are many skills and processes used in studying history, which are also useful in the study of mathematics. These skills can give those who study history new insights into their own learning of mathematics. For example, since history uses logic, reason and various forms of evidence to justify interpretations of past events, these ideas can be seen as analogous to the processes of seeking justification for mathematical statements. The teacher alert to these parallels is in a stronger position to make links and encourage student discussion and understanding.

Conversely, mathematical thinking may support students in several ways when studying history. The investigation of primary evidence and the process of deciding which are the key results, factors or connections in historical events; the identification of causes and effects from perceived patterns, and making conjectures from evidence, are all activities which can be enriched by the skills learnt in mathematical problem solving. Researching secondary evidence on mathematical topics in books, encyclopaedias and articles and using original texts, instruments and materials to replicate the process of doing mathematics from another time or place can help to deepen one's understanding of historical periods and the ways in which people of the time tackled everyday problems.

These connections suggest that while the contexts and intentions of the two subjects may be quite different, there are not only cognate skills that are being used in the practice of both history and mathematics, but that making them explicit may help learners to recognise and develop these transferable processes.

2.4.3   History of mathematics linking topics within mathematics

Until relatively recently, different strands of mathematics have been developed by mathematicians who were acquainted with most, if not all, areas of mathematics of their time. Although this is now probably impossible, mathematicians make increasing links within the subject to try to prevent fragmentation. Davis and Hersh (1982) show a list of present and past mathematical topics to illustrate the rapid expansion of the subject over the past 100 years. The history of mathematics is the ideal context where students can be shown how interdependent the different areas of mathematics are today, and how they have been steadily becoming more and more interrelated over time.

This is true from almost as far back as we have records. The popular conception of Euclid's *Elements,* for example, is that it is a geometry text. However, anyone who spends even a short time studying the different books of the *Elements* comes to realise that this is a synthesis, by mathematicians of one particular culture, of a large part of the mathematics that has gone before, and that it links together a wide range of different mathematical ideas within its formal geometrical context. Understanding and recognising the links between different areas of mathematics can be approached from a relatively elementary background. We only have to mention here Euclid, Al-Khwarizmi and Descartes to indicate how an inventive teacher with historical resources can demonstrate how arithmetic, algebra and geometry are related in the work of these mathematicians and how the relational ideas deepened and developed over time.

During the nineteenth century, we see more examples of the synthesis and consolidation (as Wilder calls it) of old with new mathematics. The concentration on the processes of the different branches of mathematics, and the consequent development of and relationships between these processes (as general properties of

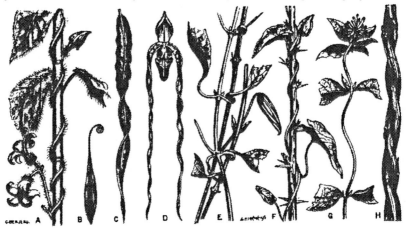

Figure 2.1: Symmetry and spiral formations in plants, from J. Bell Pettigrew, Design in nature, *1908.*

areas of mathematics), produced hierarchies of abstract structures which became successively more inclusive of the mathematics that had gone before. Unexpected alliances have emerged between different branches of mathematics. Algorithm theory which began in the first part of the twentieth century has its roots in logic which was one of the most profound studies of the ancient Greeks (Chabert 1999; Schreiber 1994). A later development, model theory, is responsible for the return of infinitesimals, in the non-standard analysis of Abraham Robinson.

An impressive example of links between different parts of mathematics is the notion of symmetry, which had its first appearance as a geometrical notion (Weyl

1952). In the late eighteenth and early nineteenth century it began to be recognised that algebraic permutations had the same structure as some geometric symmetries and later, as the theory of algebraic group structure developed, this was applied to classify different types of geometric transformations. In this way, geometry and algebra became inextricably linked through the recognition of common properties which emphasised higher order procedures and operations. It is these procedures and operational generalisations which we focus on when we are discussing the deeper nature of mathematics, and it is precisely these structural aspects that we try to help our students develop when we are teaching, even at elementary levels (see Ch. 5). The recent proof of Fermat's Last Theorem by Andrew Wiles shows how a problem unsolved for many years, tantalising mathematicians thereby, has over its history considerably enriched the body of mathematics known as number theory. The solution of this old problem was achieved using one of the most recent notions in mathematics, modular elliptic curves. This is just one example of how the conjunction of the old and the new is a commonly occurring event in the development of mathematics today.

2.4.4 History of mathematics linking mathematics with other disciplines

(i) The physical and biological sciences

The link between mathematics and the teaching of physics has a long and well-established tradition, and history offers many examples of problems and alternative solutions. Thinking about a problem from an historical context can make the learning of physics and mathematics more meaningful. A good source of examples covering many areas of the science curriculum can be found in journals like *Science and Education* where discussions range over the epistemological bases for scientific beliefs, the nature of evidence, the processes of scientific method, and the sense in which pupils' concepts may or may not be like those of earlier scientists. The naming of a concept such as *force* identifies a general phenomenon, while the formation of equations describing the relationships between concepts is a way of modelling them in measurable terms. However, naming and establishing relationships between concepts are theoretical activities, and through these theoretical constructions we may be led to believe in the reality of the objects we ourselves have created. In discussions such as these, it can be seen that not only may mathematics be used and developed as a tool to solve problems, but also that the epistemological bases of the concepts involved call into question the ways in which these ideas are symbolised in mathematics itself. Some examples of such situations follow.

Floating, sinking, Archimedes' principle and relative density

The familiar story of how Archimedes investigated whether the King's jeweller had cheated him, by using a proportion of silver in a crown supposed to be made of solid gold, can be used with quite young children.

A familiar sight in some primary classrooms is that of children experimenting to determine which objects sink and which float and asking why they do this. By using a bowl of water with graduations on the outside it is possible to make a reasonable estimate of an object's volume. Many things can be learnt in this way about the basic techniques of measurement, and about relative densities, perhaps unconsciously re-enacting experiences similar to those of our ancestors. We know now that the density of silver is less than that of gold and might refer to the anecdote where Archimedes "ran naked through the street shouting *eureka, eureka*" (Heath 1921 ii, 19). The reason for his enthusiasm was the discovery of the principle of displacement, but the thrill of making a discovery is a very emotional event and an opportunity can be taken by teachers to help pupils share this kind of excitement. Hitchcock (1996) has provided many useful ideas for the dramatisation of scientific events, and plays such as Brecht's *Galileo* (1952) or Whitemore's *Breaking the code* (1987) can be used to explore the emotions and the scientific and political contexts of discovery (see §7.4.10 and §10.2.1 for fuller discussion of plays in the classroom).

Dynamics, velocity, acceleration and energy

Roll a marble down each of two inclined planes. The planes have the same height but different slopes. Which marble has the greater velocity when it reaches the bottom of its slope?

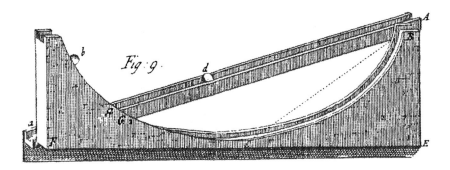

Figure 2.2: 18th century model illustrating Johann Bernoulli's 'brachystochrone problem': the marble b that falls along the cycloid passes G before the marble d that falls along a straight line (from Desaguliers, Experimental philosophy, *1734)..*

Frequently students limit themselves by only searching for a formula to solve the problem, which is a straightforward task in itself but restricts the students' thinking about the physical possibilities. By drawing the velocity and acceleration vectors, the forces acting on the marbles can be described, and by graphing the velocities of the marbles, the teacher can introduce a practical example of one of the problems which motivated the development of the calculus. Although the ideas used in the concept of energy conservation are difficult, they were already implicit in some, partially metaphysical, considerations of Galileo, developed later through Huygens' experiences of collisions, generalised by Leibniz, and further generalised by Johann Bernoulli, finally reaching Joule's formulation half way through the nineteenth century. Students' own learning might follow a similar path helping them to gain greater insight into both physics and mathematics.

Invariance, non-Euclidean geometry and relativity

A teacher alert to the history of mathematical and scientific ideas has a rich set of resources for illuminating pupils' studies. For students beginning to study twentieth century physics, the teacher can outline the contribution to these developments of some interesting and accessible mathematics. The beginnings of the concept of invariance, so important in the physics of the twentieth century, can be found in the projective geometry of Desargues and Pascal, with the ideas of projection and section. The techniques of projection and section are intuitively appealing and were developed by Chasles, and later by Poncelet and other nineteenth century geometers into a method for proving theorems within the new geometry. In a different context we find Lagrange developing ideas of invariance from purely arithmetical and algebraic problems. By the late nineteenth century, Cayley, J.J. Sylvester and Gordan had developed the invariant theory of algebraic curves to a high degree of complexity. Also other non-Euclidean geometries had been developed and the concept of axiomatic systems was beginning to emerge. Hilbert's consolidation of these different geometries by their generalisation as groups with certain invariant properties under specific transformations led to startling new ways of conceptualising problems. Hilbert's systematic study of theoretical physics, in close collaboration with Minkowski, led to Minkowski's early work on relativity theory. Without the tools of Riemannian Geometry and the theory of invariance Einstein's general theory of relativity and gravitation could not have been stated.

(ii) Geography and economics

Eratosthenes measured the radius of the earth, based on the knowledge that Syene, a town at a distance of 20 000 stadia to the south of Alexandria was on the same meridian. At noon on the summer solstice, a vertical gnomon cast no shadow in Syene while at the same time in Alexandria an upright gnomon ('pole') cast a shadow corresponding to an angle of one fiftieth of the circle (Heath 1921 ii, 106). Using the telephone or the Internet, schools in two different cities of known coordinates could liaise. Their students could try Eratosthenes' method to find the circumference of the Earth by comparing the angle of the sun in each place at the same time (Ogborn, Koulaides & Papadopetrakis 1996).

This example shows how a mathematics teacher can collaborate with colleagues in the geography department. The next examples indicate themes that would be suitable for collaboration with the economics teacher.

In 1485 the *Treviso arithmetic* was published as a manual demonstrating the power of the new Hindu-Arabic notation for arithmetic, and the ways in which this made many calculations easier (Swetz 1987). This was largely motivated by the expansion of commerce and banking in the nearby Italian cities and the growing trade with Northern Europe. In the next decade, in 1494, Pacioli devoted three chapters of his *Summa de arithmetica* to trade, bookkeeping, money, and problems of exchange. Finding ways of handling money efficiently and accurately has led to a number of technical developments, each embodying innovative mathematical concepts.

In our own century electronic computers have become indispensable in business; the development of these machines engaged some of the most brilliant minds in mathematics and physics. Conversely, we can point to a variety of operations and notions which come into mathematics directly from the experience of money or are reinforced through these means. Notions of expectation and risk, which originated in gambling, later became essential in life insurance, as part of the science of statistics. Gambling also led to the theory of probability and this now finds applications in the most important areas of the theoretical sciences. Derived from these classical theories are the modern theories of mathematical economics.

Another aspect of the links between geography, economics and mathematics are the voyages of discovery. The motivations range from curiosity to the expansion of empires but the essential needs are the same: accurate maps and ways of finding

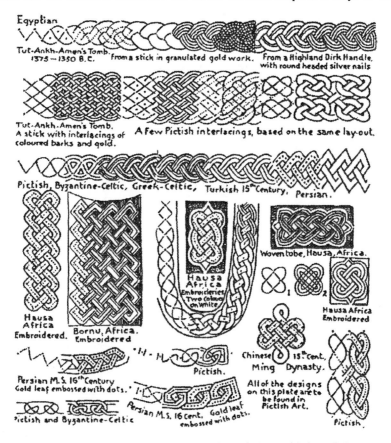

Figure 2.3: Interlacings through time and through the world, from G. Bain, Celtic art, 1945

one's exact position on an empty ocean. Looking at popular accounts of adventurers does not bring these problems immediately to mind, but the history of navigation is full of dramatic stories and the slow but sure development of instrumentation and calculational devices which enabled sailors to find their way across the sea, by day or by night. Here, astronomy and mathematics are the key to success.

(iii) Mathematics, art and music

We are often tempted to look for mathematics in artistic creations. A common example is that of the Islamic tile patterns which are so often taken out of their original context and treated just as examples of plane symmetry groups. While we can engage in this on one level as an example of applied abstract algebra, do we take time to wonder at the significance of these patterns in the contexts in which they were created? If we mathematise artistic creations in this way, rather like regarding Renaissance painting merely as examples of the development of perspective, we may stand in danger of decontextualising and dehumanising them. On the other hand an understanding of Renaissance painting or Islamic patterns which includes

 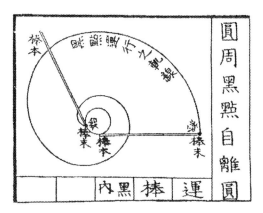

Figure 2.4: Links between mathematics, biology and history are found in the investigation of spirals, which goes back to at least the time of Archimedes; here in the shell of a nautilus and in a Japanese image of 1866

recognition of the creators' mathematical skills is all the richer thereby.

Finding and identifying pattern is generally recognised as a mathematical activity, but the sources and examples of pattern in nature and in artefacts are, in some sense, works of art in themselves. So, Islamic patterns, friezes and wallpaper patterns world-wide; wrought iron work in gates and fencing in Mozambique, Europe and the Americas; weaving patterns in fabrics, baskets and carpets across the continents should be regarded as cultural products in their own context, but can also be seen as examples of ethnomathematical activity. As people have developed their own mathematics, so have they also developed artistic traditions and artefacts. By helping students and others to see and participate in artistic projects, which overtly use mathematics from some time and place, their understanding of the underlying mathematical concepts and skills can be heightened.

A sense of rhythm is one of man's universal and basic perceptions. Building on this, the human voice or the sound of a musical instrument and the variety of musical creation is infinite. Each culture has its own aesthetic values and so different styles of music arise. However, underlying all of these are rhythm and structure, the basic elements which can be found in music lessons from primary school to the conservatoire. The theory of music is full of applied mathematics, which can be found over most of musical history, from the description of the musical intervals in the different scales to the compositions and writings of Karl Orff and Herbert von Karajan.

(iv) Ritual, religion and philosophy

Mathematics in history is inextricably linked with both emerging and developed philosophies and religions. There is considerable evidence for the early development of counting and geometry in ancient rituals, traces of which are found in Egypt, Babylon, India, China, and Greece. The circle and square were sacred figures with special significance and their properties were studied by the priests. The observation that the square on the diagonal of a right triangle was equal to the sum of the squares on the other two sides was widely known, and the dissection and rearrangement of areas found both ritual and practical applications in the building of altars and the construction of temples. Also, in exploring the properties of numbers we observe how religious practice may be affected by mathematics in number symbolism and mysticism.

Mathematics is also a science of the infinite. Hermann Weyl speculated that the presence of the infinite in mathematics runs parallel to religious intuition. Be that as it may, metaphysical speculations are present in mathematics as in many other human activities, and it is just these metaphysical notions that create the problems with the understanding of many modern day mathematical concepts. At a somewhat deeper level of cultural influence we can see how notions of mathematical proof have influenced theology. Such was the search for certainty that Spinoza, in the seventeenth century, employed the Euclidean formulation of elementary notions, axioms and theorems in his proofs of the existence of God.

In a situation where both mathematics and philosophy are part of the curriculum, there are opportunities to enrich both disciplines with historical examples. Following step by step the birth of western rationality, by integrating the knowledge that students possess from their philosophy course with that acquired during their mathematics lessons, can enable them to achieve a deeper level of knowledge about the cultural role of mathematics. Until the eighteenth century, no clear distinction existed between philosophical and mathematical thinking. 'Natural philosophy' was the name given to scientific activities until relatively late, so the aspects of western rationality that are revealed during mathematical activities can be regarded as the counterpart of those explored during the philosophy class.

2.5   Conclusion

This chapter takes into account the metacognitive level, the level at which didactics and history of mathematics meet each other.   An exploration of educational theory which goes over curricular planning, and a history which goes over a story of events can try to understand the how and the why of what is happening, in history and in the classroom.   This metacognitive request pushes the didactician firstly to investigate the nature of mathematics, and then the surrounding philosophical, multicultural and interdisciplinary issues.   The analysis we have done leads us to see even more clearly that school mathematics has to reflect, in a way it has not always done hitherto, some aspects of mathematics as a cultural activity.

–   from the philosophical point of view: mathematics must be seen as a human activity, with its cultural and creative aspects.

–   from the interdisciplinary point of view: when mathematics is linked with other subjects, the connections must be seen not only in one direction.   Students will find their understanding both of mathematics and their other subjects enriched, through the historical liaison, sympathies and mutual aid between the subjects.

–   from the cultural point of view: mathematical evolution comes from a sum of several contributions.   Mathematics can be seen as having a double aspect: an activity both done within individual cultures and also standing outside any particular culture.

References

Atiyah, Michael. *et al.* 1994.   Responses to A. Jaffe and F. Quinn, 'Theoretical mathematics: toward a cultural synthesis of mathematics and theoretical physics', *Bull. Amer. Math. Soc.* **30**, 178-207

Begg, A., *et al*, 1996.   'Mathematics and culture in Oceania', Paper presented to the Mathematics and Culture WG at ICME-8, Spain, July 1996.

Bishop, Alan J. 1995 .'Mathematics education between technology and ethnomathematics: should it be common? Does it make sense?', *Proceedings of CIEAEM* **45**, 53-62.

Brecht, Bertold 1952.   *Galileo*, London: Indiana University Press

Cajori, Florian 1980.   *A history of mathematics* (3rd edition), New York: Chelsea

Chabert, Jean-Luc *et al.* 1999. *A history of algorithms: from the pebble to the microchip,* Berlin: Springer

Davis, Philip J., Hersh, R. 1982. *The mathematical experience*, Boston: Birkhäuser.

Gillies, Donald 1992.   *Revolutions in mathematics*, Oxford: Clarendon Press

Grugnetti, L., Speranza, F. 1999.   'General reflections on the problem of history and didactics of mathematics', *Philosophy of mathematics education newsletter* **11**.

Joseph, George G. 1991. *The crest of the peacock*, London: I.B. Tauris

Heath, Thomas L. 1921.   *A history of Greek mathematics*, Oxford: Clarendon Press

Heyting, A. 1956. *Intuitionism: an introduction,* Amsterdam: North-Holland

Hitchcock, Gavin 1996. 'Dramatizing the birth and adventures of mathematical concepts: two dialogues', in R. Calinger (ed.), *Vita mathematica: historical research and integration with teaching*, Washington: Mathematical Association of America, 27-41

Hofstadter, Douglas 1979. *Gödel, Escher, Bach: an eternal golden braid,* New York: Basic Books

Høyrup, Jens 1996.   'Changing trends in the historiography of Mesopotamian mathematics: an insider's view', *History of science* **34**, 1-32

Karajan, H. von 1985. *Musik and Mathematik,* Salzburger Musikgesprach 1984 unter Vorsitz von Herbert von Karajan, Springer-Verlag

Kitcher, Philip 1983. *The nature of mathematical knowledge,* Oxford: University Press

Kuhn, Thomas 1962. *The structure of scientific revolutions,* Chicago: University Press

Lakatos, I. 1976. *Proofs and refutations: the logic of mathematical discovery,* Cambridge: University Press

NCTM 1969. *Historical topics for the mathematics classroom,* Reston, Va: National Council of Teachers of Mathematics (31st NCTM Yearbook, reprinted 1989)

Ogborn, J., Koulaides, V., Papadopetrakis, E. 1996. 'We measured the Earth by telephone', *SSR Science Notes,* 87-90.

Schreiber, Peter 1994. 'Algorithms and algorithmic thinking through the ages' in: I. Grattan-Guinness (ed.), *Companion encyclopedia of the history and philosophy of the mathematical sciences,* London: Routledge, 687-693

Seidenberg, A. 1962. 'The ritual origin of counting', *Archive for history of exact sciences* 2, 1-40

Sobel, D. 1995. *Longitude,* Penguin Books.

Speranza, F., Grugnetti, L. 1996. 'History and epistemology of mathematics', *Italian research in mathematics education 1988-1995* (Malara, Menghini, Reggiani eds.), Roma, 126-135.

Swetz, Frank 1987. *Capitalism and arithmetic: the new math of the 15th century,* La Salle: Open Court

Thurston, W. P. 1994. 'On proof and progress in mathematics', *Bull. Amer. Math. Soc.* 30, 161-167.

Waerden, B.L. van der 1961. *Science awakening I,* Oxford: University Press

Weil, A. 1978. 'History of mathematics: why and how', *Proc. International Congress of Mathematicians,* Helsinki 1, 227-236.

Weyl, Hermann 1949. *Philosophy of mathematics and rational science,* Princeton: University Press

Weyl, Hermann 1952. *Symmetry,* Princeton: University Press

Whitemore, Hugh 1987. *Breaking the code,* Oxford: Amber Lane Press

Wilder, Raymond L. 1950. 'The cultural basis of mathematics', in *Proc. International Congress of Mathematicians,* Cambridge, Ma USA.

Wilder, Raymond L. 1968. *Evolution of mathematical concepts: an elementary study,* New York, Wiley

Wittgenstein, Ludwig 1956. *Remarks on the foundations of mathematics,* Oxford: Blackwell

Chapter 3

Integrating history: research perspectives

Evelyne Barbin

with Giorgio T. Bagni, Lucia Grugnetti, Manfred Kronfellner, Ewa Lakoma, Marta Menghini

Abstract: *The question of judging the effectiveness of integrating historical resources into mathematics teaching may not be susceptible to the research techniques of the quantitative experimental scientist. It is better handled through qualitative research paradigms such as those developed by anthropologists.*

3.1 Introduction

Over the past twenty years or so there has been a growing interest in history by teachers and educators. What consequences may this interest have for mathematics education? And how can we judge its effectiveness? A great many articles have appeared in increasing number over this time, including educational reports, reflections of teachers and accounts of teaching experiences. This material gives different arguments in favour of including a historical dimension in the teaching of mathematics, and often contains reasons for why the teacher believed it to be effective. We also can identify through this material different ways in which it is effective, depending for example upon whether the presence of history is implicit or explicit in the teaching situation; and whether the use of history is local, being used for a particular topic, or global—that is, characterising the didactic strategy or the way the mathematics is taught.

The two most commonly presented reasons for the inclusion of a historical dimension are that history of mathematics provides an opportunity for developing our view of what mathematics is; and that it allows us to have a better understanding of concepts and theories. In each of these there is a sequence of developing understandings: the history of mathematics can first change the teacher's own perception and understanding about mathematics, then it will influence the way mathematics is taught, and finally it affects the way the student perceives and

John Fauvel, Jan van Maanen (eds.), *History in mathematics education: the ICMI study*, Dordrecht: Kluwer 2000, pp. 63-90

understands mathematics.   We can evaluate the effectiveness of introducing an historical dimension into the teaching of mathematics through an examination of this process.

The breadth of the arguments is such that we cannot approach the question of using history of mathematics in a quantitative or piece-meal fashion.   We offer examples later of case-study evaluations that use a holistic and qualitative approach. Our approach should not be seen as prescriptive: we do not propose models or programmes.   On the contrary, a view of the whole of the process suggests we should be cautious; there are limits and risks attached to an approach that takes too simplistic a view of the significance of history in mathematics education.

The change which this may bring about in the image of mathematics held by the teacher can be presented as a contrast between a formal presentation of mathematics and a heuristic approach provided by history.   This difference corresponds to a contrast in pedagogic style: that of the traditional teacher, where knowledge is handed out by the teacher, and a learning process based on mathematical activity by the student.   The heuristic view is associated with a constructivist view of mathematics in which knowledge is constructed step by step and concepts are clarified through solving new problems.   History here is not only a revelation but also a source of reflection for the teacher, as is shown in the examples given in sections 3.3 and 3.4.

The historical dimension encourages us to think of mathematics as a continuous process of reflection and improvement over time, rather than as a defined structure composed of irrefutable and unchangeable truths.   The latter view is one that may be held by the teacher fresh from college or university and without experience of research.   Thinking about mathematics as an intellectual activity, rather than as a finished product, means thinking of problems to be solved, of the importance of conjectures and the value of intuition.   In this sense, the pupil in mathematics and the mathematical researcher are engaged in the same activity.   The historical dimension here can bring about a global change in a teacher's approach, whether or not the historical element is explicitly present in the classroom.   Historical knowledge helps the teacher to understand stages in learning as well as to propose problems inspired by history.   It is interesting to note that teachers in some countries are tempted to contrast the image of mathematics which history presents with that given by the 'modern mathematics' reforms which were popular in the 1960s.   Under modern mathematics reforms the teaching of mathematics began with the most recent formulation of concepts of mathematics, which is the exact opposite of the historically-informed approach.

Historical awareness also leads teachers to change the way they think about their students.   As shown in sections 3.5 and 3.6, the responses students make to an historical problem take on a new character when they are compared with the responses made by mathematicians through the ages.   Historical and epistemological analysis helps the teacher to understand why a certain concept is difficult for the student and can help also in the teaching strategy and development.   This has two particular consequences for how the teacher can use the historical dimension effectively.   First, the teacher can adopt a constructive attitude towards the errors the students make.   Secondly, the teacher can focus on producing a variety of responses to a given problem, relate them to what the students know or to the connections within their present knowledge. The historical dimension leads to the idea that mathematics is no longer a sequence of discrete chapters (in geometry, algebra or analysis), but is an activity of moving between different ways of thinking about mathematical concepts and tools.

When we learn about the historical development of mathematics it affects how we think about the time our students spend in developing mathematical understanding. If it took several centuries for mathematicians to be able to make explicit our current concept of a limit, for example, it is going to take a considerable time for our students as well. There is time needed, also, to deal with the epistemological problems inherent in manipulating the infinite. And then it takes time to move from the idea of the limit as a tool for solving problems to the idea of the limit as part of an integrated body of mathematical knowledge linked to other concepts, such as that of real number or set. We should note, however, that even if students are led to construct their knowledge in a way that parallels the historical development, it does not mean that there will be an exact match between the student's construction and the historical sequence. After all, obstacles encountered by mathematicians in history may not be those that face the student of today. Nonetheless, learning that there were obstacles is in itself beneficial.

If the teacher decides to introduce history explicitly in class, it can be done either as part of a global approach in terms of a didactic strategy or in a local way, in the context only of teaching a particular topic. In addition to the points made above, the teacher may wish to provide a cultural context for mathematical knowledge by locating this knowledge within the history of mankind and ideas. Where explicit use of history is concerned, there are limitations and risks. It is seen in section 3.7 that it can be difficult to understand the procedure used by a mathematician of ancient times if it is not set within the historical context. There is a difficulty here for the teacher to resolve, well before it becomes one for the student (this raises the question of the training of teachers). At least two types of danger can arise when using history explicitly. First, using piece-meal historical illustrations can give a false and truncated view of what mathematics, and indeed history, was really like historically. Alternatively, in trying to present a global historical view, we could be in danger of ending up with an education in mathematics history quite independent of the needs of mathematics education. At worst, one could fear that mathematics might one day be replaced by a teaching of its history.

It is therefore a question of integrating history *within* the teaching of mathematics, and that is why teachers talk of a historical dimension, a historical style, or a historical perspective in mathematics education. These terms describe, in a general way, the teacher's active mobilisation of all his or her historical and epistemological reflections. In evaluating the effectiveness of using history in mathematics classrooms, we have to consider all the aspects of a historical dimension. It is possible to appreciate the effectiveness of using history through an ethnographic approach to examples of practice. Section 3.2 suggests we should proceed by an analysis of case studies, using the observations of participants and interviews with students and teachers, and drawing on existing written accounts, in particular on articles where teachers explain why and how their historical approach to mathematics has changed the way they teach.

3.2   The historical dimension: from teacher to learner

Evelyne Barbin

A good place to start an analysis of the effectiveness of using a historical dimension in the teaching of mathematics is to ask 'does it work?'. First we need to establish what the 'it' is: that is, we need to determine the nature of teacher's objectives when they use history as part of their mathematics teaching. Only after that would we be able to seek an answer to our question. But the question remains of how an answer might be reached. It is tempting to ask for a 'scientific' study of the problem. Unfortunately, there exist no successful studies where the impact of an historical dimension can be measured by using a battery of tests for determining the competences of students, nor comparative experiences between classes where an historical dimension was or was not used. The reason for this is that the attainment of objectives claimed for using history cannot be measured by assessments (Rogers 1993). Objectives such as interest or understanding of a concept cannot be measured in a quantitative way. It is even less appropriate to use quantitative methods for trying to measure the impact on mathematics education when history is used in a global way. Such attempts as have been made to formulate and pursue such studies have failed on methodological grounds.

We shall attempt a response to our initial question in another way, through the use of a *qualitative* analysis of the changes that can occur when history has a place in the teaching of mathematics. In particular we shall look at the way in which a change in the teacher brings about a change in the teaching, which in turn leads to a change in the student. The methodology used here is *ethnographic*, which is often used in educational research (Eisenhart 1988). We shall consider nine articles written by teachers of mathematics and intended for other teachers. These articles are of significant interest, in that they present case studies by teachers of work in their own classrooms, and are spread over a sufficiently long time span for us to be able to assess agreements and differences. They include at the same time some introspection by the teachers on their own conceptions and intentions, and personal observations of the effects of the outcomes. In their choice of aims, we can also read about the role the teachers played in the reported cases and whether, in their view, the observed changes could be generalised to other classes. The nine articles

John Fauvel, Jan van Maanen (eds.), *History in   mathematics education: the ICMI study*, Dordrecht: Kluwer 2000, pp. 66-70

surveyed were all published in France between 1991 and 1998 in the review *Repères IREM*: namely, Bühler 1998, Farey and Métin 1993, Friedelmeyer 1991, Friedelmeyer 1993, LeGoff 1994, M:ATH 1991, Métin 1997, Nouet 1992, and Stoll 1993.

These articles all concern mathematics teaching in the lycée (15-18 years) and are addressed to the same readership. Six of them deal with an explicit use of history, one using problems from the history of mathematics and the others using the reading of historical texts. The other three articles deal with an implicit use of history, one of which is local (a problem inspired by history) and the other two global. In overview, three of the authors use history primarily so as to bring about a change in the way mathematics is viewed, for five of them the aim is to improve the learning of mathematics, and one of them uses history as a way of aiding the mental construction of mathematical concepts.

In the articles, taken as a whole, we can identify five results of using history. It can bring about a change in

− the teacher's mathematical conceptions
− the student's mathematical conceptions
− the role of the teacher
− the way students view mathematics
− the students' learning and understanding

These five types of effect are not all discussed in each of the articles and we need to consider the different ways the authors articulate them.

In three of the articles, the authors write about how the study of historical texts has changed their own mathematical conceptions and of the changes they perceive in their students. The group M:ATH writes that

The confrontation with mathematical texts changes the view of mathematics for both teacher and student. Mathematics becomes alive, it is no longer a rigid object. It is the object of enquiry, controversy, contains mistakes and uses methods of trial and error.

They add: "reading old texts excites the curiosity of the students and encourages them to question" (M:ATH 1991). For Monique Nouet, "the prime objective attached to the history of mathematics concerns one's view of the discipline: it is possible to show that mathematics is a science on the move", and one of her final class students (17 years) wrote that "mathematics has for me passed from the status of a dead science to that of a living science, with an historical development and practical applications" (Nouet 1992). Jean-Marie Farey and Frédéric Métin wanted to share a surprise with their students (Farey and Métin 1993):

The image that we give of our specialism through teaching is too often that of a frozen world, merciless and hardly human [...]. Some people turn to the history of mathematics: astonishment and wonder! [...]. We are no longer dealing with a finished product but with something in continuous evolution; it is no longer a case of accepting a discipline of divine nature, but of understanding tools, methods and concepts.

Their article describes the different reactions of 15 year old (2nd class) students when presented with a text by Ben Ezra (12th century) which explains the method of double false position.

Some of the authors describe the change in their attitude towards teaching that comes from a new understanding of the nature of mathematics. André Stoll finds the formal definition of the integral given to students of the final class (17 years) quite unacceptable given that "two thousand years were required to bring the infinitesimal calculus to fruition", and he suggests a sequence of problems inspired by the works of Archimedes, Ibn-Qurra and Fermat. The resultant effect on his teaching was to set up a link between integral calculus and differential calculus and to introduce a definition of integral in a 'natural' way (Stoll 1993). Jean-Pierre Friedelmeyer explains that it is not a question of expecting our students to follow the same evolutionary process that took place historically, but that an understanding of history helps the teacher "better to understand certain difficulties that the student has and to construct a shortened path whereby the difficulties are confronted with a full awareness of the causes of those difficulties" (Friedelmeyer 1991). This is the approach adopted in his later article where he explores the root cause of the difficulties of the current teaching of analysis which, he claims, lie with the concept of numerical continuity, "since the student's intuition is based on a long-standing idea of geometric continuity". A historical perspective provides the opportunity of entering into "times when understanding was closer to the intuition held by our students, and this aids us in managing the stages by which the concepts and fundamental tools of analysis are constructed, and to set the notions of meaning and rigour in context" (Friedelmeyer 1993).

Their new perception of mathematics also radically alters the view the teachers have of their students' learning processes. Nouet considers that the most important aspect of the history of mathematics for her students is to raise the question of the time needed to deal with a topic: she allows time for her students to construct their ideas slowly and to identify moments of misunderstanding. During the course of the school year adjustments take place, and the way the students express their ideas improves as the teacher delves more deeply into the topic. In this way the students are reassured and some regain confidence in themselves (Nouet 1992). History encourages the teacher to see the student as a thinking and inquiring being, and to take a fresh attitude towards the work the student produces. Frédéric Métin's purpose in presenting his 2nd class (15 years) students with a text by Legendre on the approximation of π was "to encourage them to talk about the way they thought about numbers and approximations". In addition to questions about the mathematics, he asked them to comment on what they found awkward in the notation, why Legendre wrote 'equals', and what they thought about it (Métin 1997). Martine Bühler set her students the famous problem of sharing out winnings when a game of chance is interrupted *(le problème des partis)* and she analyses the seven methods invented by her students. Her own knowledge of the solutions proposed by Pascal, Fermat and Huygens helped her in encouraging the students to follow through their different ideas (Bühler 1998).

Most of the nine articles deal with examples where history is explicitly introduced into a mathematics lesson and comment on its effects on learning. None of the authors attempted to quantify their results. In order to assess whether the use of history has an effect on the learner, they did not feel it appropriate to carry out a standard evaluation of the students' abilities in solving some exercise or other. Judgements about the effects of the historically-based teaching strategy rests on other grounds. It is important to make the point that when students become better at understanding, it has a significant effect on their learning (Kieran 1994). The conviction that the use of history improves the learning of mathematics rests on two assumptions about the process of learning: the more a student is interested in mathematics, the more work will be done; and, the more work that is done the greater will be the resulting learning and understanding. It may be added that the interest provoked by the use of history goes beyond its being just a motivating factor. The work which the students are asked to do involves real mathematical activity and the learning does not consist solely in diligently working through exercises (Barbin 1997).

Historical knowledge enriches the mathematical culture of the teachers. This has important consequences for the way it is taught, and also how the role of the learner is perceived. In proposing that their pupils read a problem from an historical source, Farey and Métin (1993) adopt a new attitude:

The teacher does not adopt the position of the person who animates the classroom, but voluntarily steps back [...]. We are not wanting the students to find a method of solution, nor to carry out a simple application of a method; the students therefore react in ways that are different to how they usually behave.

Jean-Pierre LeGoff writes that a teacher is also a researcher, and above all an intellectual, who can find through the history of mathematics the pleasure of teaching. The pleasure which the history of mathematics offers the teacher can also benefit the student through the wealth of knowledge the teacher gains, something that can be summarised by analogy with painting and the neat comment "the more colours an artist uses, the richer will be his touch" (LeGoff 1994).

In illustrating and defending the use of an historical dimension in teaching mathematics, many of the authors point out to the sort of historical training their teaching rests on. Two of the articles make a different point too, indicating limits to the use of history in teaching mathematics, and drawing attention to potential risks. Métin concludes his article on a pessimistic note with respect to the reading of historical texts; it seems to have most advantage for the better students, and he indicates that he is turning now towards a more global use of history through introducing cultural and historical aspects into all of his teaching (Métin 1997). LeGoff distrusts any historical dimension which would be dictated by an official curriculum, he fears that such a move would only propose a 'historical veneer', or a teaching of history which would create a screen in front of the mathematics, or perhaps an historical introduction to texts which would present them retrospectively as superceded by later knowledge (LeGoff 1994).

LeGoff's reflections raise the question of whether an historical dimension should be incorporated into the official mathematics curriculum, which is dealt with more fully elsewhere in this book. Up to the present, the French curriculum mentions history in connection with mathematics teaching only as a possibility: the teacher is completely free to use history or not. If some form of general recommendation were to be imposed, we face two difficulties. First, it would be hard to implement, unless teachers were to have a specific and solid training. Second, a too specific curriculum could have the perverse effect of making the historical aspect rigid, or it might separate the history from the mathematics. These difficulties would multiply with developmental changes to course models or to teaching methods imposed on teachers under the pretext of efficacy or because of a new view of the science of teaching. These points should be borne in mind when considering the political context of educational reforms. To return to the overall point of this section, however, the articles examined above show considerable agreement between the different authors and we have identified qualitative similarities in the changes in the attitudes of teachers and students. This suggests that the experiences described are not exceptional, but can be generalised to all mathematics education, provided we take account of the particular nature of the specific examples we have mentioned.

References for §3.2

Barbin, E. 1997. 'Sur les relations entre épistémologie, histoire et didactique des mathématiques', *Repères-IREM*, n°27, avril, 63-80

Bühler, M. 1998. 'Un problème de dés en terminale', *Repères IREM*, n°32, 111-125

Eisenhart, M. 1988. 'The ethnographic research tradition and mathematics education research', *Journal for research in mathematics education*, **19** (2), 99-114

Farey, J.-M., Métin, F. 1993. 'Comme un fruit bien défendu', *Repères IREM*, n°13, 35-45

Friedelmeyer, J.-P. 1991. 'L'indispensable histoire des mathématiques', *Repères IREM*, n°5, 23-34

Friedelmeyer, J.-P. 1993. 'Eclairages historiques pour l'enseignement de l'analyse', *Repères IREM*, n°13, 111-129

Kieran, C. 1994. 'Doing and seeing things differently: a 25-year retrospective of mathematics education research on learning', *Journal for research in mathematics education* **25**, 583-607

LeGoff, J.-P. 1994. 'Le troisième degré en second cycle: le fil d'Euler', *Repères IREM*, n°17

M:ATH 1991. 'Mathématiques : approche par des textes historiques', *Repères IREM*, n°3, 43-51

Métin, F. 1997. 'Legendre approxime π en classe de seconde?', *Repères IREM*, n°29, 15-26

Nouet, M. 1992. 'Histoire des mathématiques en classe de terminale', *Repères IREM*, n°9, 15-33

Rogers, L. 1993. 'The assessment of mathematics: society, institutions, teachers and students', Didactics of mathematics, *Erasmus ICP-92-G-2011/11*, 603-613

Stoll, A. 1993. 'Comment l'histoire des mathématiques peut nous dévoiler une approche possible de l'intégrale', *Repères IREM*, n°11, 47-62.

3.3   The indirect genetic approach to calculus

Manfred Kronfellner

The 'New Math' of the 1960s and 1970s aimed to introduce in school from the very beginning a university level of rigour.   The obvious difficulties experienced by pupils in the process of learning and understanding of these concepts led educators in the 1970s to develop alternative approaches, such as 'simplified analysis', while still maintaining the demand for exactness.   But in practical mathematics education these revisionist proposals did not succeed.   In reaction to the problems posed by New Math, in the 1970s the genetic method was reinvented (or rediscovered). Roland Fischer, for example, proposed the idea of a heuristic approach with 'subsequent exactification' (Fischèr 1978).   In contrary to the New Math ideology he argued for teaching the essential concepts at a heuristic ("naive") level initially, then to apply the concepts, theorems and algorithms on this low level as far as possible, and to increase exactness and rigour only afterwards.

In the case of differential calculus this method leads to the following teaching strategy.   Do not define the concept of limit, at first, in the usual formalistic way, but use only a heuristic idea such as 'unlimited approximation' (Kronfellner & Peschek 1991, Bürger *et al.* 1991, Kronfellner 1998, 76ff).   The symbol $\lim_{z \to x}$ is in this phase of the teaching strategy not a well defined mathematical concept, but only an abbreviation for the phrase "when z approaches (unlimited) to x".   Other theoretical concepts, such as continuity, are also avoided in this phase.   Based on these intuitive conceptions some rules of differentiation, restricted to polynomial functions, are derived and applied to those tasks usually treated in school mathematics.   After this period, when the need occurs for rules to treat further types of function, the necessity and the advantages of a more exact definition of the concept of limit will be elaborated, and subsequently used for more exact proofs of the rules already used, as well as for proving additional theorems.

This approach has epistemological potentialities.   It can be characterised as genetic, more precisely 'indirect genetic' in the sense of Otto Toeplitz (1927), although in his original proposal Fischer was not motivated by historical goals.   The 'indirect genetic method' means that there is no need to mention historical details explicitly.   The historical development only acts as a guideline.   It shows the teacher (or the textbook author) the crucial way forward: namely, that those aspects of a concept which historically have been recognised and used before others are probably more   appropriate   for   the   beginning   of   teaching   than   modern   deductive reformulations.   Newton,   Leibniz,   Euler   and   others   of   the   early   calculus   era contributed successfully to the development of mathematics and its applications

John Fauvel, Jan van Maanen (eds.), *History in   mathematics education: the ICMI study*, Dordrecht: Kluwer 2000, pp. 71-74

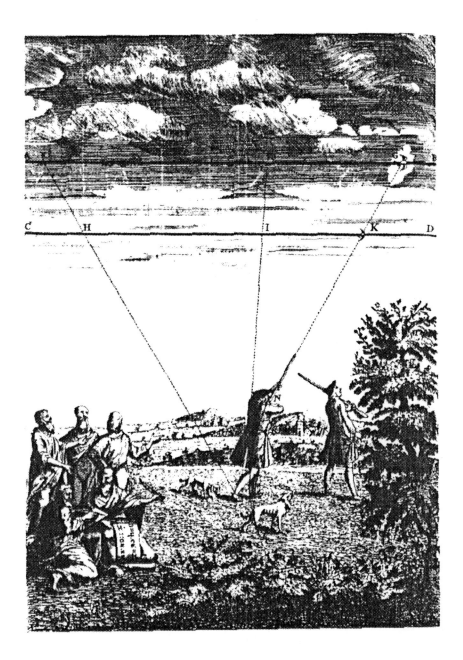

Figure 3.1: These gentlemen, who seem to be on the verge of inventing the pigeonhole principle, introduced Newton's calculus to English readers by showing its benefits for traditional country pursuits. From Newton's Method of fluxions (1736).

without an exact concept of limit, and without the concept of continuity. This can be used as an epistemological argument to avoid exaggerated exactness in too early a stage of the teaching process, to proceed to contextual interpretations and applications in order to make these (pre-) concepts meaningful to the pupils, and only after that to elaborate more exact definitions.

The historical development also shows that the community of great mathematicians needed a long time, more than one century, to build the conceptual basis of the subject. This underlines once more the intrinsic difficulty of these insights and reminds us to be patient with our pupils and not to assault them too early with such difficult mathematical concepts and too high levels of abstractness and exactness. Teachers sometimes feel guilty or dissatisfied when they teach a subject on a lower level than they are familiar with from their university study. Such teachers can hopefully be reassured by pointing to the historical development. Is the mathematical level of Euler and others really too low for our pupils?

The indirect genetic method has broader potentialities. According to Toeplitz one advantage of the approach consists in the possibility of the teacher making parallels to the historical development visible, and supplementing the teaching with additional historical details without being forced to change the order of succession of teaching units. When, for example, the teacher rephrases "approaches unlimited" as something like "infinitely close to", (s)he has an opportunity to speak about infinitesimals, differentials, their dubiousness, and their recent interpretation in Robinson's 'non-standard analysis'. When starting the phase of subsequent exactification (s)he can report explicitly about the famous criticism of Bishop Berkeley, the long lasting development from Newton, Leibniz, to Cauchy, Weierstrass and the reason for the search or need of an exact philosophical and conceptual basis (in geometrical style, according to the axiomatic method of Euclid's *Elements*). The need for additional concepts can be underlined by reporting about Bolzano and the concept of continuity which he needed for an exact proof of the Intermediate Value Theorem. (The search for a proof here shows also the process of historical exactification, given that the theorem appears geometrically evident and was used already earlier by Euler and Gauss without scruples (Hairer & Wanner 1996, 205). Further historical details which can easily be built in are Fermat's maximum method, Descartes' or Fermat's tangent method, remarks on the 'priority dispute' and its political background, on the use of different notions and symbols and their influence on the further development, on ancient roots of infinitesimals and infinity (actual versus potential infinity), and so on. In this way some brief asides about the development of calculus or of mathematics in general can contribute to an appropriate image of mathematics as a dynamic and developing science, contrary to some public opinion, and as an important part of our culture.

In spite of these advantages the indirect genetic approach contains also limitations and risks. It may be that this method needs more time to teach than a straight-forward and more formalistic one, even if no additional historical details are explicitly mentioned. Furthermore the approach may look somewhat long-winded or fuzzy, though only for those who are already experienced in using mathematical formalism; so it is an impression which teachers have, rather than their pupils. The latter, by contrast, feel rather confused by unfamiliar abstract symbols and concepts.

Certainly, teachers have to be convinced that this additional need of time is a fruitful investment and not wasted, even when this cannot be verified by indubitable empirical results. They should be encouraged to make their own experiences with this method, and to compare the results and their feeling with previous experiences. It is especially important that teachers should learn confidence, not to have guilty consciences when doing mathematics on less precise level than that of their university studies.

Another limitation consists in the very real difference between the learning pupil and the great mathematician of the past. Pupils do not necessarily feel a lack of rigour, whereas the great mathematicians had a more subtle perspective. It is well known that, from Newton onwards, several versions of the calculus were explored from the perspective of validity and rigour, a fact underlined by Bishop Berkeley's famous criticism.

On the other hand, there is the converse danger, when teachers set out to teach according to the indirect genetic method, of underestimating the advantages of modern notation and exactness. Although too high a level of rigour and formalism in an early phase of the teaching seems to be obstructive, it is still an important goal of mathematics education to show the merit of precise and exact formalism and to teach students to recognise and use this advantage. This goal has to be taken into account, but mainly in a later phase of teaching. Similar arguments hold for the possibilities and advantages of modern technology, in computer algebra systems.

References for §3.3

Bürger, H., Fischer, R., Malle, G., Kronfellner, M., Mühlgassner, T., Schlöglhofer, F. 1991. *Mathematik Oberstufe 3*, Wien: Hölder-Pichler-Tempsky

Fischer, R. 1978. 'Die Rolle des Exaktifizierens im Analysisunterricht', *Didaktik der Mathematik* 6, Heft 3, 212-226

Hairer, E., Wanner, G. 1996. *Analysis by its history*, New York: Springer

Kronfellner, M., Peschek, W. 1991. *Angewandte Mathematik 3*, Wien: Hölder-Pichler-Tempsky

Kronfellner, M. 1998. *Historische Aspekte im Mathematikunterricht*, Wien: Hölder-Pichler-Tempsky

Toeplitz, O. 1927. 'Das Problem der Universitätsvorlesungen über Infinitesimalrechnung und ihre Abgrenzung gegenüber der Infinitesimalrechnung an höheren Schulen', *Jahresberichte DMV* 36, 90-100

3.4 Stochastics teaching and cognitive development

Ewa Lakoma

We can distinguish two essential models of how mathematics education takes place. In the traditional model, the teacher plays the main role and gives to students ready-made, already-existing, independent knowledge. The main style of teaching is a

John Fauvel, Jan van Maanen (eds.), *History in mathematics education: the ICMI study*, Dordrecht: Kluwer 2000, pp. 74-77

lecture, during which the teacher shows students definitions of mathematical notions and typical examples of applications of these notions. The main task of students is to become acquainted with these definitions and applications and to use them in typical exercises. This model takes account only of logical connections between mathematical concepts, treated synchronically (that is, irrespective of any development over time).

An alternative model, which has arisen more recently, is one in which students learn mathematics in a more active way and construct, step by step, their own mathematical knowledge. In this active style the role of the teacher is quite different. Here the teacher plays the role of tutor, advisor, observer and helper, helping students to work in the direction and manner appropriate to their abilities. This model has arisen through recent research in mathematics education which shows that the epistemological structure of mathematics, in the matter of students' cognitive development, differs from that presupposed in the scientific, synchronic model (Freudenthal 1983; Sierpinska 1996). Thus one of the main aims of didactics of mathematics is to gain the knowledge necessary to create a new style of mathematics teaching. In this approach it is necessary to recognise the structure of mathematics not only from the logical, formal point of view but also from the diachronic perspective, which takes into account the historical development of mathematical concepts.

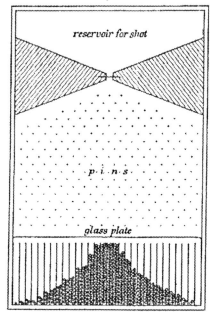

Figure 3.2: Galton's Quincunx (1873), a model for a probabilistic experiment.

In the case of probability and statistics, which for brevity we call stochastics, understanding the mathematics is not enough to work out an approach to teaching. The didactical structure must differ essentially from the scientific one, which is founded on Kolmogorov's axiomatics (Hacking 1975; Lakoma 1992). The high-level probabilistic concepts are too abstract and too far from the real context to be understandable by students who are not becoming mathematicians. In education we need another, non-axiomatic, structure of probability which shows the domain as a mathematics alive and in the process of being developed.

The dual character of the probability concept, described from the historical point of view by Ian Hacking in his work *The emergence of probability* (1975), has become an inspiration for research into how the probability concept develops in today's classroom (Lakoma 1990). Today's environment is certainly different from what it was some centuries ago, but the main research hypothesis is that the dual character of probability —laws of chance versus degrees of belief—

plays as important role today in the process of learning probability and statistics as it did in history. One of the main aims of these research studies in stochastics education is to explore and to understand the process of forming in the student's mind such mental objects (Freudenthal 1983) as can be given mathematical structure through the concept of probability. Many individual cases of students learning probability and statistics at secondary level were carefully analysed (Lakoma 1990, 1998, 1999b, 1999c) which led to the didactical hypothesis that the process of learning probability concepts has a strong interactive nature. Various forms of interactions among students seem to stimulate the process of forming probabilistic notions, in a way which respects a student's natural cognitive development.

Hacking's dual characterisation of probability arose from his historical analysis of its emergence in the seventeenth century (Hacking 1975, 12):

It is notable that the probability that emerged so suddenly is Janus-faced. On the one side it is statistical, concerning itself with stochastic laws of chance processes. On the other side it is epistemological, dedicated to assessing reasonable degrees of belief in propositions quite devoid of statistical background.

Analysis of early probabilistic reasonings shows that both these aspects became intertwined, starting from about the time of Pascal. The history suggests that in order to acquire the probability concept it is necessary to accept consciously its dual nature. Therefore in the process of probability teaching it is necessary to create such conditions that will make possible to form in students' mind the dual probability concept.

Working out the historical phenomenology of probabilistic concepts (Lakoma 1992)—in the sense of Freudenthal (Freudenthal 1983)—led to what has been called the Local Model's approach to probability and statistics teaching (Lakoma 1990, 1996, 1998, 1999b, 1999c). This approach gives students an opportunity to learn probability in a way which respects their individual cognitive development. The fundamental idea of this approach is to use forms of teaching which stimulate the student's initiative. Natural activities in the process of probability learning are involved by discovering and formulating problems which arise from them, and searching for solutions, even partial ones, according to the individual students' abilities. These activities allow students to develop both aspects of probability and keep them in balance. The methodology of the process, which comes directly from Isaac Newton, may be described having the following steps: discovery of a problem; formulation of a problem; construction of a model representing the 'real' phenomenon; analysis of this model; confronting the results obtained from the model with the 'real' situation. At the early stages of education students build models which just fit to the concrete phenomena. These are local models. At the more advanced levels these models become more general, appropriate to the whole class of phenomena and much more sophisticated mathematically. What is important is the explanatory value of a local model.

The history of probability is used for two important purposes: not only to elaborate a didactical approach to stochastics teaching but also to understand students' ways of probabilistic thinking (Lakoma 1990, 1998, 1999b, 1999c). By observing when students are able to use in their arguments both aspects of probability we can recognise whether they are already fluent with the mature, dual, probability concept or if they still need to make efforts to form in their mind the essential duality of the concept. So, it is worth stressing that knowledge of the historical development of probabilistic concepts serves us as a tool to evaluate a degree of maturity of students' probabilistic knowledge and understanding. It thus serves us as *a tool for measuring the effectiveness of the didactical approach to stochastics teaching.*

Thus, the example of stochastics teaching shows that using the history of mathematics in mathematics education can be effective in :

– creating a didactical approach to mathematics teaching which takes account of the student's cognitive development;

– recognising the student's ways of arguments as corresponding with past problems, and encouraging their responses to real situations similar to those known from the history of mathematics;

– organising the process of learning mathematics according to the student's actual abilities.

For assessing the effectiveness of using history of mathematics in mathematics education, a qualitative analysis seems to be more useful than a quantitative approach. Using history of mathematics is found to be effective when we try to recognise general mathematical competencies in the performance of students rather than particular skills. It is possible to evaluate this kind of effectiveness after some years in which they have been learning mathematics in active style, by observing students' progress and actions in real situations when they use their knowledge.

References for §3.4

Freudenthal, H. 1983. *Didactical phenomenology of mathematical structures*, Dordrecht: Reidel

Hacking, I. 1975. *The emergence of probability*, Cambridge: University Press

Lakoma, E. 1990. *Local models in probability teaching* (in Polish), doctoral thesis, Warsaw University

Lakoma, E. 1992. *Historical development of probability* (in Polish), Warsaw: CODN-SNM

Lakoma, E. 1998. 'On the interactive nature of probability learning', *Proceedings of CIEAEM-49, Setubal*, 144-149.

Lakoma, E. 1999a. 'On the historical phenomenology of probabilistic concepts – from the didactical point of view', in A. Boyé, F. Héaulme and X. Lefort (eds), *Contribution à une approche historique de l'enseignement des mathématiques*, Nantes: IREM des Pays de la Loire, 439-448

Lakoma, E. 1999b. 'The diachronic view in research on probability learning and its impact on the practice of stochastics teaching', *CIEAEM-50 Proceedings*, Neuchatel, 116-120

Lakoma E. 1999c. 'Del calculo probabilístico al razonamiento estocástico: un punto de vista diacrónico', in R. M. Guitart (ed), *Uno-revista de didactica de las matematicas* **22**, 55-61

Sierpinska, A. 1996. 'The diachronic dimension in research on understanding in mathematics – usefulness and limitations of the concept of epistemological obstacle', in Jahnke, H.N., Knoche, N., Otte, M. (eds.), *History of mathematics and education: ideas and experiences*, Göttingen: Vandenhoeck & Ruprecht, 289-318

3.5   Ancient problems for the development of strategic thinking

Lucia Grugnetti

One of the risks in introducing history of mathematics in mathematics education is the anachronism which consists in attributing to an author knowledge that he never possessed,   There is a vast difference between recognising Archimedes as a forerunner of integral and differential calculus, whose influence on the founders of the calculus can hardly be overestimated, and seeing in him, as has sometimes been done, an early practitioner of the calculus. If the risk of *anachronism* is a big one for historians, it is not smaller in doing history of mathematics in mathematics education. So, when a past mathematician or other scientist is introduced in the classroom, it is desirable to outline the political, social, economical context in which he lived. In this way it is possible to discover that facts and theories, studied in different disciplines, are concretely related (Grugnetti 1994).   The interaction between history and didactics of mathematics must, however, be developed taking into account the negative influences that each can have on the other (Pepe 1990). A possible negative influence of history on didactics is the creation of a domain with interesting and curious references which are, in effect, not essential and are felt to be irrelevant. But the history of mathematics does offer several examples which gain by an interdisciplinary approach (Pepe 1990) such as, for example, the number systems of the ancients; Galileo, the mathematisation of the physical world and the experimental method; Descartes and the analytical method.

When ancient problems are used, teachers and pupils can compare their strategies with the original ones (Grugnetti 1994). This is an interesting way for pupils to be led to understand the economy and the power of present mathematical symbols and processes. And another point: observing the historical evolution of a concept, pupils can remark that mathematics is not fixed and definitive.

An example of way that the history of mathematics can foster an interdisciplinary approach, generating material across several school subject areas, is given by the *Liber Abaci* (1202) of Leonardo Pisano (known as Fibonacci). This provides a source of problems which concern different teachers and subjects, such as:

– Italian and Latin: what kind of language is that of the *Liber Abaci*?
– history: the development of the Middle Ages in Europe and Islam
– geography: the West, the Middle East, the Islamic world
– mathematics: pupils' strategies for solving some problems
– Fibonacci's strategies: why did he solve his problems in the way he did?

John Fauvel, Jan van Maanen (eds.), *History in mathematics education: the ICMI study*, Dordrecht: Kluwer 2000, pp. 78-81

An important component of this approach is the possibility for the students to compare their strategies with the ancient ones. The students can, for example, understand the economy and the effectiveness of modern algebraic processes compared to the ancient methods. The activity of recognising and comparing strategies is one of the most important aspects to develop in mathematics learning. Only once students become able to compare different strategies (for solving problems, but also for proving theorems), can the process of generalisation evolve.

It is interesting to ask 13/14 year old pupils to try to translate the following problem from the *Liber Abaci*:

In quodam plano sunt due turres, quarum una est alta passibus 30, altera 40, et distant in solo passibus 50; infra quas est fons, ad cuius centrum volitant due aves pari volatu, descendentes pariter ex altitudine ipsarum; queritur distantia centri ab utraque turri.

When this was done in class, in Italy, several translations (into Italian) were discussed, of which the final version was reached, which in English may be rendered as follows:

Two towers, the heights of which are 30 paces and 40 paces, have a 50 paces distance. Between the two towers there is a font where two birds, flying down from the two towers at the same speed will arrive at the same time. What is the distance of the font from the two towers?

Figure 3.3: The two towers problem, here from the Calandri manuscript (1491), as it appeared on the cover of the Mathematical Gazette of March 1992, an issue that was especially devoted to history in mathematical education.

The 13/14 year old pupils then solved the problem, using the Pythagorean theorem and solving an equation.

The real interest of this problem was that of analysing and discussing Fibonacci's strategy in which arithmetic writing of operations is not given and in which the Pythagorean theorem is implicitly used. This is a literal translation of Fibonacci's text:

If the higher tower is at a distance of 10 from the font, 10 times 10 is 100 which added to the higher tower times itself is 1600, which gives 1700, we must multiply the remaining distance times itself, which added to the lower tower times itself, i.e. 900, gives 2500. This sum and the previous one differ by 800. We must move the font away from the higher tower. For example by 5, i.e. globally by 15, which multiplied by itself is 225, which added to the higher tower times itself gives 1825, which added to the lower tower times itself gives 2125. The two sums differ by 300. Before the difference was 800. So, when we added 5 paces, we reduced the difference of 500. If we multiply by 300 and we divide by 500, we have 3, which added to 15 paces gives 18 which is the distance of the font from the higher tower.

Pupils had to interpret Fibonacci's sentences and translate them into mathematical symbolism. This activity was done in small heterogeneous groups. In modern symbolism Fibonacci's procedure can be written as:

$$10^2 + 40^2 = 100 + 1600 = 1700 \text{ and}$$

$$(50 - 10)^2 + 30^2 = 40^2 + 30^2 = 1600 + 900 = 2500$$

(Fibonacci says: "this sum and the previous one differ by 800")

$$15^2 + 40^2 = 225 + 1600 = 1825 \text{ and}$$

$$35^2 + 30^2 = 1225 + 900 = 2125$$

(Fibonacci says: "the two sums differ by 300"). He now uses the diagram:

and his last sentence could be written as:

$$(5 \times 300): 500 = 3;$$

$$3 + 15 = 18.$$

The discussion brought to the class's attention the method of 'false position', one of the oldest ways to solve problems (which was used also by the ancient Egyptians).

For the students it was an occasion for understanding that it is more economical to solve this problem using a simple algebraic equation, which Fibonacci could not use. The class discussion centered on the reasons why Fibonacci could not use algebra in our sense. In this way a historical example could contribute also to give to students the opportunity to compare arithmetical and algebraic procedures.

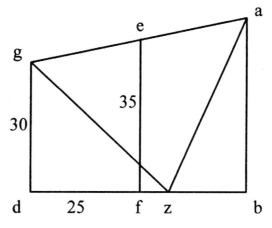

Figure 3.4: Fibonacci's diagram

The Fibonacci problem did not finish here! In fact, Fibonacci considers a second strategy to solve it. After having explained that the triangle *agz* (where *z* is the font) is isosceles on the base *ag* (with *ae* = *eg*) by construction, Fibonacci adds (see Figure 3.1):

40 and 30 is 70; the half is 35, in fact the line *ef*. The lines *df* and *fb* have 25 in length, the difference between 35 and the lower tower is 5, which, multiplied by 35 is 175, which divided by the half of the distance between the two towers, in fact 25, gives 7 (the line *fz*). Therefore *dz* is 32 and it remains 18 for the line *zb*.

It would be interesting to discuss with students some aspects of Fibonacci's procedure which, as we can see, is based on the similarity of triangles *efz* and *ghe* where *h* is the intersection point of *ef* and the parallel to *df* which contains *g*. The open raising of hypotheses by the students, and the subsequent discussion about them, are important elements in this didactic strategy.

Through this kind of activity it is possible to introduce history of mathematics in mathematics education avoiding the risks and reaching the aims mentioned at the beginning of this section.   There is certainly prima facie evidence that it was effective for the students.

References for §3.5

Grugnetti, L. 1994. 'Relations between history and didactics of mathematics', *Proceedings of PME XVIII*, Lisbon, 121-124
Pepe, L. 1990. 'Storia e didattica della matematica', *L'educazione matematica* 3 (1-2), 23-33

3.6   Difficulties with series in history and in the classroom

Giorgio T. Bagni

The history of mathematics provides a collection of useful examples for assisting in the learning of mathematics, which can be used by the teacher in a number of ways (Fauvel 1990, 1991; Pepe 1990; Barbin 1991; Grugnetti 1992; Furinghetti 1993; Furinghetti and Somaglia 1997). This section examines some topics in the history of infinite series which help us to understand better the difficulties faced by today's pupils.

The study of infinite series is an important topic of the mathematical curriculum of the upper secondary school. For several centuries it has played a central role in the study of analysis (on which see Boyer 1969; Edwards 1994), as well as provided a number of counter-intuitive obstacles for the learner. A sum of infinitely many addends, for example, is often considered by pupils to be 'infinitely great'. In this instance the history of mathematics can both help the teacher to understand the pupil's difficulty and suggest what to do about it.

A time-honoured problem in this area is Zeno of Elea's paradox of 'Achilles and the Tortoise'. This concerns a convergent geometric series. Pupils may experience difficulty in absorbing the difference between convergent, divergent and indeterminate series, and this can cause problems and inconsistencies in their minds. Let us consider directly a famous indeterminate series, the one consisting of +1 and -1 in alternation. In 1703, the Italian mathematician Guido Grandi stated: "From 1-1+1-1+... I can obtain 0 or 1. So the creation *ex nihilo* is quite plausible" (Bagni 1996, II) (we may note here the theological motivation for the argument, which may interest some pupils). Grandi's argument was based on bracketing the series in two alternative ways

$$(1-1)+(1-1)+(1-1)+(1-1)+... \ = 0+0+0+0+... = 0$$

$$1+(-1+1)+(-1+1)+(-1+1)+... \ =1+0+0+0+... \ = 1$$

The sum of the alternating series was considered 1/2 by many mathematicians in the 17th century. According to Grandi, this can be justified by considering the sum of the geometric series

$$\frac{1}{1+x} = \sum_{i=0}^{+\infty}(-x)^i = 1-x+x^2-x^3+...$$

Then putting $x = 1$ into the series we should have:

John Fauvel, Jan van Maanen (eds.), *History in mathematics education: the ICMI study*, Dordrecht: Kluwer 2000, pp. 82-86

$$\sum_{i=0}^{+\infty}(-1)^i = 1-1+1-1+... = \frac{1}{2}$$

We may note that there is a simpler way to reach the same (false) conclusion. The same result can be achieved by the following procedure

$$s = 1-1+1-1+... \Rightarrow s = 1-1(1-1+1-1+...) \Rightarrow s = 1-s \Rightarrow s = \frac{1}{2}.$$

It can be interesting for some pupils to ask them what is the fallacy in this argument. (It is because the argument works only once you have established that the series does indeed have a sum which is a number 's' like any other; but that is what you are trying to establish. Nowadays we accept such a geometrical series as having a sum only if $|x| < 1$.)

Gottfried Wilhelm Leibniz, too, studied Grandi's series, and he wrote to Jacopo Riccati summarising the argument mentioned above:

I do not know if Mr. Count Riccati, and Mr. Zendrini have seen about the question whether 1-1+1-1 etc. is 1/2, as R. P. Grandi stated, someway correctly. In fact $1/(1+x)$ is $1-x+xx-x^3+x^4-x^5$ etc. so if x is 1, we have 1/(1+1) = 1-1+1-1+1-1 etc. = 1/2. It seems that this is clearly absurd. In the *Acta Eruditorum* from Leipzig I think I have solved this problem.

(this letter was probably written in 1715; see Michieli 1943, 579). In fact Leibniz studied Grandi's series in some letters to Christian Wolff, where he introduced an interesting probabilistic argument that influenced Johann and Daniel Bernoulli too. Leibniz noticed that if we stop the series 1-1+1-1+... at some finite stage, taken at random, it is possible to have 0 or 1 with the same "probability". So the most "probable" value is the average between 0 and 1, so 1/2 (Leibniz 1715). This argument was accepted by some later distinguished mathematicians, notably Joseph Louis Lagrange and Siméon Denis Poisson.

Later in the 18th century, Leonhard Euler wrote in his textbook on differential calculus *Institutiones calculi differentialis* (1755): "We state that the sum of an infinite series is the finite expression by which the series is generated. From this point of view the sum of the infinite series $1-x+x^2-x^3+...$ is $1/(1+x)$ because the series arises from the development of the fraction, for every value x". Euler considered infinite series as a part of algebra of polynomials (Kline 1972, 537). So series were considered to be polynomials that can express the original function, without any convergence control. As we shall see, this situation can be important in the educational field.

Jacopo Riccati (Grugnetti 1985, 1986) criticised the convergence of Grandi's series to 1/2 in his *Saggio intorno al sistema dell'universo* (Ricatti 1754/1761, 87), he wrote:

Grandi's argument is interesting, but it is wrong, because it causes contradictions. [...] Let us consider $n/(1+1)$ and, by the common procedure, let us obtain the series $n-n+n-n+n-n$ et.cet. $= n/(1+1)$. If we remember that $1-1 = n-n$, or $1+n = n+1$, we have that either in this series or in Grandi's series there are the same number of 0.

The contradiction involving "the same number of 0" was reached in this way. Having written $1/2 = 1-1+1-1+...$ "by the common procedure", Riccati introduced the series:

$$\frac{n}{2} = n-n+n-n+...$$

Let us compare the considered series, we can write:

$$s = 1-1+1-1+1-1+... = (1-1)+(1-1)+(1-1)+... = 0+0+0+...$$

$$s' = n-n+n-n+n-n+... = (n-n)+(n-n)+(n-n)+... = 0+0+0+...$$

So Riccati concluded that Grandi's procedure is incorrect. His argument cannot be accepted (notice that it is based upon the "common procedure", which is not correct for an indeterminate series), although his conclusion is clear and correct (Riccati 1754/1761, 86):

The mistake is caused by [...] the use of a series from which it is impossible to get any conclusion. In fact, [...] it does not happen that if we stop this series, the following terms can be neglected in comparison with preceding terms, this property is verified only for convergent series.

Educational aspects

Let us turn now examine some educational aspects. This issue was raised with *Liceo Scientifico* students in Treviso (Italy) who did not know infinite series, although they had been introduced to the concept of infinite set. The following question was given to them (45 pupils 16-17 year olds and 43 17-18 year olds—88 pupils in all):

In 1703 the mathematician G. Grandi studied the addition 1-1+1-1+... (addends, infinitely many, are always + 1 and -1). What is your opinion about it?

Pupils answered as follows:
26 pupils (29%) said the answer is 0
18 pupils (20%) said the answer can be either 0 or 1
5 pupils (6%) said the answer does not exist
4 pupils (5%) said the answer is 1/2
3 pupils (4%) said the answer is 1
2 pupils (2%) said the answer is infinite
30 pupils (34%) gave no answer.

First of all, notice that the greater part of the pupils interpreted the question as an implicit request to calculate the 'sum' of the series. Only 5 students (6%) explicitly stated that it is impossible to calculate the sum of Grandi's series. Note too that a fifth of the pupils suggested the possibility of two answers.

The students were interviewed about their answers. Some of them used, in effect, similar arguments to those found in the eighteenth century. "If I want to add always 1 and -1, I can write (1-1)+(1-1) so I can couple 1 and -1: so I am going to add infinitely many 0, and I obtain 0." (Marco, 3rd class, and 15 other pupils). And those students who stated that the sum of the series is 1/2 justified it by arguments similar to the probabilistic argument of Leibniz. "If I add the numbers I have 1, 0, 1, 0 and always 1 and 0. The average is ½." (Mirko, 4th class).

So students' justifications are remarkably similar to some we find in the history of mathematics. In particular, we can recognise, explicitly or implicitly, that some students felt as did several mathematicians in the 17th and 18th centuries, that an infinite series can be always considered a polynomial: the notion of convergence, not considered before the work of Gauss, had not entered the Italian pupils' heads yet either . This seems to bear out in this instance the view of Piaget and Garcia (1983), that historical development and individual development are parallel.

Didactic reflection

Anna Sfard states that in order to speak of mathematical *objects,* it is necessary to make reference to the process of concept formation, and supposes that an *operational* conception can be considered before a *structural* one (Sfard 1991, 10). As regards infinite series, the passage from an operational conception to a structural one is hard, because of the necessity of some basic notions (for example the limit concept).

As regards the *savoir savant,* the historical development of mathematical concepts can be considered as the sequence of (at least) two stages: an early, intuitive stage, and a mature stage; several centuries can pass between these stages. In the early stage the focus is mainly operational, the structural point of view is not a primary one. For example, in the early stage of working on infinite series (that is, at least until Gauss's works) main questions of convergence were not fully considered. From the educational point of view, a similar situation can be pointed out (Sfard 1991): of course, in an early stage pupils approach concepts by intuition, without a full comprehension of the matter. Then the learning becomes better and better, until it is mature.

There is a clear analogy between these situations. And the experimental results given above show that in the educational passage from the early stage to the mature one we can point out, in our pupils' minds, some doubts and some reactions that we can find in the passage from the early stage to the mature one as regards the *savoir savant,* too. Of course, processes of teaching and learning take place nowadays, after the full development of the *savoir savant,* as regards either early stage, either mature stage. So the *didactic transposition,* whose goal is initially a correct development of intuitive aspects, can be strongly based upon the results achieved in the mature stage, too, of the development of the *savoir savant.*

Moreover the process of teaching-learning and the *didactic transposition* must consider that, as we previously underlined, pupils' reactions are sometimes similar to corresponding reactions noticed in several great mathematicians in the history of mathematics. This correspondence can be a very important tool for the teacher in developing the effectiveness of history as a resource base, but it needs a clear epistemological skill.

References for §3.6

Bagni, G.T. 1996-7. *Storia della Matematica,* I-II-III, Bologna: Pitagora

Barbin, E. 1991. 'The reading of original texts: how and why to introduce a historical perspective', *For the learning of mathematics* **11** (2), 12-13

Boyer, C. 1969. 'The history of the calculus', in: Hallerberg *et al.* (ed.), *Historical topics for the mathematics classroom,* Reston: NCTM 31st yearbook, 376-402

Edwards, C.H. Jr. 1994. *The historical development of the calculus,* Berlin: Springer

Fauvel, J. (ed.) 1990. *History in the mathematics classroom: the IREM papers,* Leicester: The Mathematical Association

Fauvel, J. (ed.) 1991. *For the learning of mathematics* **11** (2), special issue on history in mathematics education

Furinghetti, F. 1993. 'Insegnare matematica in una prospettiva storica', *L'educazione matematica,* **3-4**, 123-134.

Furinghetti, F., Somaglia, A. 1997. 'Storia della matematica in classe', *L'educazione matematica,* **18**, V

Grugnetti, L. 1985. 'Sulla vecchia ed attuale equazione di Riccati', *Rendiconti del Seminario della Facolt adi Scienze dell'Universita di Cagliari* **55** (1), 7-24

Grugnetti, L. 1986. 'L'equazione di Riccati: un carteggio inedito tra Jacopo Riccati e Nicola II Bernoulli', *Bollettino di Storia delle Scienze Matematiche,* **6** (2) 45-82.

Grugnetti, L. 1992. 'L'histoire des mathématiques: une expérience interdisciplinaire fondée sur l'histoire des mathématiques', *Plot* **60**, 17-21

Kline, M. 1972. *Mathematical thought from ancient to modern times,* New York: Oxford University Press, reprint: 1991

Leibniz, G.W. 1715, 'Epist. G.G.L. ad V. claris. Ch. Wolfium' (Letter by Leibniz to Wolff), *Acta Eruditorum Supplementum* **5** (1711-1719)

Michieli, A.A. 1943. 'Una famiglia di matematici e di poligrafi trivigiani', in Riccati, I., *Atti del Reale Istituto Veneto di scienze, lettere ed arti,* Cll, II.

Pepe, L. 1990. 'Storia e didattica della matematica', *L'educazione matematica* **3** (1-2), 23-33

Piaget, J., Garcia, R. 1983. *Psychogenèse et histoire des sciences,* Paris: Flammarion

Riccati, Jacopo 1754/1761, *Saggio intorno al sistema dell'universo,* in: *Opere,* 1, Lucca 1761

Sfard, A. 1991. 'On the dual nature of mathematical conceptions: reflections on processes and objects as different sides of the same coins', *Educational studies in mathematics* **22**, 1-36

3.7   On potentialities, limits and risks

Marta Menghini

Let us try to think about *potentialities, limits* and *risks* (Grugnetti 1994) in connection with determining the effectiveness of using history of mathematics in the classroom.

Consider first the case in which one asks history of mathematics for help in teaching an argument. The simplest way is to use history *implicitly*: to take *ideas*. This means that history is not an aim for itself, but a teaching itinerary is constructed which must utilise suggestions from various sectors, always keeping in mind the *didactic* aims. So, in speaking of the implicit use of history, we are not referring

John Fauvel, Jan van Maanen (eds.), *History in   mathematics education: the ICMI study,* Dordrecht: Kluwer 2000, pp. 86-90

here necessarily to the "indirect genetic method" of O. Toeplitz (1927) discussed in section 3.3 above. The indications which derive from history can be very slight. This way of using history is growing as more research is done on learning difficulties. One tries, in fact, to go back to the origins of a concept, to conceive it in a different situation, to return to the instant in which the theory "branched out". More generally, one enters into the area of *didactic transposition*, that attempts to calibrate the didactic operation in relation to the conceptual difficulties and complexities of a given topic.

In a first broad classification, we can say that in those cases the *limits* are those of the classroom, like always in teaching; the *potentialities*, of course, consist in a better understanding of the topic. The *risks* are involved with a lack in didactical transposition; more precisely, the risk is to follow too much the real historic path. A didactical competence is needed, more than a historic one: a historian could easily be drawn to criticise the somewhat distorted and adjusted interpretations given to a certain event for pedagogic purposes.

A more detailed example is that of teaching analysis, discussed in section 3.3, in which history is used to help pupils better understand a topic which is known to cause difficulties for pupils. Another case would be to give, using history, some additional knowledge, in order to help in understanding a more general topic. Let us explain this by an example.

In his *Conic Sections* of the 3rd century B.C. (Apollonius 1923, Apollonius 1952) Apollonius describes a geometric procedure for the sectioning of a cone with a plane in order to determine what we today would call the equation of the parabola, and then of the other two conic sections (Mancini Proia and Menghini 1984, Menghini 1991). In this procedure one determines a relationship between a segment of the axis of the parabola and a segment perpendicular to it, such that the segments correspond to what we today would call the coordinates of a point. The procedure is not common enough in schools to be considered a mathematical fact 'without history'. Nevertheless, in teaching conic sections, one can follow different paths which make more or less explicit use of the history of the subject.

In an *implicit use*, one presents the topic along classical lines but without mentioning Apollonius and his era. There are notable advantages (*potentialities*) for the teacher in this strategy: an application of 3-dimensional geometry, an optimal connection between synthetic geometry and analytical geometry, and above all, the possibility to connect, using their equations, the definition of the conics as sections of a cone to that of a conic as a locus of points. For pedagogic reasons one can simplify the original procedure (already rather simple) in many parts: notation, use of x-y coordinates, limitation to a right cone, limitation to the parabola. In particular, it will be necessary to modify one passage. Given a relation of the type $a{:}b = c{:}d$ (where a, b, c, d are segments), in order to substitute, in a further expression, the value $a = b{\cdot}c/d$, Apollonius must introduce a new parameter (which requires a complicated geometric explanation), since he cannot rely on the algebraic procedure, i.e., substituting the segments with their lengths. This "simplification" is a deep modification, it must be stressed with teachers. As we said earlier, the *limits* of such a use of history are connected with the mathematical difficulty of the topic. This treatment is interesting and deep, but it is not easier than the usual treatment of conic sections. The *risks* (in this case, the lack in didactical transposition) are connected with the "simplifications" we mentioned above: the teacher must be able to translate and to reinterpret certain passages into a language which is comprehensible for the pupil.

Another teaching aim could be that of 'giving to mathematics a cultural vision'. By immersing a topic in a historic period, we can obtain, within this aim, a changing of point of view, an understanding of a mentality, which are important *potentialities* in a mathematics curriculum. The principal *risks* consist in pointing out the historical aspect more than the mathematical one, to have a sort of history-course with a low mathematical level. A *limit* of this aim is that it demands a certain historic competence from the teacher.

Let us take up again the example of Apollonius, which earlier we dealt with *implicitly*. The historical context of Apollonius' procedure can be explicitly addressed, for suitable classes, and this offers other interesting aspects in addition to those of the implicit treatment. One can, for example, compare the definitions of Apollonius with those of his predecessors, analyze some of the properties known to Apollonius, observe this 'precursor' of the cartesian plane, see how to move from one definition to the other, and watch how the concept of conic sections has changed over the centuries with a definite 'cultural' growth.

As to the simplifications, it is obvious that one has to rely on algebraic procedures in an 'implicit' scholastic treatment, since the requirements of Greek geometrical handling are so intense. But the algebraic procedure is preferable *even* if one chooses the second path, in which history is used in an *explicit way*. Only in this case (in fact, to avoid a 'misstatement') one must clearly tell the students that at this point one must substitute the lines of Apollonius with algebraic statements. In helping the students to understand the difference the teacher can again underline different styles in the treatment of a topic.

In the explicit case the treatment becomes longer than in the first one, so, in addition to the *limits* of the classroom, we have also the limits of time: it is difficult to treat topics in this way more than two or three times in a year. The *risk* is that, underlining the historical aspect, the teacher tells students about Apollonius, the Greeks, the synthetic methods, and that Apollonius found a brilliant method to determine a sort of equation of the parabola by sectioning a cone, but doesn't explain this method mathematically. So the story becomes a too long anecdote, which can be boring and reduces its effectiveness, especially for gifted students.

This risk, in connection with the same aims, can be observed too in the case of the reading of original texts. To read in the classroom passages from the original works of famous mathematicians is a simple and realistic way to introduce the history of mathematics into teaching. From the viewpoint of the historian, it is a useful way to begin to develop an interest in history. From the didactic point of view, it is one of the first efforts to move out of the usual canons of the teaching of mathematics, provided that one evaluates carefully the interest, the relevance and the cultural contribution of the chosen text (Barbin 1991). A text which highlights a fundamental moment in the history of mathematics (a change in language, an innovative idea, a 'rich' problem) is interesting also from a didactic point of view. Once in a while, for example, a passage is appealing because the language used is different from the usual one, because the problem discussed is interesting, because in its solution one uses for the first time an actual technique of demonstration (or on the contrary, because one uses a technique different from the one commonly used).

But we have to pay attention to the use that a teacher can make of the text. When trainee teachers were asked if they would use the original text of Apollonius (Bottazzini *et al.* 1992) with their pupils, referring to the passage mentioned before, many of them said yes, because one can explain to pupils that today's notation is simpler and more elegant. But the same conclusion can be drawn from the reading of many passages! For example in Italy it sometimes happens that 16th century algebraists (who write more or less in Italian) are read in the classroom, and the same conclusion is drawn. A better historic competence could help in the discovery of a deeper significance in the various passages (see section 3.4). Another *risk* connected to the lack of historical competence is given by an autonomous bibliographical research. Someone not expert in history can hardly know if a chosen passage is effectively representative of a certain historical period or of a way of thinking. The choice of a work by an unknown and too original author isn't helpful. It is possible that it presents a unique problem to propose in class, but it is not appropriate (and may be distracting) to underline its historical aspect.

Another problem, already hinted at, is that sometimes the reading of original texts does not really interest the good students of mathematics but is more meaningful to the students with greater interest in humanistic topics. Is it true that some of them understand the mathematical problem only through the reading of the passage (Lit and Siu 1998)? Or do they like this kind of classroom activity because one doesn't need to do real mathematics?

In the case of an *explicit use* of history the objectives of the educator are quite different from those of an *implicit* use. The intention may be, again, to intervene in the conquest of a concept, but above all, one wants to describe a historical period, to show the evolution and the stages in the progress of mathematics. In this case, even with the necessary simplifications, the emphasis *is* on history.

But we can do something more, we can go inside history maintaining an attitude as open as possible to historical investigation. This is the case in which we want to stress even more that mathematics is something that is developed and can be constructed, when we want to stress the creative side of mathematics more than the cultural one. Here we need the pupil to become an actor: *implicitly* by posing problems taken also from history, *explicitly* letting him follow the path of the development of a certain piece of history. The *risk* in this case is to 'teach' this development. Among the *potentialities*: the pupil has the courage to discuss something done 200 years ago, even by a great mathematician (while he has not the courage to discuss what his teacher says). All this, as we saw in sections 3.5 and 3.6, is strictly connected to the way the teacher acts.

References for §3.7

Apollonius de Perge 1923. *Les coniques; oeuvres traduites pour la première fois du grec en français avec une introduction et des notes par Paul Ver Eecke*, Bruges: Desclée de Brouwer
Apollonius of Perga 1952. *Conics*, tr. R. Catesby Taliaferro, Chicago

Barbin, E. 1991. 'The reading of original texts: how and why to introduce a historical
    perspective', *For the learning of mathematics* **11** (2), 12-13

Bottazzini U., Freguglia P., Toti Rigatelli L. 1992. *Fonti per la storia della matematica*,
    Firenze: Sansoni

Grugnetti, L. 1994. 'Relations between history and didactics of mathematics', *Proceedings of
    PME XVIII*, Lisbon, 121-124

Lit C.-K., Siu M.-K. 1998. 'A research project on the effect of using history of mathematics
    in the school classroom', *Report for the ICMI Study in Luminy*.

Mancini Proia, L., Menghini, M. 1984. 'Conic sections in the sky and on earth', *Educational
    studies in mathematics* **15,** 191-210.

Menghini, M. 1991. 'Punti di vista sulle coniche', *Archimede* **41**, 84-106.

3.8   Suggestions for future research

At the beginning of this chapter, we mentioned the many articles appearing in recent
years about the use of history in teaching mathematics. For pursuing an investigation
on the effectiveness of history in the classroom, it seems desirable to collect and to
study two kind of materials:

1. to collect experiences of teachers who use history. The purpose is to study their
    aims, their steps, the problems they meet in teaching, the advantages and the
    disadvantages in their eyes.
2. to collect questionnaires and interviews of teachers and pupils about
    mathematics. The purpose is to study their approaches to mathematical
    concepts, such as the infinite, and mathematical ideas, such as mathematical
    rigour.

The optimal way to explore all these materials is, as we explained earlier,
necessarily qualitative, recognising that ethnographic methods are appropriate to
explore the question of effectiveness. But we have to make precise what these
methods mean for the specific question of the relations between history and
mathematical teaching.

   If we think about the future, we have to take into account how the teaching of
mathematics will evolve and what problems may arise in the next years. A major
point is the interest and enthusiasm towards mathematics found in educational
circles. This question has two levels. One is the pertinence of mathematics in the
curriculum. In some countries, there is a trend to reduce the quantity of
mathematical teaching or to orient it towards applied subjects. This is linked, in
particular, with the use of new technologies. Secondly, the difficulties of
interesting pupils themselves in mathematics. This point is very important if we
think that it is not possible to engage in real mathematical activity without an
enthusiasm or intellectual interest for mathematics. The introduction of history of
mathematics can play here a decisive role. History is a source to define perennial
knowledge, that is, knowledge which permits us to understand the world. But more
than that, we can find in history the meanings of mathematical knowledge, to
understand what it is for and what are the problems mathematics helps to solve. All
this reinforces the image of mathematics for teachers and pupils, so for history to
respond effectively in this regard is essential.

Chapter 4

History of mathematics for trainee teachers

Gert Schubring

with Éliane Cousquer, Chun-Ip Fung, Abdellah El Idrissi, Hélène Gispert, Torkil Heiede, Abdulcarimo Ismael, Niels Jahnke, David Lingard, Sergio Nobre, George Philippou, João Pitombeira de Carvalho, Chris Weeks

Abstract: *The movement to integrate mathematics history into the training of future teachers, and into the in-service training of current teachers, has been a theme of international concern over much of the last century. Examples of current practice from many countries, for training teachers at all levels, enable us to begin to learn lessons and press ahead both with adopting good practices and also putting continued research efforts into assessing the effects.*

4.1 Earlier views on history in teacher education

Almost since the beginning of internationally coordinated mathematical activities, the importance of a historical component in the training of future mathematics teachers has been stressed by historians of mathematics and by mathematics educators and has been backed by the mathematical community. Already in 1904, the third International Mathematical Congress, held in Heidelberg, adopted a motion recommending the introduction of a historical component (*IMC* 1904, 51):

Considering that the history of mathematics nowadays constitutes a discipline of undeniable importance, that its benefit—from the directly mathematical viewpoint as well as from the pedagogical one—becomes ever more evident, and that it is, therefore, indispensable to accord it the proper position within public instruction.

The Congress wished to see established, on an international level (*ibid.*, 51):

. . . that the history of the exact sciences be taught at the universities, by introducing lecture courses for the four parts: 1. Mathematics and Astronomy, 2. Physics and Chemistry, 3. Natural Sciences, 4. Medicine.

John Fauvel, Jan van Maanen (eds.), *History in mathematics education: the ICMI study*, Dordrecht: Kluwer 2000, pp. 91-142

The motion was proposed by expert mathematicians, historians of mathematics and mathematics educators, including David Eugene Smith (USA), Paul Tannery (France), Anton von Braunmühl, Emil Lampe, Max Simon, Paul Stäckel, and Ernst Wölffing (Germany), and Gino Loria (Italy). They made reference to earlier similar motions passed by the International Congress for Comparative Historical Research (Paris 1900) and by the International History Congress (Rome 1903).

Since mathematics students were at this time (and in the following decades) almost exclusively studying for a teaching licence, the effect of this motion was to recommend the introduction of mathematics history into teacher training. The motion even recommended, additionally, the proposal "to introduce the elements of the history of the exact sciences into the curriculum of the particular teaching disciplines of the high schools" (*ibid.*, 51 sq.)

The readiness to agree to such appeals has probably not diminished since that time. The problem, however, lies not simply with putting such appeals into practice. Nowadays, we can also see a profound shift in the motivations and justifications advanced for such claims and, consequently, decidedly different forms of practice from those intended by the proponents of the use of history at the beginning of this century.

The changes that have occurred during the last two or three decades can best be illustrated by considering a characteristically traditional position as presented by the Dutch teacher and historian of mathematics Eduard Jan Dijksterhuis. The position he adopted is highly revealing since it was published in an ICMI study directly preceding our present one, namely the Dutch contribution of 1962 to an international ICMI study on the state of teaching mathematics. Contrary to earlier periods, Dijksterhuis here made a distinction between two different career orientations: the profession of mathematics teacher and the career of mathematician.

As regards the latter, Dijksterhuis expressed his conviction that "the history of mathematics does not form an essential part" of the study of mathematics—at best forming a complement serving some historical or cultural curiosity. In justification, Dijksterhuis (1962, 34) claimed that:

present-day mathematics has [...] adopted and preserved all (from older mathematics) that was valuable and discarded the rest. There is not the slightest reason for occupying oneself with this rest once more.

For the other career pattern, that of mathematics teacher, he proposed a historical component as an essential core of the study course (*ibid.*, 34 sq.):

An entirely different situation presents itself for those who are qualifying for the profession of mathematics teacher in a secondary school. Their principal task will be to hand on mathematical knowledge to the new generation and, if possible, to engender love and admiration of man's achievements in this field through the centuries. For those students a knowledge of the historical evolution of the science is an asset which is not only valuable, but downright indispensable, and which alone, naturally in combination with a good command of present-day mathematics, will enable them to perform their duties satisfactorily. They are constantly concerned with phases from the development of mathematics which have long ago become a thing of the past and they have to make those phases clear and attractive to adolescents who in this way have to be trained in mathematical thinking.

It is striking how relatively implicit Dijksterhuis's justification remains for such a strong claim. He bases his claim exclusively on the value of history in the mathematics classroom of the future teacher, saying nothing about the use of history as a means of developing the teacher's own knowledge. He refers to the motivational function of history in the classroom and to the function of mental discipline and training which could be exerted without emphasising particular mathematical subjects (but probably intending classical mathematical topics).

And it is likewise striking how sure he was about the content structure for the historical component. In eight pages he set out "what historical topics *have to be* considered important for prospective mathematics teachers". (*ibid.*, 35; our emphasis) Greek mathematics is depicted as the "principal subject", "a thorough knowledge" of it is considered as "absolutely indispensable": it provides the conceptual and methodological guidelines. (*ibid.*, 36 sq.). The mathematics teacher whom Dijksterhuis has in mind is one for secondary schools, and presumably for their upper grades, as he himself had been for much of his career. Clearly, among the variety of types of secondary schools, only classically orientated 'grammar schools' (in Germany and the Netherlands *Gymnasium*) are considered: schools emphasising a historical approach by their entire curriculum and spirit, and thus supporting such an orientation in mathematics teaching.

Contrary to this traditional position from the 1960s, almost all of its assumptions—explicit as well as implicit—about the aims, functions and methodologies of a historical component have now changed, at least in general. No longer is the historical component of only indirect use for the trainee teacher's later classroom experiences; no longer is Greek mathematics regarded as the key field of historical knowledge; no longer is there a clear consensus about the content or structure of school mathematics courses; no longer is the historical component restricted to teachers of secondary schools—teachers in primary schools are now seen to be helped by historical resources as well.

On the other hand, the consensus about the usefulness of mathematics history courses which was apparent in the 1904 ICM motion can no longer be supposed to be shared by the entire mathematical community. This much is already implied in Dijksterhuis's view that history is inessential for 'general' mathematical studies, and remains a widespread view today.

References for §4.1

Dijksterhuis, E. J. 1962. 'The Place of History in the Training of a Mathematics Teacher', in: L.N.H. Bunt (ed.), *The Training of a Mathematics Teacher in the Netherlands.* Report of the Dutch ICMI Subcommittee, Groningen: Wolters, 34-43.

IMC 1904: *Verhandlungen des Dritten Internationalen Mathematiker-Kongresses 1904,* Leipzig 1905

4.2 International overview

Over the last two decades, the number of persons trained and competent in the history of mathematics has considerably increased in many countries. Some of these graduates have entered the teaching profession. The use of history in the mathematics classroom has become more common and we notice an increase in the number of courses in the history of mathematics in teacher training institutions. Nevertheless, these courses represent largely individual initiatives; the issue is how such courses may be more widely established and how to ensure a more stable and official status for a historical component in teacher education.

In what follows, we describe the current state of teaching history of mathematics to future mathematics teachers in a number of countries which, taken as a whole, will provide a fairly representative picture.

One of the characteristic trends is that practising a historical component in teacher training is no longer restricted to those countries with an extended tradition in mathematics history and a considerable mathematical community. We find a growing number of countries at the 'periphery' where, comparatively recently, historians of mathematics, or mathematics educators with a strong interest in mathematics history, have achieved an academic position where they are able to introduce mathematics history courses into teacher training after having qualified themselves by specialised research, mainly at one of the metropolitan centres.One example of this trend is **Morocco**. Historical information is present in its mathematics textbooks and we even find some simple activities based on this information, as Abdellah El-Idrissi remarks. In the past, teachers used to avoid these passages because they felt they lacked sufficient knowledge or were not convinced of the value of such an approach. Only comparatively recently do we find historians of mathematics and interested mathematics educators developing mathematics history at the universities and at the ENS (*Écoles Normales Supérieures*), responsible for teacher training. Research seminars are used to establish an infrastructure for communication, and courses in mathematics history are being offered. At the moment, such courses are offered at two of the four ENS. Up to now, these courses have been entirely optional and without an official or general status. The principal source material—both for the information given in school textbooks and for teacher education courses—comes from the prevailing cultural heritage, that is from the history of Arab mathematics.

Another example is presented by **Brazil** where in a few universities historians of mathematics have become established as university professors in recent years. Evidently, there is not yet an official status for mathematics history within teacher education, but at these universities courses are offered for future teachers. At several universities, graduate programmes in the history of mathematics have a formal status, usually in connection with mathematics education. The first initiative in Brazil for generally introducing mathematics history was taken by the Brazilian Society of Mathematics, remarkably, which suggested history courses as a component of mathematical studies as long ago as 1979. As a result of national meetings and seminars, there is now a considerable community of mathematics teachers actively concerned with the relation between mathematics education and mathematics history. This lends support to the use of a historical component in teacher training and the introduction of history into the classroom. This remarkably strong movement is particularly inspired by a new vision of mathematics history known as *ethnomathematics*. This has been promoted not only by the Brazilian scholar Ubiratan d'Ambrosio and others (e.g. Paulus Gerdes, Marcia Ascher) who have developed it internationally as a historically oriented research field, but it also features in the work of the Brazilian mathematician Eduardo Sebastiani Ferreira. The attractiveness of the historical dimension of ethnomathematics resides in its emphasis on a culture's own historical roots—in the Brazilian case on the unravelling and appreciation of mathematical elements of earlier, indigenous cultures in Latin-America.

The last example in this group is provided by **Hong Kong**. Due to the long-standing and successful research and teaching of Man-Keung Siu, a key person in the mathematics education community of Hong Kong, most teacher education courses there include some elements of history and many teachers are interested in historical issues, as Chun-Ip Fung reports. There are no official regulations requiring courses in mathematics history for mathematics teachers; yet at two of the universities in Hong Kong such courses are regularly offered. The courses at both universities provide us with the first example of another new trend: the extension of the history of mathematics to primary education. While one of the universities (Chinese University of Hong Kong) is exclusively concerned with the initial training of primary school teachers, the other one (University of Hong Kong) has courses for primary and secondary school teachers and also includes mathematics history in in-service training courses. It is also interesting to note an emphasis on a balanced account of the contributions of different cultures to the development of mathematics (thus avoiding a possible tendency towards Sino-centrism). In these courses the time spent on history ranges from a few hours to over forty hours. Course objectives range from simply opening up the historical dimension for teachers to highlighting the development of school mathematics, instructional use of historical materials or even, if time permits, an introduction to the world history of mathematics.

The next group of examples is from countries where there exists a longer tradition of research and teaching in mathematics history but where, for various reasons, a historical component is relatively poorly established.

In **Italy**, a considerable tradition of research in mathematics history has existed since the nineteenth century, but the development of mathematics as a school discipline took place in a manner quite different from that in other European countries (*cf.* Schubring 1996, 377sq.). Here no differentiation between study courses for mathematics teachers and those for mathematicians has emerged: mathematicians and teachers take the same final academic examination, the *laurea*. There are courses in mathematics education or in mathematics history at many universities, although they do not constitute a necessary part of the *laurea* examination. Both kinds of course are taught by the holder of a post in 'complementary mathematics' (*matematiche complementare*), a position which can be filled by experts from the fields of either mathematics education, mathematics history, or mathematical epistemology.

There is another peculiar element in the Italian system. In addition to the academic *laurea*, there is a national examination of mathematics teachers. This is organised by the state in a centralised way, and is the same for the whole of Italy. Although the examination confers the *abilitazione*, the qualification needed to be employed as mathematics teacher, the examination takes place rather infrequently (the last time, after an interval of more than six years). According to the programme for these oral exams, knowledge about "the most important moments of the history of mathematics" can be a subject for examination—which implies a rather traditional understanding of this discipline. Teachers qualifying from the university at a time between two national exams can only be employed temporarily. For a permanent position they have to pass the next national exam. There is now (October 1998) a plan to establish a 'Scuola di specializzazione per insegnanti'. This would provide a specialisation for teachers over two years, for those who have passed the *Laurea* exam, and the *abilitazione* would then be automatically conferred.

In the **Netherlands**, with a shorter research tradition but with a specialised centre for the subject at Utrecht, history of mathematics is taught at five of its twelve universities as an optional part of the study course, not specifically concerned with the training of mathematics teachers. While the universities confer the 'first degree' to teachers, i.e. the ability to teach in all grades of secondary schools, the polytechnics confer the 'second degree', qualifying for teaching in the lower secondary grades. In relation to the latter qualification, the teaching of mathematics history at the various institutions shows a broad spectrum. Some just have scattered information within the mathematics courses while others have formal historical courses (van Maanen 1995).

In **France**, research and teaching in mathematics history used to be performed within the discipline of philosophy, while the mathematical departments, in general, took no interest in history. Consequently, the history of mathematics was almost universally absent in the training of mathematicians and mathematics teachers. The situation began to change following the establishment of Institutes for Research into Mathematics Education (IREM) in 1969, whose task it was to provide in-service training for mathematics teachers. When, in 1975, the *Commission Inter-IREM d'Histoire et d'Epistémologie des Mathématiques* was created, it began to organise in-service training in the history of mathematics, with the aims of promoting the introduction of mathematics history into the classroom and of enriching teachers' understanding of mathematics. In particular, this second aim is characterised by concerns for the epistemological dimension of mathematics. This perspective, sketched in (Barbin 1995), is quite dominant in the French approaches to history, and can be seen as a reflection of the original official position of the history of mathematics as part of the discipline of philosophy. The *Commission inter-IREM* itself, as well as organising working groups at numerous local IREMs, has published an enormous number of pertinent papers and books which together constitute the richest source of historical material anywhere available (see §11.10.2).

The restriction of historical training to a voluntary component of in-service mathematics teacher training looked ready to change in 1989 with the establishment of the IUFM, *Instituts Universitaires de Formation des Maîtres*, the first time that higher education institutes had been established for teacher training in France. Primary school teachers had previously been trained at teacher training *écoles normales.* Secondary mathematics students can enter the IUFM after having obtained a *licence* in mathematics (requiring three years of university studies). While the IUFM provide, a professional form of teacher training for secondary schools, the mathematics part of the courses is to a large degree provided by professors connected with an IREM. In particular, those among them who are committed to the work of the *Commission Inter-IREM* are active in ensuring a historical component in courses for future mathematics teachers. Sometimes, there is opposition from colleagues who are against such an innovation. At one IUFM, for example, an attempt to make the history component compulsory for secondary school teachers failed and was retained as only an optional subject. At the present time, history of mathematics is not identified as part of the IUFM's official curriculum programme. In a recent list of competences, established by the *Ministère de l'Education,* history of mathematics is mentioned as a topic to be studied, but not to be assessed. Nonetheless, in several of the 29 institutes there are either optional courses in history in the second (final) year or the option of choosing this subject for the concluding so-called professional thesis, as Éliane Cousquer reports.

Following the setting up of the IUFM, a recent development is for some universities to offer 'pre-professional' modules in order to prepare students better for teacher training. These modules include some history of mathematics and history of science. Another new development is the introduction of 'culture générale' modules into the first years of university studies for science and mathematics students and these also often contain either lectures or taught courses in the history of mathematics.

In **Germany** the situations in the former German Democratic Republic and in the pre-1990 Federal Republic used to be different. In the Federal Republic (known as West Germany), which we consider here first, a decisive break occurred in the late 1960s and the 1970s. The cultural values of those social classes which had until then dominated the aims and visions of the educational sector—the so-called *Bildungsbürgertum*—lost ground, as a consequence of, first the student movement, and later on the process of radical individualisation. The key pattern of the former established set of cultural values had been *historicism*, that is referring actual values to supposed or real historical roots, preferably in the 'Christian Occident' or in classical Antiquity. From the 1970s these values could no longer be regarded as socially shared. The effect of this radical break with tradition

was that all elements of school curriculum content reminiscent of historicism were removed. In language teaching, the classical texts by poets like Goethe and Schiller were replaced by non-literary texts ('Gebrauchstexte') and even history instruction itself was in danger of being replaced by the study of social processes. In the same way, all allusions to mathematics history which used to be present in mathematics textbooks—mainly for the upper grades of the socially and culturally high status *Gymnasium*—were eliminated. The *Gymnasien* themselves were dismantled in 1972 and replaced by secondary schools no longer emphasising historicism and classical values (with the exception of Bavaria). New mathematics textbooks produced since the 1970s are practically void of any historical references. Only recently has a new textbook series been published which includes some elements of history—this time no longer restricted to the upper grades or selected schools.

For teacher education at universities, i.e. for future *Gymnasium* teachers, there used to be some rather isolated and rare historical lectures, while teacher education at the pedagogical colleges for primary and lower secondary schools contained no history at all. This situation has gradually changed since the 1970s. Firstly, as a consequence of the student movement, reflection on mathematics and its social function—and this included history—became enhanced so that at some universities history became integrated into the curriculum for lecture courses, as an optional subject (although this had little effect in practice). However, graduates of the newly expanded history of science centres at the universities of Hamburg and Munich would soon make it possible for other universities to offer courses in the history of mathematics. And, following the integration of the pedagogical colleges with the universities in most of the federal states during the 1980s, it is now also possible for future primary teachers to take such courses. Further details are given in a forthcoming article by Schubring.

The development in Germany is *ad hoc* and has no official support. This is illustrated by events following the integration of the GDR into the Federal Republic in 1990. As in other fields, the new federal states immediately adopted the regulations of the old Federal Republic. The compulsory teaching of history as part of mathematics education was mostly abolished and the centre for the history of science at Leipzig was dismantled. Nowadays, in the majority of the new states, mathematics history does not figure at all in the curriculum for teacher education. In two of the new states some minimal history is again prescribed, but this involves just three universities.

In general, looking at teacher examination regulations issued by the federal states, one can detect a certain progress. In half of the current 16 states, mathematics history is mentioned either as an optional subject of studies (but not of exams) and usually grouped together with reflection on foundations and logic or, in four states, as a compulsory subject of studies for future secondary school teachers (but rarely as a subject of examination). Even in the latter cases, there is not much emphasis on history as such—'insight' into the development of mathematics is expected—and in both cases a bare minimum of study time is prescribed, seldom more than two weekly hours in a one-semester course. The two 'western' states who prescribe these studies are those housing the two specialised history of science centres, the small state of Hamburg and Bavaria, which is the only state to have held on to a considerable part of the *Gymnasium* traditions. It is also worth mentioning that in a small number of states, in Baden-Württemberg and Brandenburg for example, regulations provide an opportunity for future primary school teachers to study the history of mathematics.

Whatever the regulations, the practice is quite different. Whereas occasional history courses may be offered in all states, including those not mentioning history in their regulations, and local curricula for teacher education at a number of universities include mathematics history, regular courses outside the Munich and Hamburg centres, and now Berlin, are rare. Where lecture courses are offered, these are initiatives by enthusiasts, and there is no coordination or common structure. Unsurprisingly, no specific teaching material is available for the courses. Finally, it is worth noting that the history of mathematics education and of mathematics teaching are both mentioned in the curriculum regulations of several states for trainee teachers. This itself opens up a number of possibilities, at least potentially.

The four European countries we have just considered have a rather poor record for the teaching of the history of mathematics, despite the strengths of their long-standing mathematical tradition and a strong mathematical community. Some other European countries, with smaller communities of mathematicians or a less impressive mathematical tradition, fare rather better in this respect.

Austria, for instance, unlike its neighbours Germany and Italy, includes the history of mathematics as a recognised component of teacher training for secondary schools. Although Austria is a federal state, there are common national regulations for the examination of secondary school teachers throughout Austria, as Manfred Kronfellner reports. Future mathematics teachers are required to take an oral examination in either philosophical aspects or historical aspects of mathematics, and this necessitates prior study. At all universities, there are regular courses in the history of mathematics to enable students to prepare for the examination. In one university, the Vienna Technical University, the study course in history is to become compulsory. It is worth noting that there are in Austria two history textbooks specifically for mathematics teachers (Kaiser/Nöbauer 1984; Kronfellner 1998). These textbooks emphasise the *Problemgeschichte*, i.e. the evolution of mathematical ideas.

In **Poland**, the teaching of mathematics history is widely practised, and this practice was unaffected by the political changes in Poland around 1989. Most universities offer a course in the history of mathematics, as Ewa Lakoma was able to establish by a questionnaire sent to about one hundred persons teaching at the various teacher training institutions who are interested in or active in the history of mathematics. These courses may be either compulsory or optional. Up to now, there are no general regulations concerning the curriculum for future teachers. Each university runs its own programme. The Ministry of Education is, however, preparing regulations to certificate university programmes for teacher education. In July 1998, only a few universities did not have history of mathematics in the curriculum for mathematics teachers. The courses usually comprise 30 to 60 hours per year and are given as lectures to students of the third, fourth, or fifth year of the five year study course. The history of mathematics lecture course often has supplementary exercises, demanding usually 30 hours per year.

The history of mathematics lecture courses in different universities have a relatively common structure, namely:
1. The first traces of concepts of number and shape in ancient times
2. Empirical mathematics in ancient Egyptian and Babylonian times
3. Greek mathematics before and after the time of Alexander the Great
4. Mathematics in the East: China, India and Arab countries
5. European mathematics in mediaeval times and during the Renaissance

6.   The development of the calculus and probability in the 17th and 18th centuries
7.   Algebra from the 17th to 19th centuries
8.   Set theory in the 19th century
9.   Geometry in the 19th century; the development of non-Euclidean geometries
10. The Erlangen programme and the Hilbert programme
11. Hilbert's problems
12. The Polish school of mathematics
13. Notes on the law of parallelism in the history of mathematics teaching; how to benefit from historical knowledge in mathematics education.

While the structure here seems to reflect the general pattern of certain mathematics history textbooks, topic 12 shows the extent to which course content can be related to the cultural history of one's own country. Ewa Lakoma reports:

The tradition of the Polish school of mathematics is so strong that the history of mathematics, in a natural manner, is a matter of interest to mathematicians and to students. In fact, Professor Andrzej Mostowski, a great mathematician, also gave lectures on the history of mathematics.

Topic 13 in the list is also significant, showing the new approach which mathematics educators take to the history of mathematics. History can serve as a source of reflection or, often in too simplistic a way, as a direct guideline for the practice of teaching.

As with Poland, the strength of the cultural roots of mathematics in **Portugal** seem to have inspired the establishment of a historical component within mathematics teacher education when a specific diploma for mathematics teachers and a related curriculum was set up in Portugal in 1972. The basic components of this study course are didactics, methodology, and psychology of learning. Initially, a one semester course used to be devoted to mathematics history, but this can now be extended to two semesters. This course is offered at all universities and appears in all programmes for the training of teachers for upper secondary schools. For the training of teachers for primary schools and lower secondary grades (1 to 6), a history of mathematics course is not generally included; there would not be enough time, since these students also have to study other disciplines (according to a communication by Jaime Carvalho e Silva, and Amaro 1995).

The structure of the course 'The history of mathematics' at Coimbra University shows the importance attached to the cultural history of the nation, inspired by the achievements of Portuguese mathematicians in the period of the 'voyages of discovery' in the 15th and 16th centuries, and the consequences of the educational reforms of 1772. It has four components:

–   History of analysis from Archimedes to Weierstrass
–   History of geometry
–   History of numerical analysis
–   History of mathematics in Portugal.

The last of these components discusses the following issues:

–   Why study history of mathematics and history of mathematics education?

- The university reform of 1772
- The *Libro de Algebra* of Pedro Nunes (1567)
- The *Principios mathematicos* of Anastacio da Cunha (1790)
- The mathematics education reform of Sebastião e Silva (1962-1973).

Another course, taught at the Universidade do Minho, at Braga, emphasises the mathematical contribution of antiquity (Babylonians, Egyptians, Greek mathematics) and for modern times just two aspects: Hilbert and the history of mathematics in Portugal (Amaro 1993, 456).

The purpose of teaching the history of mathematics, as stated by Portuguese mathematics educators, primarily at the pedagogical colleges and teacher training institutes for grades 1 to 6 ('Escolas Superiores de Educação'), is to enhance the mathematical understanding of future teachers and to develop methodological reflection about teaching practice. For example, at the Pedagogical College at Castelo Branco students in their final year of a degree course in mathematics or science take a course whose declared aims are, among others:

- to construct a basic knowledge about the development of mathematical thinking with respect to numbers, numeral systems, early computing, fractions, and geometry;
- to foster an understanding of how mathematics is used and why it is needed in society;
- to develop an understanding of the nature of mathematics;
- to develop teaching and learning skills, based on the study of specific aspects of mathematics history. (*ibid.*, 457)

In order to participate successfully, the student teachers have to choose a topic from the syllabus of the 5th or 6th grade and are expected to devise a plan for a learning unit which connects mathematics history and learning activities, based on an exploration of available historical literature. (*ibid.*, 457 sq.)

National seminars on the history of mathematics take place, the eleventh of which was in 1999. These seminars promote cooperation between secondary school teachers and university professors. Recently, a number of students have obtained a master of education degree in which the history of mathematics was a major component.

The term 'cultural identity' could be used for the cases of Poland and Portugal in order to describe the specific 'rooting' of mathematics within their respective cultures and societies. These types of socially shared values can be described as arising 'from below'. We can also identify examples of the introduction of cultural values through the educational sector 'from above'. In these cases, ideological judgements are responsible for shaping the content and structure of the educational system, as with certain centralised state policies. (Ideology' here is not intended to carry an *a priori* negative character: the term. 'idéologie' evolved in France around 1800 as the science of ideas and of their emergence and development.) This is of relevance for the institutionalisation of mathematics history since, to take recent well known examples, socialist countries used to integrate mathematics history into Marxist philosophy. We will discuss the impact of this for teacher education for some particular cases.

Our first example is **China**. Although China has one of the oldest traditions in mathematics, the subject fell into decline after 1600 and from the middle of the nineteenth century Western mathematics became dominant. According to the accepted view within the People's Republic of China, the traditions of Chinese mathematics fell into oblivion and became only revalued and reassessed when the Republic was established in 1949. As Dianzhou Zhang reports:

Once the Chinese people won their real independence in 1949, the government launched a movement of patriotism, and asked mathematical educators to foster pupils' patriotic thought by means of incorporating more knowledge of Chinese history of mathematics. This led to researches into the ancient history of mathematics being conducted. As a consequence, when Chinese historians of mathematics were invited to compile new textbooks, a number of mathematical results were then renamed after, or more correctly attributed to, Chinese authors.

As an example, Zhang mentions the replacement of Pythagoras theorem by 'Gou Gu theorem', Pascal's Triangle by 'Yang Hui Triangle', and Cavalieri's principle by 'Zu Geng principle'. It may be noted here that in recent Arab textbooks Pascal's triangle is referred to as 'Ibn Munim's triangle'. As regards teacher training, Zhang reports that

in normal colleges and universities there is supposed to be an optional course on the history of mathematics (45 classroom hours). However, because of the lack of mathematical historians to teach the subject, many universities are unable to offer a course of mathematics history when the students elect to do it.

In actual fact, it appears that research into the history of mathematics only rarely transfers into teaching.

In the former **Soviet Union**, mathematics history was cited in support of the case for the validity of the Marxist thesis that the development of scientific ideas is determined by social conditions (the famous 'externalist position') by Boris Hessen in his seminal and ground breaking paper on the social roots of Newton's *Principia* given to the 1931 International Congress of History of Science (Hessen 1931). This had an important influence on mathematics teacher training. For students of mathematics at the pedagogical institutes, future secondary school teachers, a course in the history of mathematics became compulsory. The course programme, valid for the entire country, was prepared and supervised by a committee comprising the experts in this field. It would appear that the course was well taught at the better of the pedagogical institutes. In fact, the number of specialised textbooks, many translated into other languages, show that this course was well established (examples are the books published by G. P. Boev in 1956; by Rybnikov, 3 editions between 1960 and 1994; by G. I. Gleizer, 3 editions between 1964 and 1983; by I. Ya. Depman in 1965 and by B. V. Bolgarskii in 1974). A particularly popular book was, and still is, a textbook especially prepared for the pedagogical institutes: A.

Youschkevitch *et al.* (eds.), *Khrestomatiya po istorii matematiki* (Source book on the history of mathematics), Moscow 1976-77.

In **Russia** today, the situation is clearly more variable and there is no longer a centrally prescribed programme for teacher training. Institutions can devise their own programmes. It is clear, however, that at many pedagogical institutes (now renamed pedagogical universities) history courses continue to be a component of the training of mathematics teachers and in some instances may even be compulsory. (This information is kindly provided by Sergei S. Demidov, Moscow.)   At the Rostov Institute, a course on the history of mathematics teaching at Russian schools has been developed (see Poljakova 1993, 1997).

A revealing case study, described in a forthcoming article by Hans Wußing, is provided by the former **German Democratic Republic**. While the history of mathematics was a research area and was taught at a modest level, though without any particular official support, in the first two decades of the GDR's existence, the government rather suddenly, and without being urged to do so by the discipline, declared in 1969 that the study of mathematics history was to be a required component of teacher training for secondary schools. Analogous decisions were taken for physics, biology and chemistry. In general, it proves easier to introduce mathematics history into textbooks than to change the practice of teacher training. In the case of the GDR, however, the reverse was true.   While historians of mathematics and publishers of textbooks were unhappy with the quality of those parts of school mathematics instruction materials that related to history, the situation for teacher training was much better. A training programme was established so that after a certain time almost all the universities and pedagogical colleges were provided with professors or lecturers competent to give courses in the history of mathematics. Suitable teaching materials and textbooks were developed, in particular Hans Wußing's successful and much translated *Vorlesungen zur Geschichte der Mathematik* (Lectures on the History of Mathematics).

We have seen that there are situations where individual initiatives have succeeded in introducing a historical component into teacher training and we have raised the issue of whether such individual initiatives might become more widely adopted. We have also seen that there are cases of centralised, directed programmes decided by a ministry or state which might well include elements of the history of mathematics. It is interesting to note that there are some fortunate cases in small countries, with just one university or teacher training institution, where individual actions become, in effect, official measures.

One such case is **Latvia**, one of the Baltic states, formerly part of the Soviet Union and now an independent country, with only one university, in the capital Riga. Daina Taimina reports that when she began to lecture at Riga University (now the University of Latvia) in the late 1970s she was able to establish a course in the history of mathematics which had not previously been offered. Presumably, establishing this course was facilitated by the then current state policy of support for history of mathematics in mathematics education. The course became well accepted and a part of the regular study programme for mathematics teachers. Eventually, in

1990, she was able to publish a textbook for this history course, being the first textbook in the Latvian language on the history of mathematics. The course comprises 25 hours in one semester and is taken as one of the last courses a prospective teacher takes before finishing. Former students rated the history course as the finishing touch to all the other mathematics courses they had studied. In the new 1998 regulations for teacher training in Latvia, the mathematics history course is prescribed as a standard course of two hours per week in the last, ninth, semester within a study field of 'educational issues'.

A somewhat analogous success has been achieved in **Cyprus**, where there is only one university in the capital, Nicosia. Thanks to the energetic activity of individuals, mathematics history has been introduced into the training of future primary school teachers. This training is given at the Department of Education of the University of Cyprus. This component in Cyprus is therefore also an example of the new trend to bring some mathematics history into the training of the primary school teachers. This particular innovation is all the more remarkable since these future teachers are trained as 'generalists', required to teach almost all subjects of the primary school syllabus, unlike their secondary colleagues who are usually trained for teaching just one or two subjects. A further difficulty for initiatives of this sort lies in the fact that primary students, both here in Cyprus and in other countries, may not be well prepared in mathematics from their own secondary school studies and may even have dropped the subject. We present in the next section (§4.3.1.2) the approach being adopted in Cyprus to use history to improve the attitude of these students towards mathematics and even to enhance their mathematical competences.

This trend to create a historical component for future teachers is also represented, albeit rather patchily, in **Britain**. Teacher training institutions are no longer quite as autonomous as they were, but a number of universities include some aspects of the history of mathematics in courses for future school teachers, either as taught components or as study topics. The British Society for the History of Mathematics (BSHM) established an education section in 1990 (HIMED) which organises annual conferences and promotes the use of history in mathematics education. Both the Mathematical Association and the Association of Teachers of Mathematics promote the use of history of mathematics in teaching through journal articles and conference activities. Among universities offering history of mathematics courses, whether for mathematics or mathematics education students, the Open University, a distance learning university with a large number of students, is prominent.

The presence of more historical components is related to changes in social conditions, in particular in the composition of school populations. Many industrialised countries are becoming ever more 'multi-cultural' because of growing migration from the so-called Third World. This is particularly the case with former colonial powers. These social changes are clearly reflected by new claims being made for the teaching of mathematics history in schools, which in turn influence the context in which history is presented to trainee teachers.

One argument for including history in mathematics courses is that it helps to 'humanise' mathematics. While this may seem a traditional motivation for the subject, a new reason being proposed is that it helps to overcome mathematics anxiety or mathematics avoidance. It is argued for instance that many girls in Britain do not continue with mathematics learning beyond the age of sixteen because mathematics is seen as being about things, not people. Probably the strongest support for including history as a part of mathematics is a new claim that history helps to emphasise the subject's multicultural inheritance and the culturally dependent nature of the subject. It is argued that including the history of mathematics is particularly important in a rapidly growing multicultural and pluralistic society, and that in Britain it helps to counter still prevalent eurocentric or colonialist views. A revealing presentation of this claim is given as a rationale for a history course at the University of Greenwich (Sheath, Troy and Seltman 1996):

Finally, we intend that students be aware of the issues inherent in interpreting the mathematics of other times and cultures from the viewpoint of our own. It may be argued that deep in the consciousness of the West is the assumption of cultural superiority, the assumption that almost all gains in human civilisation, and certainly mathematics, have originated in WASP (White Anglo-Saxon Protestant) culture. Such ethnocentricity urgently needs to be tempered by knowledge of the contribution of all humanity to present-day mathematics, which is itself global in character. The hierarchical view whereby some contributions are considered superior to others, as if there were some quantitative measure, has to be tested.

Such challenges to 'WASP' cultural hegemony have been formulated even more radically in the **United States**. There have resulted profound changes in the declared rationales of the educational system and in the content and structures of syllabuses, with the intention of replacing the cultural values and curriculum representative of exclusively 'dead white males' by consideration of the contributions of women, of minorities and of other cultures.

As with all federal states, it is quite difficult to give a fair general description of the educational scene for the whole USA and, in particular, as to the acceptance of a historical component for trainee teachers. There is the added complication that US requirements for teacher certification are defined not only by individual Boards of Education but also by other organisations. Probably the most influential of such organisations is NCATE (National Council for Accreditation of Teacher Education), one of the two accrediting bodies in the USA for teacher education programmes. Each individual university or college decides for itself whether to have particular programmes, such as teacher education, accredited and to which standard. Having courses accredited ensures the employability of the graduates.

A survey carried out by Victor Katz (1998) shows that in the majority of the US states certification requirements for teachers at secondary schools require the study of a course in the history of mathematics, whether this is for mathematics teacher education programmes or for individuals presenting themselves for accreditation. Traditionally 'neutral' formulations of competences in, for example 'foundations and history of mathematics' (Maine) are the exception. The minimal expression of mainstream programmes is to require "studies of the historical and cultural significance of mathematics" (Pennsylvania). More explicit and typical is the requirement in Montana that:

for the prospective teacher, the programme shall . . . include experiences in which they . . . explore the dynamic nature of mathematics throughout history and its increasingly significant role in social, cultural and economic development.

The trend of multi-culturalism is tellingly explicit in the California state requirements. Standard 14 of the Commission on Teacher Credentialing states, of programmes in teacher education:

History of Mathematics: Each programme requires students to have a foundation of knowledge about the history of mathematics, and a historical perspective regarding the development of mathematics.

The rationale for standard 14 is that "a foundation in the history of mathematics enables students to gain a rich understanding of the origins of mathematical concepts". Reviewers who judge whether a programme meets this standard are expected to consider the extent to which

(i)   The programme requires students to understand the chronological and topical development of mathematics.
(ii)   The programme requires students to understand the contributions of historical figures, including individuals of various racial, ethnic, gender, and national groups.
(iii)   The programme requires students to understand the contributions of mathematics to society, and its impact on society.
(iv)   The programme provides opportunities for students to be exposed to the mathematical discoveries that have affected the course of civilisation.
(v)   The programme has other qualities related to this standard that are brought to the reviewers' attention by the institution.

The requirements of the NCATE emphasise most explicitly the new trend for multi-culturalism and the consideration of minorities. The process of meeting NCATE Standards is fairly rigorous. For example, to meet the requirements for accreditation for grades 7-12 mathematics teachers, a programme must require students to meet a long list of 'outcomes' specified in three broad areas: mathematics, teaching preparation and field-based experiences. A given programme must state, for each particular outcome, how it is met, whether by a specific course or by experiences over several courses or in other ways. Outcome 1.7 for Grade 7-12 mathematics teachers, for example, states that

Programmes prepare prospective teachers who have a knowledge of historical development in mathematics that includes the contributions of underrepresented groups and diverse cultures.

This quotation is from the NCATE Curriculum Guidelines for mathematics which were prepared by the National Council of Teachers of Mathematics and are (for grades 7 to 12 mathematics teachers) from the 1993 revision of these guidelines (p. 429). Identical quotations could be made for Kindergarten to grade 4 teachers (p. 417) and for grades 5 to 8 teachers (p. 423).

Although primary trainee teachers should also study history of mathematics, according to NCATE, primary teachers in the USA are not certified in an academic field. Most of them have a very minimal background in mathematics so that there is not yet in practice a drive towards introducing a historical component for them.

One can conclude that the majority of prospective teachers for secondary schools are exposed to history of mathematics in some form, and that it is delivered with a dominantly multi-culturalist perspective.

A further new trend in the justification for including a historical component in mathematics teaching is a reflection of changes occurring in the school system, at least in a number of industrialised countries. For whatever reasons, there is an increasing tendency for students in secondary schools to show a marked distaste for mathematics. This is related to the deep structural change of those schools which prepare students for entrance to the universities. These schools were once for a (relatively small) minority of students from a social élite. These schools now receive students who want to proceed to a university education from a much broader proportion of the population, perhaps 30, 40 or even 50 percent of an age group instead of a small minority, yet the curriculum has not been correspondingly modified. The mathematics curriculum, for instance, is regarded by students as particularly boring. This social pressure against mathematics as a main school discipline is felt most strongly in Scandinavian countries.

In **Denmark**, for instance, mathematics history is included for its humanising qualities so that students see better the attractiveness of mathematics. A first step was the Ministry's new syllabus of 1988 for the *gymnasium* (i.e. grades 10-12), according to which mathematics should be taught with due respect to three aspects: its history, its inner structure, and its applicability. Eventually, an entirely revised syllabus of 1994 for the *folkeskole* (i.e., grades 1-10) demanded that mathematics history should be included in the teaching of mathematics and the importance of mathematics for the development of the society should be illustrated, thus giving the measure a clear social perspective.

As Torkil Heiede reports, in Denmark the history of mathematics is now therefore in some sense obligatory, and for all school grades. This was, in fact, only possible because the history of mathematics has a long and unbroken tradition in mathematics education in Denmark. At Copenhagen University the tradition goes back about hundred years (especially to the two experts in Greek geometry, the mathematician H.G. Zeuthen and the philologist J.L. Heiberg). History of mathematics was given a new impetus there in the 1930s, during the residence of Otto Neugebauer, the expert in Babylonian and Egyptian mathematics. Furthermore, at the four newer universities, particularly in Aarhus, the history of mathematics has been developed, with the result that mathematics teachers in Danish gymnasiums (all of whom have a master's degree in mathematics, or perhaps in physics or chemistry) have always had the opportunity of attending a course in the history of mathematics as part of their training.

The new syllabuses have not only encouraged the production of new mathematics textbooks with integrated aspects of mathematics history but the historical component in teacher training for secondary schools has changed to become an important subject for future teachers. At the universities the number of students attending undergraduate courses in the history of mathematics has increased considerably. For instance at the University of Copenhagen, the mean audience used to be 20 students per year, but in the year after the new regulations for the gymnasium had been published, the number swelled to 40 and in the following year to 80 as older students realised that it was necessary for them to take this course if they were to become gymnasium teachers. Now the numbers have stabilised again at a level of 50 per year. Unfortunately, the larger attendance combined with departmental adjustments between the undergraduate study subjects has forced a reduction in both the size and scope of the course. Also at the teacher training colleges which educate the future *folkeskole* teachers, history of mathematics now has a more prominent place than ever before.

As regards in-service training for teachers at gymnasiums, the number of courses in the history of mathematics has been increased in recent years. In-service training for *folkeskole* teachers is better structured, with courses in the history of mathematics frequently occurring, usually as part of more general mathematics courses (Heiede 1996b).

The last new trend to be presented here is, from a structural viewpoint, highly innovative. It primarily concerns countries which had been formerly subject to colonialism and where their own cultural traditions had been not only overlaid by the colonial power's own culture and values but intentionally suppressed. Since these traditions were oral, they would inevitably become lost through lack of use.

The best example of a new evaluation of these traditions is that of **Mozambique**, a former Portuguese colony (Gerdes 1998). Here, the unravelling of hidden or suppressed ethnic Black African traditions in developing and practising geometry and arithmetic has not only developed into a research programme for mathematics history and mathematics education (thus constituting a novel approach to ethnomathematics) but has also become the rationale for mathematics teacher education. This programme of ethnomathematics, called 'mathematics in history', fulfils the function of permitting the trainee teachers to establish an intrinsic relationship towards mathematics and enables them later to teach mathematics to their students as a meaningful subject rooted in their own culture. When the Mathematics Teacher Education Programme began after independence, "few students ... actually liked mathematics; many spoke frequently about mathematics as 'the beast with seven heads', apparently having no utility in society and no roots in Mozambiquan and African cultures."(Gerdes 1998). Mathematical traditions and practices of daily life which have survived colonial rule can be incorporated into the school curriculum. We have already noted the value other countries place on a link between the mathematics that is studied and the cultural history of the society, but this Mozambique programme relies on a much more intrinsic relationship.

The ethnomathematical programme 'mathematics in history' was inaugurated at the Pedagogical University of Mozambique, training teachers for primary and secondary schools and has now built up considerable experience. The aim of the programme is (quoted from Gerdes 1998, 41):

to contribute to a broader historical, social and cultural perspective on and understanding of mathematics. The first theme 'Counting and Numeration Systems', gives a good start, because the students can begin to analyze and compare the various ways of counting and numeration they learned in their life. After they have discovered the rich variety at the national level, they then are brought into contact with systems both from other parts of Africa and the world, and from other historical periods.

The aims of this course permeate the entire curriculum but special emphasis is given in a course during the fourth of the ten semesters. Mathematics in history is, moreover, a subject of later specialisation, for a thesis and examination. As Gerdes emphasises, "mathematics is a universal activity; that is, it is a *pan-cultural* and *pan-human* activity", going on to stress that the development of mathematics is not *unilinear* but *multilinear* (*ibid.*, 47).

In concluding this section, we can sum up by saying that the scope, function and vision of mathematics history in teacher training programmes are undergoing profound changes. Earlier idealistic views, focusing on a standard canon of Western, particularly Greek, mathematics are largely fading away, as is the view that it is only suitable for students of the upper grades of secondary schools for the social *élite*. Everywhere in the educational system, there is evidence of systemic changes, even crises, and the introduction of mathematics history responds in different ways to these crises and changes. In many industrialised countries, we find a widespread aversion of students to mathematics or, at least, an avoidance of it. Here, mathematics history is seen as a way of combating this distaste for mathematics by presenting mathematics as a living, 'human' subject. This might also apply to social subgroups, minorities or populations hitherto excluded from higher learning.

While in many of the European States, with eminent traditions in mathematics, new visions on the role of mathematics history within the teaching of mathematics have not yet widely emerged, we find that countries on the periphery, as it were, have been more successful in this respect. By broadening the cultural perspectives, in particular in the centres of former colonial powers or in countries where racism has been rife, mathematics history has achieved a novel and important function in helping to create a multicultural vision.

References for §4.2

Amaro, G. 1995. 'The use of mathematics history and epistemology in mathematics education of teachers', *Proceedings 1. UEE Montpellier 1993*, 453-459

Barbin, E. 1995. 'L'histoire des mathématiques dans la formation des enseignants de mathématiques en France', 491-492

Gerdes, Paulus 1998. 'On culture and mathematics teacher education', *Journal of mathematics teacher education*, 1, 33-53

Heiede, Torkil 1996b. 'History of mathematics and the teacher', in: R. Calinger (ed.), *Vita mathematica*, Washington DC: M.A.A., 231-243

Katz, Victor 1998. 'History requirements for secondary mathematics certification', report.

Maanen, J. van 1995. 'The place of the history of mathematics in teacher training. The situation in the Netherlands', *Proceedings 1. UEE Montpellier 1993*, 495-496

Poljakova, T. S. 1993. 'Programma kursa po istorii otechestvennogo skol'nogo (...), v 1992-1993 gg.', *Matematika v skole*, **3**, 32-34

Poljakova, T.S. 1997. *Istoria otechestvennogo skolnogo matematicheskogo obrasovanja*, Rostov

Schubring, Gert 1996. 'Changing cultural and epistemological views on mathematics and different institutional contexts in 19th century Europe', in: C. Goldstein, J. Gray, J. Ritter (eds.), *L'Europe mathématique - mythes, histoires, identités. mathematical europe - myths, history, identity*, Paris: Éditions de la Maison des Sciences de l'Homme, 361-388

Schubring, Gert (forthcoming). 'Mathematik-Geschichte in Mathematik-Unterricht und Ausbildung der Mathematik-Lehrer in der alten Bundesrepublik', in: *Proceedings of the conference 'Komparative Forschung zur Entwicklung des Mathematikunterrichts und der Mathematik-Didaktik in der BRD und der DDR 1945 bis 1990'*, Bielefeld/Ohrbeck 1996

Sheath, G., Troy, W. and Seltman, M. 1996. 'The history of mathematics in initial teacher training', *Proceedings 2nd UEE Braga 1996*, ii, 136-143

Wußing, Hans (forthcoming). 'Mathematikgeschichte als Bestandteil der Mathematiklehrer-ausbildung in der ehemaligen DDR', in: *Proceedings of the conference 'Komparative Forschung zur Entwicklung des Mathematikunterrichts und der Mathematik-Didaktik in der BRD und der DDR 1945 bis 1990'*, Bielefeld/Ohrbeck 1996

4.3 Examples of current practice

We can summarise the grounds for including a historical component in the training of teachers, as these purposes have been developed over the last decades, as setting out to achieve four main functions:
1. letting teachers know of the past of mathematics (the direct teaching of the history of mathematics);
2. enhancing teachers' understanding of the mathematics they are going to teach (methodological and epistemological function);
3. equipping teachers with the methods and techniques of incorporating historical materials in their teaching (use of history in the classroom);
4. enhancing teachers' understanding of the evolution of their profession and of the curricula (history of mathematics teaching).

The following examples of practice in a number of countries show how these functions are currently achieved in teacher education.

4.3.1 Current practice in initial teacher training

4.3.1.1 Hong Kong: On finding a place for history in primary mathematics teacher education

Chun-Ip Fung

Setting objectives is troublesome for brief history courses of some 15 hours duration. Unfortunately, this was what I have been confronted with during the past

John Fauvel, Jan van Maanen (eds.), *History in mathematics education: the ICMI study*, Dordrecht: Kluwer 2000, pp. 110-113

seven years when teaching history of mathematics to both pre-service and in-service teachers, who have not more than a high school graduate level in mathematics and who are supposed to teach until Grade 9. The main struggle is to resolve two possibly conflicting aims: whether you want students to acquire a rough picture of what has happened in mathematics in the past, or whether you want students to be able to capture and organise historical material for instructional use. Is it better to develop a historical viewpoint during the study of historical materials or a pedagogical one?

During the past seven years of experimentation, I have had first-hand experience of the struggle between these aims. In an attempt to enhance students' historical knowledge, a brief overview of the world history of mathematics was given, important historical events being highlighted. To provide summative assessment, an open-book written examination was set (this was for a group of serving teachers and also a group of pre-service teachers). Examples of the questions are:

1. Right-angled triangles were studied in the *Elements*, the *Zhou Bi Suan Jing*, and the *Jiu Zhang Suan Shu*.
   a) Give evidence from each.
   b) Comment on the difference of their achievements.
2. Euclid's *Elements* were famous for rigour and deductive reasoning. In Book 1, for example, propositions were built on definitions, postulates and common notions.
   a) Give an example to illustrate the above description.
   b) Can you see any exception? Please comment.
3. Some people say that the idea of limit existed at the time of Euclid. Discuss this issue with reference to Proposition 1, Book 10 of the *Elements*, which reads:

Two unequal magnitudes being set out, if from the greater there be subtracted a magnitude greater than its half, and from that which is left a magnitude greater than its half, and if this process be repeated continually, there will be left some magnitude which will be less than the lesser magnitude set out.

This approach concentrated on building up students' historical knowledge, at the expense of the pedagogical dimension. It was not particularly successful, even on the level of the acquisition of factual knowledge. There are several possible explanations. Firstly, the students' mathematical knowledge was sketchy to begin with. This exerted great pressure on their reading of mathematical texts, which are generally not written for readers of their mathematical background. Secondly, not having enough time to adjust to a more historical approach, students often had a high anxiety level. Thirdly, the fact that students often lack general historical awareness for cultures.

An alternative approach was tried, which requires students to locate and organise historical materials for teaching purpose. Owing to the unavailability of original sources, students are only able to consult secondary sources. For most of the time, students preferred to read books containing short popular accounts. Traditional history of mathematics texts were shunned. This may be due to the fact that most history of mathematics texts presuppose a certain knowledge of mathematics and a certain familiarity with Western history, one or both of which was absent for these

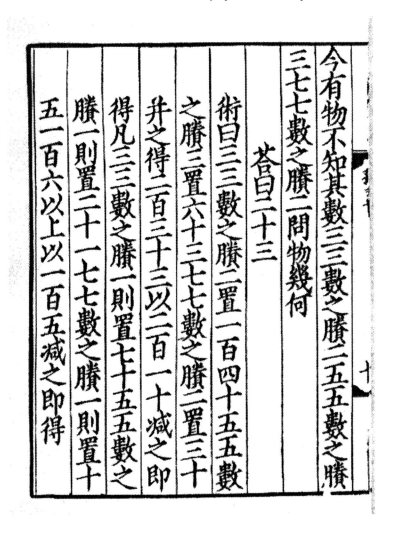

Figure 4.1: The Chinese remainder theorem from SUN ZI SUAN JING (Master Sun's Mathematical Manual), c. 4th century. For Hong Kong student teachers this type of source presents difficulties.

students. A problem which this led to was the accuracy of quoted historical facts. Since students made use of sources which focused more on arousing interest than on accuracy, the historical events and information about characters given in these texts might be based on unconfirmed or even erroneous information. Without a critical training, students often muddle facts with rumours, since the mathematics story books on which they so heavily relied do not normally distinguish fact from hearsay. At the end of the course, I was assigned the difficult task of evaluating the project, using assessment criteria which called for evaluating the student's work according to (i) the accuracy of historical materials used, and (ii) the extent to which the curriculum structure enhanced mathematics learning. Apparently, criterion (i) is essential to ensure that the student's work has something to do with history. However, it is practically impossible for me, as a non-historian, to check the accuracy of material case by case. The difficulty which this illustrates is that an instructor who is a pedagogue may be reluctant to play the role of a historian, whether or not a lack of qualification is taken into account.

My experience seems to suggest that a compromise between the two intended aims is necessary. This could be done by moving away from the idea of a history course, and developing in its place a mathematics course having some historical connections. To this end, technical mathematics would be the prime object of study. Examples from history are selected and presented in modern language. Choices, both of mathematics and of history, are made according to relevance for school mathematics and to technical complexity. Teaching develops via the instructor's explaining a segment of mathematics, with the students reading assigned historical materials, and finally students presenting their instructional designs as to how to incorporate the segment into school teaching. This mode was tried out and met with relatively smaller resistance.

4.3.1.2   A pre-service programme for primary teachers implemented in Greece and Cyprus

George Philippou and Constantinos Christou

In this section we present a primary pre-service programme based on history of mathematics, developed in two universities during the last nine years. The programme proved to be effective in terms of changing students' attitudes toward mathematics.

Entering characteristics   In Greece and Cyprus, the programme of study for primary teachers caters for the 'generalist teacher', that is, for a person capable of teaching all subjects in the primary school curriculum. A typical such programme consists of a broad set of areas (science, literature, psychology, sociology, learning theories, research methodology etc.) including one or two compulsory mathematics courses and one methodology course. These courses are intended to enhance the students' mathematical understanding and their ability to transform mathematical knowledge into didactical situations.

Mathematical teacher preparation   Pre-service programmes provide for growth in content knowledge and pedagogical content knowledge. The former is the amount of knowledge 'per se' and its organisation in the mind of the teacher, and the latter includes useful forms of representation ideas, analogies, illustrations, examples, explanations, etc. That is, what makes a topic easy or difficult for the learner (Grouws and Schultz 1996).

The question of what mathematical content is most appropriate for the primary teacher is intriguing. Traditionally, it was taken for granted that the primary teacher

John Fauvel, Jan van Maanen (eds.), *History in   mathematics education: the ICMI study*, Dordrecht: Kluwer 2000, pp. 113-117

needs to be better versed in general pedagogy rather than in mathematics as a discipline. Recently, however, the emphasis is rather on the mathematical world view than on the content. Prospective teachers need mathematical experiences that challenge old and foster new dispositions, leading to self-confidence, developing the ability to apply mathematical methods and symbolism, viewing mathematics as a study of patterns and relationships, and opening a perspective on the nature of mathematics through historical and cultural approaches. The potential of history of mathematics to enhance mathematical understanding, to motivate the learner to make necessary connections, and to realise the continuity of human culture has been repeatedly advocated by those experienced in the field.

4.3.1.2.1   Teachers' beliefs and teacher education

Content knowledge and pedagogical content knowledge is translated into practice through the filter of one's philosophy of mathematics and its learning (Swafford 1995). Thus, apart from knowledge and abilities, prospective teachers are expected to develop positive attitudes and beliefs related to the task. Teacher education should enable trainee teachers to transform and enhance their beliefs in relation to classroom actions. Such change is expected to improve teacher classroom behaviour, though we cannot assume that changes in beliefs will necessarily be translated into changes in practice.

Beliefs and attitudes are mental states organised around an object or situation through experience, predisposing one to respond in a favourable or unfavourable way. Beliefs are propositions that are accepted as true by the individual; they constitute the individual's subjective knowledge about self and the environment, physical or mental. Richardson (1995) identifies beliefs as the teacher's own theories, which are sets of interrelated conceptual frameworks tidily connected with action; they are a kind of knowledge-in-action. Beliefs are thought to drive action, but experience and reflection on action may lead to modified beliefs i.e., there is an interactive process between the two variables. Attitudes include motivation, interest, confidence, perseverance, willingness to take risks, tolerance, and resistance to premature closure (Reynolds 1992).

4.3.1.2.2   The teacher preparation program

A pre-service primary teacher mathematical programme can rely on an overall grasp of the nature and significance of the subject, an 'advanced literacy' in the fundamental concepts and methods and a competence in mathematical thinking. A guided journey through the history of mathematics would enable students to construct mathematical meanings and support their new conceptions about mathematics by changing their beliefs and attitudes towards mathematics and its teaching. The specific programme considered here was based on selected works and paradigms from the history of mathematics, exploring the cultural environment of the genesis of these works and ideas. How mathematical thinking evolved, seen by following the solution of some major problems that intrigued and inspired the leading mathematical minds from the classical Greek world until modern times, was expected to function as a strong motivation and aid. Coming to know some of the successes, and understanding some of the failures, of well-known mathematicians would offer students an insight concerning the nature and the significance of mathematics. Hands-on experience together with the incentive to follow the steps of major characters was assumed to free students of some misconceptions, fears, and negative attitudes.

The journey started with 'pre-Hellenic mathematics', proceeded to Greek mathematics, passed through Islamic and Hindu contributions, elements of the mathematics of the mediaeval and the Enlightenment period, and culminated with six rather lengthy units, selected from contemporary mathematics (for a list of topics see Philippou & Christou 1998b).

At the University of the Aegean, Rhodes (UA), the programme comprised one content course and one method course, while at the University of Cyprus (UC) it comprised two content courses and one method course. In both cases the courses were structured so as to facilitate active learning. The three credits were divided into two hours lecturing and a one and a half-hour activities session. The students were led to construct their own meanings and draw conclusions by working on tasks and examples from the history of mathematics.

4.3.1.2.3   The programme evaluation

After being run for four years (1988-1992) at the UA, the programme was assessed in terms of its effectiveness in improving students' attitudes toward mathematics. A questionnaire was administered concurrently to students at entry (E1) and the end of the programme (E2) (Philippou 1993). For comparison purposes, the same set of questions was administered to comparable samples in two other rural Greek Departments of Education. At the UC a longitudinal assessment process was adopted (1992-1995); namely, the same set of questions was administered before the commencement of the programme (Ph1), after the first course (Ph2) and at the end of the three courses of the programme (Ph3) (Philippou & Christou 1998a).

The questionnaire consisted of three complementary scales. The Dutton scale comprised 18 statements ranging from highly negative attitudes toward mathematics, e.g., 'I detest mathematics and avoid using it at all times', to the most favourable e.g., 'mathematics thrills me, it's my favourite subject'. The liking-disliking scales comprised ten items each requesting the subjects to choose the reasons of liking or disliking the subject, and the self-rating scale was an eleven point linear scale on which the subjects were expected to locate their feelings with respect to mathematics.

Several statistical tests were applied. At the UA we used the t-test for each item on the Dutton scale and the liking-disliking scales to test for differences between E1 and E2, whereas for UC the χ^2-test was used for Ph1, Ph2 and Ph3. The points of the self-rating scale were grouped into four categories: highly negative, negative, neutral, positive and highly positive attitudes. The Median Polishing Analysis was also applied on the responses of the three phases at UC. To this end, the Dutton scale was partitioned into three parts reflecting feelings of *satisfaction, anxiety,* and *appreciation* of the usefulness of mathematics. In addition, ten semi-structured interviews were carried out, to elicit the views of those interviewed and their feelings with regard to the programme.

4.3.1.2.4   Results and discussion

The analysis revealed an alarmingly high proportion of students bringing along extremely negative attitudes. For instance, 26% and 24% of the students in UA and UC respectively endorsed the statement "I detest mathematics and avoid using it at all times". Similar proportions endorsed the statements "I have never liked mathematics", "I have always been afraid of mathematics" and "I do not feel sure of myself in mathematics". The same pattern of responses also appeared in the self-rating scale, in which 36.9% and 33.5% of the subjects located themselves in the range 1-5. Students liked mathematics mainly because "it develops mental abilities" (58%, 47%) and "it is practical and useful" (48%, 39%), while they disliked mathematics primarily because of "lack of understanding" (31%, 24%) and because of "lack of teacher enthusiasm" (32%, 25%).

Changes in attitudes were observed in both universities, but it was greater in the case of UC. In the UA, the *t*-test indicated significant differences at the 0.01 level in six items, indicating improvement of attitudes, while no such difference was observed in the control group. In the UC, the χ^2-test revealed significant differences in attitude on 14 out of 18 statements of the Dutton scale. For instance, the proportion of students who 'detest mathematics' dropped from (26%, 24%) to (16%, 12%) and of those who 'never liked mathematics' from (36%, 28%) to (32%, 18%) in UA and in UC, respectively. Conversely, the proportion of those who 'enjoy working and thinking about mathematics outside school' went up from 18%, 20% to 27%, 40% in UA and in UC, respectively. The proportion of subjects who detest mathematics also dropped, according to responses on the self-rating scale from (14.6%, 14.3%) to (5.9%, 3.1%).

The Median Polishing Analysis showed a positive change throughout the three phases in all three sub-scales. The overall effect was found to be low (34%, 21%, and 41%, for the three scales, respectively), indicating a rather low level of endorsement of the ideas portrayed by the items. Attitude change, however, was shown by Row Effects to be remarkable in all three sub-scales. That is,

– in the satisfaction scale, a positive change: – 14.5% → 3.5% → 3.5%;
– in the anxiety scale, a steady negative change: 3% → 0% → –3%; and
– in the usefulness scale, a steady improvement: – 4.5% → 7.5% → 9.5%.

In brief, the programme was found to be effective in improving prospective teachers' attitudes. It produced attitude change as evidenced by different instruments in a variety of situations. According to students' evaluations, their introduction to history of mathematics played a major role in this development, though some related variables, such as instructors' enthusiasm, have not been ruled out.

References for §4.3.1.2

Grouws, D.A., K. A. Schultz 1996. 'Mathematics teacher education', in: J. Sicula (Ed.). *Handbook of research on teacher education,* London: Prentice Hall, 442-458

Philippou, George N. 1994. 'Misconceptions, attitudes and teacher preparation'. in: *Proceedings of the Third International Seminar on Misconceptions and Educational Strategies in Science and Mathematics,* Ithaca NY: Cornell University.

Philippou, George N., Christou, Constantinos 1998a. 'The effects of a preparatory mathematics programme in changing prospective teachers' attitudes toward mathematics', *Educational studies in mathematics,* **35,** 189-206

Philippou, George N., Christou, Constantinos 1998b. 'Beliefs, teacher education and history of mathematics'. *Proceedings of PME 22* (Conference of the International Group for the Psychology of Mathematics Education, **4,** 1-9

Reynolds, A. 1992. 'What is a competent beginning teacher? A review of the literature', *Review of educational research,* **62 (1),** 1-35

Richardson, V. 1996. 'The role of attitudes and beliefs in learning to teach', in: J. Sicula (Ed.). *Handbook of research on teacher education,* London: Prentice Hall, 102-119

Swafford, J.O. 1995. 'Teacher preparation', in: I. M. Carl (Ed.), *Prospects for school mathematics,* Reston VA: NCTM, 157-174

4.3.1.3   UK: A new dimension in educating mathematics teachers

David Lingard

4.3.1.3.1   Context

Routes to Qualified Teacher Status (QTS) at both primary and secondary levels have undergone considerable change in recent years. At secondary level these could now include a 2-year or 3-year BSc Honours undergraduate course (in Mathematics and Education), or a 2-year or 1-year postgraduate (PGCE) course. At primary level, the routes could include a 3-year BA Honours undergraduate course, or a 1-year postgraduate (PGCE) course.

As part of the inevitable accompanying curriculum review, these changes have seen the introduction of some form of history of mathematics unit into the majority of these courses at a number of universities offering these QTS routes. The only students who miss out are usually those on the 1-year PGCE courses for whom, sadly, the pressures on time are already enormous. Some of these postgraduates may however have followed a history unit as part of their undergraduate mathematics degree course.

4.3.1.3.2   Rationale

The Mathematics Education Centre at Sheffield Hallam University is perhaps typical of the institutions which have embraced the history of mathematics as an integral

John Fauvel, Jan van Maanen (eds.), *History in   mathematics education: the ICMI study,* Dordrecht: Kluwer 2000, pp. 117-122

component of QTS courses for both primary and secondary student teachers. We believe that some knowledge and understanding of and immersion in the history of mathematics is an important ingredient in the education of mathematics teachers because it helps them to:
1. make more sense of mathematics,
2. humanise the subject in the school classroom,
3. emphasise the continuous and continuing development of mathematics, and
4. appreciate the multi-cultural inheritance and culturally dependent nature of mathematics.

To this, George Sarton would have added: "The study of the history of mathematics [...] will enrich their minds, mellow their hearts and bring out their finer qualities." (Sarton 1936), with which we would concur.

4.3.1.3.3   Content

In theory, these history units form a 35-hour or 70-hour taught course, for primary and secondary students respectively. In practice, about one third of that time is handed over to the students for individual enquiry and research, and to group and individual tutorials to support and guide this.

The topics for the taught sessions are designed to :
a) give students an overview of the history of mathematics,
b) focus upon a number of key events, discoveries, developments and publications (e.g. The Rhind Papyrus, Greek geometry and proof, the quest for a value for π, analytical geometry and the calculus, the history of algebra, Chinese mathematics and the Nine chapters, etc.),
c) examine in more detail the life and work of one or two significant mathematicians (examples so far have included: Pythagoras, Archimedes, Al-Khwarizmi, Newton, Germain, Euler, Kovalevsky and Ramanujan),
d) consider some related themes (e.g. the contribution of women, the effect of religious patronage and persecution, the translation of texts, collaboration and plagiarism),
e) look at some of the 'unsolved' problems that have fascinated many mathematicians over the years and which have been responsible for the development of new mathematics (e.g. the three problems of antiquity, the Goldbach conjecture, the Riemann hypothesis, the four-colour map, Hilbert's problems, and Fermat's last theorem), and throughout
f) set the history of mathematics into the wider context of world history.

The selection of topics, themes and mathematicians etc. may vary from course to course and from year to year. Choices are dictated partly by student interest, perceived relative importance (historically and/or mathematically) and, inevitably, the personal interest, experience and knowledge of the tutors!

4.3.1.3.4   Teaching and learning styles

The teaching and learning styles adopted by the tutors are of particular importance. These are intended to reflect and role model good classroom practice. There are no lectures. Classes are taught in groups of up to 25 students, and the taught sessions will include a mixture of direct teaching (exposition), small group work, paired work, individual tasks and plenary discussion. More specifically these include:

THE GREAT HISTORY
BALLOON DEBATE.

Figure 4.2: The Sheffield Balloon Debate (Student poster reproduced in the BSHM Newsletter #35, Autumn 1997)

(i) Problem solving, e.g. the Egyptian method for finding the volume of a truncated pyramid (*cf.* The Moscow Papyrus, problem 14),

(ii) Reading and research,

(iii) Debate (see for example 'The Sheffield balloon debate', Hicks 1997, and figure 4.2)

(iv) Field trips (to support work on Newton, for example, students might spend a day visiting his birthplace, Woolsthorpe Manor, the King's School in Grantham where he was a pupil, and the Grantham museum),

(v) Individual 'work in progress' presentations, in order that peers may benefit from the research and study undertaken for assessed assignments,

(vi) The use of television, video and audio programmes (see e.g. §4.3.1.3.5 below),

(vii) Group exercises, including for example the production of a wall poster to illustrate the development of mathematics over a particular period of time,

(viii) Quizzes designed to promote research (and enjoyment !).

4.3.1.3.5   A typical session
To illustrate the above in more detail, there follows a summary of a recent 3-hour taught session on the BSc Secondary course (Year 1, 17 students):

Topic : the contribution of women to the history of mathematics.

Preparation : Students were given a booklet of key readings (extracts from Mozans 1913, Burton 1986, Osen 1974 and Downes 1997) one week before the session and asked to study these.

a) Groups of 3 or 4, each group given a large envelope containing over 50 'clues' (pieces of evidence) relating to four famous contemporary mathematicians. Object: to sort the evidence and identify the four. The adjacent resources centre was available for limited research (the mathematicians were Gauss, Germain, Lagrange and Poisson).

b) Listen to audio cassette 'Who is Sophie Germain ?' (Cassette M006/B, Maths Miscellany, Open University, 1994). Students also follow the dialogue with a tape transcript.

c) Plenary discussion, including how and when this might be used in the school classroom, with which pupils and with what additional accompanying resources/activities.

d) Closer examination of the correspondence between Germain and Gauss (referred to on the tape) about her proof concerning the primes for which 2 is a residue or non-residue. Exploration of this, mathematically, in pairs.

e) Move to adjacent PC lab. Use of Excel (spreadsheets) to extend this work. Discussion about use of this in schools and Information and Communication Technology (ICT) issues.

f) Exposition by the tutor about prejudicial attitudes towards women in mathematics and in history. Illustrated with examples from the 19th century in England, and by other 'case studies', including Hypatia, Somerville and Kovalevsky.

g) Listen to extracts from BBC radio programme 'Real women: Sophie Germain' (BBC Radio 4, broadcast on 6 March, 1998) to compare and contrast to b) above.

h) Summary and plenary discussion.

It has taken several years to develop a range of similarly inter-active sessions, and it is a constant but nonetheless enjoyable challenge to structure new ones.

4.3.1.3.6   Assessment

In some respects it has proved to be problematic to find the best mode of assessment. What we have settled for at present is a combination of a written, critical account (at least 4000 words) of some aspect of the history of mathematics, and a presentation of this work to their peers. Students are free to propose and negotiate a topic with the tutors, no two being able to pursue the same topic at any one time. In their account students are encouraged to raise and try to answer questions such as: why did this happen then? what were the catalysts for change? what were the immediate and longer term effects of this development/discovery? who plagiarised whom? whose version of events do we believe, and why? etc. Accounts must be drawn from a wide range of sources, be well illustrated and professionally produced.

The peer group presentations may last 20 minutes and should be informative, interesting, lively and inter-active (à la classroom!). They are usually done at the end of the unit, are given in historical order and often provide an enjoyable summary of the course. The current weightings are 60 % for the account, 40 % for the presentation.

The written accounts have, perhaps surprisingly, proved popular with students. Many enjoy the freedom of choice, the individual enquiry and research and the opportunity to 'publish'. Some have great difficulty in constraining their accounts to less than 5000 words, some reach 8000! The marking load for tutors is very heavy. The presentations are also time consuming, usually occupying a whole day.

We have tried a timed, written, 'open book' examination, and also a *vive-voce* with a mixture of prepared and *ad hoc* questions, but tutors and students were dissatisfied with both.

To give a flavour of the work undertaken, in addition to the more obvious (but no less valuable) choices, recent assignments have also included the lives and work of individual mathematicians: Thomas Harriot, Simon Stevin, Albrecht Dürer, Liu Hui, Eratosthenes, Heron of Alexandria, and George and Mary Boole; cultural surveys such as mathematics in China, Vedic mathematics, and mathematics and Islam; and particular topics such as the museum of Alexandria, the development of perspective, logic in the 19th century, the history of topology, the Lucasian chair at Cambridge, and the solution of equations.

4.3.1.3.7   Resources

The students rely heavily upon books and journal articles and frequently need to make use of inter-library loan facilities. Students are given a detailed booklist, currently listing over 250 popular titles available in the library, to get them started. It is interesting to note that about 80% of these have been published in the last 15 years. The proliferation of websites on the internet for the history of mathematics (well summarised and annotated by Barrow-Green 1998; and see §10.3.2) form an increasingly used source of material, especially for those with access at home, but John Fauvel's timely and cautionary article (Fauvel 1995) is prescribed reading for all students at the outset!

The Open University history of mathematics course broadcasts, their Maths Miscellany audio tapes, a variety of relatively recent television and radio programmes and even occasional coverage in the responsible press all provide further material. So too do articles in the professional journals, such as *Mathematics in school, Mathematics teaching* and the *BSHM Newsletter*.

We try to encourage students to make their own field trips, especially where these may be local for the students. So far these have included Cambridge University and the Whipple Museum, the British Museum and the Science Museum in London, George Green's mill in Nottingham, and Lincoln and Doncaster (Boole).

4.3.1.3.8   Feedback and evaluation

We were initially taken aback by the overwhelming positive feedback from the majority of students and this is documented in Lingard 1997. Colleagues in other institutions in England would seem to confirm the apparent enjoyment of such units in QTS courses. Some of these clearly relate to a desire to humanise and civilise the school curriculum, and they impinge upon the relationship between mathematics, gender and 'ways of knowing' which in turn requires a re-conceptualisation of the nature of the discipline, as argued in Povey *et al* 1999.

In the longer term, what is perhaps even more encouraging and relevant is the growing evidence, locally and elsewhere, that once in post, many student teachers are using what they have gained from the course in their classrooms. Amongst the most recent examples are: classroom murals on the history of mathematics in one school, a Millennium project on the mathematics of the last thousand years in another, and a one week study project for the whole of year 7 (11 year olds) on the life and work of four famous mathematicians. Sadly, there are as yet insufficient good classroom materials to support the work of these school teachers. This is where, in England at least, the next thrust is needed.

References for §4.3.1.3

Barrow-Green, June 1998. 'History of mathematics: resources on the world wide web', *Mathematics in school*, *27*: 4, 16 – 22

Burton, Leone 1986. *Girls into maths can go: women in mathematics – her story*, London: Holt, Rinehart & Winston Ltd

Downes, Steven 1997. 'Women mathematicians; male mathematics: a history of contradiction?', *Mathematics in school*, **26**, 3 (May), 26 – 27

Fauvel, John 1995. 'History of mathematics on the web', *British society for the history of mathematics newsletter* **30**, 59 – 62

Hicks, Lucy et al. 1997. 'The Sheffield balloon debate', *British society for the history of mathematics newsletter* **35**, 41 – 43

Lingard, David 1997. 'The role of the history of mathematics in the teaching and learning of mathematics'; appx 3: extracts from student teacher evaluation reports on the history of mathematics units at Sheffield Hallam University, 1994 – 1997, *report written for the ICMI Study*Mozans, H.J. 1913. *Woman in science*, New York/London: D. Appleton & Co.

Osen, Lynn M. 1974. *Women in mathematics*, Cambridge Mass.: MIT Press

Povey, Hilary; Elliott, Sue & Lingard, David 1999. 'How mathematics is made and who makes it: a consideration of the role of the study of mathematics in developing an inclusive mathematical epistemology', Paper presented to the Gender and Education conference, Warwick University, March 1999

Sarton, George 1936. *The study of the history of mathematics*, Cambridge: Harvard University Press

4.3.1.4   Mozambique: Mathematics in history for secondary school trainee teachers

Abdulcarimo Ismael

The programme component 'Mathematics in History' was introduced in 1990 as a compulsory course in all mathematics teachers 'licenciatura' programmes at the Universidade Pedagógica in Mozambique. This initiative found its inspiration in the ethnomathematical research, especially as it relates to didactics, which has taken place since the end of 1970s and which became organised in 1988 as the 'Ethnomathematics in Mozambique' research project.

In the teaching of 'Mathematics in History', three aspects are stressed: the origin of some mathematical/geometrical ideas; the roots of mathematics in African and Mozambiquan cultures; and the history of mathematics in Africa and in the other parts of the world. The main topics of the course are:

John Fauvel, Jan van Maanen (eds.), *History in mathematics education: the ICMI study*, Dordrecht: Kluwer 2000, pp. 122-124

- the history of multiplication: explanations of different methods of multiplication; the method of Ahmes in the Rhind Papyrus, called the 'African multiplication method'; where does the method taught in schools in Mozambique come from?
- counting and numeration systems: classification according to basis and position; survey of the numeration systems found in the students' own mother tongues and of popular counting methods (for results see Gerdes 1993 & 1994b); the binary system, calculators and African duplication systems.
- number systems: the history of natural numbers, negative numbers and associated historical issues, rational and irrational numbers (Pythagorean philosophy), imaginary numbers.
- history of algebraic equations: different methods (algebraic and geometrical) for solving algebraic equations (Egyptian, Greek, Maghreb, Babylonian).
- numerical analysis: iterative methods for solving equations (from Babylonian, Egyptian, Hellenistic, and Arab methods until the theory of Galois).

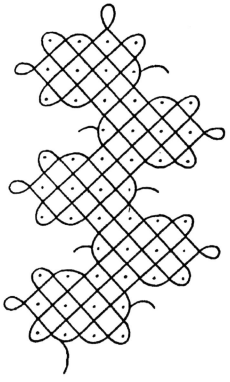

- history of geometry: geometry in African cultures; history of geometrical methods; the contributions of Euclid, Archimedes, Descartes and others.
- history of the calculus: the contributions of Leibniz and Newton; roots of the calculus in Northern Africa, India and Europe.

We imagine that most of our activities are similar to those used in other countries, but there are some aspects to be stressed which are 'innovations'. We ask students to think about cultural aspects of their own lives which can be directly or indirectly related to mathematics and probably also to its history. We require students to take an active role in the process of learning about the history of mathematics, through collecting data and by reflecting on their own counting processes (e.g. aspects of counting and spoken numeration systems), by looking for oral sources (interviewing old people and illiterate people), analysing the data

Figure 4.3: A leopard with five cubs: a characteristic example of a LUSONA, the art of sand drawing as exercised by the Tchokwe in Angola. Such SONA serve to reconstruct traditional geometrical knowledge and are used in teacher education in Mozambique and in the classroom as well. From Gerdes 1991, courtesy of the author.

they have gathered, interpreting the results and formulating their own hypotheses. In doing this we are asking them to take an active part in historiography and in reviving history. Gerdes (1990, 1995) presents many examples of ethnomathematical aspects which can serve as resources for doing history of mathematics in the classroom. His articles (1993, 1994a, 1994b) were used in the teaching of the mathematics in history course with UP-students. The articles in Gerdes 1993 and 1994b include works and article by others at the UP (Marcos Cherinda, Jan Draisma, Abdulcarimo Ismael, Abilio Mapapa, Daniel Soares).

At the end of each 'mathematics in history' course, we have carried out an anonymous evaluation. The results have shown that the students were usually surprised by what they had done; there was a high level of interest in the course, the students themselves were very motivated and convinced about the usefulness of history of mathematics in their future profession. The students also feel very confident about the potential that Africa has to offer and its contribution to the development of mathematics. Sometimes the students even go so far as to make exaggerations such as: 'all mathematics comes from Africa'.

References for §4.3.1.4

Gerdes, Paulus 1990. *Ethnogeometrie. Kulturanthropologische Beiträge zur Genese und Didaktik der Geometrie*, Hildesheim: Franzbecker Verlag.

Gerdes, Paulus 1991. LUSONA. Geometrical Recreations of Africa *[Récréations Géométriques d'Afrique]*, Maputo

Gerdes, Paulus (ed.) 1993. *A Numeração en Moçambique [Numeration in Mozambique]*, Maputo: Universidade Pedagógica

Gerdes, Paulus 1994a. *African Pythagoras: a Study in Culture and Mathematics Education*, Maputo: Universidade Pedagógica

Gerdes, Paulus (ed.) 1994b. *Explorations in Ethnomathematics and Ethnoscience in Mozambique*, Maputo: Universidade Pedagógica

Gerdes, Paulus 1995. *Ethnomathematics and Education in Africa*, Stockholm: Institute of International Education, University of Stockholm

Gerdes, Paulus 1999. *Geometry from Africa: mathematical and educational experiences*, Washington: Mathematical Association of America

4.3.1.5   Morocco: History of mathematics used in teacher training: an example

Abdellah El Idrissi

We offer here an example of using history of mathematics with secondary teacher trainees. The intention was to use history for the purposes of epistemological

John Fauvel, Jan van Maanen (eds.), *History in mathematics education: the ICMI study*, Dordrecht: Kluwer 2000, pp. 124-127

analysis.   The aim was for students to analyse mathematical reasoning and to become acquainted with the major stages of the evolution of certain concepts after having traced their development (El Idrissi 1998).

The topic used in this example is trigonometry and, in particular, the fundamental concepts of angle and the basic trigonometric ratios. The course, for about ten students, was planned to last for about twenty hours. The guiding principles for the course were as follows.

4.3.1.5.1   Guiding principles
a)   The course should be activity based with a minimum of teacher 'narration'.
b)   The topics and sessions should be based around specific problems and sources.
c)   All work should be based on original texts.   Original texts, even if they are difficult to use, provide for a better understanding and avoid imposing erroneous interpretations.
d)   In using history it is important to distinctinguish between hard facts and inter-pretations, and also to follow the tools, concepts and conventions of the period. Preferably one avoids early recourse to modern symbolism and interpretations.
e)   In teacher training historical analysis should be complemented by some teaching activities.   This will help encourage the interest of the students by providing material they will be able to use in their classrooms.

4.3.1.5.2   Course description
Trigonometry has a history of more than four thousand years and a choice has to be made of subject matter and time period. Four periods were chosen.

a The Egyptian period: *The Rhind Papyrus* (*c.* 1500 BC)

This document is one of the rare mathematical documents we possess as evidence of Egyptian mathematics.   Reading and interpreting the document is relatively recent. The *Rhind Papyrus* consists of mathematical problems together with their solutions, both of which provide useful study material.   The problems we chose concerned the 'seqt', a concept close to our idea of cotangent and which was used to determine the angle of slope of sides of pyramids. (Gillings 1972; Neugebauer 1969; Smith 1958)

b Ancient Greece: Ptolemy's *Almagest* (*c.* 150)

The *Almagest* is the oldest work that informs us of Greek ideas on trigonometry, earlier works having been lost. Two chapters of this work on astronomy are given over to trigonometry and, in particular, to the construction of a table of chords. We were interested in the underlying mathematical reasoning as well as the construction of the table. (Neugebauer 1969; Smith 1958; Halma 1813)

c The Hindu period: the *Suryasiddhanta* (*c.* 500)

Hindu trigonometry as presented in the *Suryasiddhanta* differs from Greek trigonometry in that a table of sines is constructed using the radius of the circle as the base. The other difference is the complete absence of symbols, all results being given in words. (Burgess 1858; Smith 1958)

d The Arab period: Al-Tusi's *Traité du Quadrilatère* (*c.* 1250)

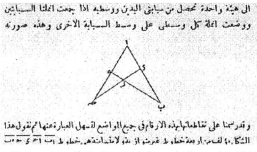

الى هيئة واحدة تحصل من مبانى البدن روسطه اذا جمت اجلنا السبابين

روشمت اجلة كل وسطى على وسط السبابة الاخرى وهذه صوره

وقدرسنا على تتاطعاتهاينه الارتابرق جميع المواضع لتسهل العبارة عنها تمنقول هذا
الشكار ٤: لقسم. او بعدخطا، با غريش؟ او نتو لاندماننتهر خطا، با رتب ٢٣

Figure 4.4: Al-Tusi's quadrilateral (from the 1891 edition)

Al-Tusi's work marks a decisive stage in the development of trigonometry in which Greek and Hindu influences are apparent. A variety of methods for the solution of triangles is given. The methods depend on different units of angle which are in fact incompatible. Resolving these differences is a challenge for the students and provides opportunities for fruitful resolution of conflicting viewpoints (Al-Tusi 1891).

Following the principles given above, pertinent extracts were selected from each source and activities based on the text were given to the students. Some activities went beyond the strict historical context in order to deal further with the mathematics or the teaching of mathematics. (A fuller discussion of original sources and their use will be found in Ch. 9).

4.3.1.5.3   Evaluation

The results of the teaching programme were analysed qualitatively. Seven statements were presented to the students. Each of the participants was asked to agree or disagree and to give reasons. It should be noted that the statements did not make explicit reference to the history of the subject. Two examples:

1. *Tangent is simply the ratio of sine to cosine. For pupils to understand tangents all that is necessary is for them to understand sine and cosine, following which we can be sure they will understand tangent.*

2. *For me trigonometrical ratios (sine, cosine, tangent, ...) seem the same as measuring angles. That is why I would plan my teaching of trigonometry as a way of measuring angles, or at least as a way of describing angles.*

The course was evaluated using semi-structured interviews. The main purpose was to get the students to explain and justify their answers. The evaluation led to three important conclusions.

First, history of mathematics helped the students to analyse mathematical concepts and, to a certain extent, to choose suitable teaching strategies. *Second*, the students became aware of difficulties that may arise if history is used to teach some mathematics or to clarify concepts. *Third*, there are cases where the history of mathematics can convey concepts, inappropriate for teaching mathematics.

References for §4.3.1.5

Al-Tusi Nasir c.1250/1891. *Traité du Quadrilatère*. tr. P. A. Carathéodory Constantinople
Burgess, E.R. 1858. 'Suryasiddhanta: a text-book of Hindu astronomy, annotated translation',
    Journal of the American Oriental Society, San Diego, *6*.

El Idrissi, Abdellah, 1998. *L'histoire des mathématiques dans la formation des enseignants: étude exploratoire portant sur l'histoire de la trigonometrie*, Ph D thesis, UQA Montréal
Gillings, R.J. 1972. *Mathematics in the time of the Pharaohs*, Cambridge, Mass.: MIT Press
Halma, M. 1813. *Composition mathématique: L'Almageste*, Paris: Henri Grand Libraire
Neugebauer, O. 1969. *The exact sciences in antiquity*, 2nd ed. New York: Dover
Smith, D.E. 1958. *History of Mathematics*, **2**, reprint New York: Dover

4.3.1.6   France: A historical module for secondary school trainees

Éliane Cousquer

In the mathematics department of the *IUFM* du Nord Pas-de-Calais, we have had a great deal of experience in teaching the history of mathematics (*IUFM* stands for *Institut Universitaire de Formation des Maîtres*). This takes place within the *IUFM* for initial teacher training, and mainly within the *IREM* for in-service teacher training. Our practice has changed and evolved over time.

Initially, a twenty-hour course in history of mathematics was compulsory, and had to be taken in either the first or the second year. This proved to be unsatisfactory in either year. In the first year, students wished to devote all their time to prepare for an examination which does not require the history of mathematics, and in the second year, the teaching practice stage and the writing up of the thesis take up most of the energy of the students who, in any case, tend to feel that the training at the *IUFM* is too diversified. Consequently, the mathematics department decided to make the history course optional and to link it to the needs of those students who are preparing for their practice stage.

Nowadays, a course is delivered in parallel to the seminars preparing for the professional thesis. The course deals with the history of a number of topics, such as algebraic equations, proof, numbers, measuring and vector calculus (see Cousquer 1998). Students can draw on these topics either for the preparation of their practice stage in schools or for the thesis. This turns out to be more satisfactory, since the students find the history of mathematics course directly relevant to their needs. Rather fewer than half group take the course but, on the whole, these fifty-odd participants find it interesting. Students gain deeper understanding through working on their thesis, which is enhanced by the work in schools teaching and by having followed the history course. The course at the *IUFM* provides a rich experience, and the participating trainee teachers wish it to continue since such courses provide meaning for the topics they had to teach during their practice stage: introduction to algebra, initiation to proofs, presentation of different kinds of numbers, vector methods in geometry. Linking the history course to the immediate concerns of the *practice stage* makes the course more valuable.

Reference for §4.3.1.6

Cousquer, Éliane 1998. *La fabuleuse histoire des nombres*, Paris: Éditions Diderot

John Fauvel, Jan van Maanen (eds.), *History in mathematics education: the ICMI study*, Dordrecht: Kluwer 2000, p. 127

4.3.1.7   Germany: A course component on the history of mathematics education and the professionalisation of mathematics teaching

Gert Schubring

The point of departure for this component is the ambiguous position of mathematics instruction and of mathematics teachers in a large number of technologically developed countries. While mathematics officially enjoys the status of a major teaching subject and is generally assigned the highest importance for developing science and technology and hence for social welfare, it is at the same time held to be accessible to only a small percentage of 'gifted' students. Hence, mathematics teachers are accustomed to be faced in practice with a disparagement of their subject by the general public and by parents. The consequences of this disparagement are either an acceptance that they can achieve only limited success with their teaching or, on the other hand, a confirmation over and again of the impression of failure in mathematics experienced by the great majority. Their training does not prepare mathematics teachers to cope with the fragile social status of mathematics instruction. Nor are they prepared for entering into a discourse legitimating the role of mathematics instruction within general education when confronted with harassing questions by parents about the value of mathematics instruction. This ambiguity is experienced by mathematics teachers even in their daily professional life.

The thinking behind this component is therefore to prepare future mathematics teachers, during their university studies, for problems of their future profession arising from the specific social and cultural resonances of the subject they teach.

As educational structures are a result of long-term processes, the component was developed as a historical one: introducing the teacher students to the history of mathematics education and in particular to the history of their profession as teachers of mathematics. The notions of profession and professionalisation are different from those used in pedagogy and in history of pedagogy. These notions are not restricted here to refer to the *social* aspect of teacher life, they rather embrace the content of teaching since a dominant element of mathematics teachers' professional identity is their own intrinsic relation to their subject, their 'love of mathematics'. Historical studies on the emergence and further development of the profession of mathematics teachers and on its field of professional activity in mathematics instruction are, thus, highly apt to contribute to instilling meta-knowledge about their subject into future teachers. Such studies will make them aware that the history of mathematics education and of their own profession is not an isolated or *internal* one, but rather a social and cultural history which relates school mathematics to the overall history of the respective countries. In fact, school knowledge is even less neutral than scientific knowledge: history of mathematics education must therefore be thought of as a part of the social history of knowledge. School mathematics develops as a

John Fauvel, Jan van Maanen (eds.), *History in mathematics education: the ICMI study*, Dordrecht: Kluwer 2000, pp. 128-131

constitutive part of the institutional history of school in the respective countries—it is subjected to social pressures on contents and methods of teaching, and its epistemology is affected by the social norms and values generally shared in a country at a given time.

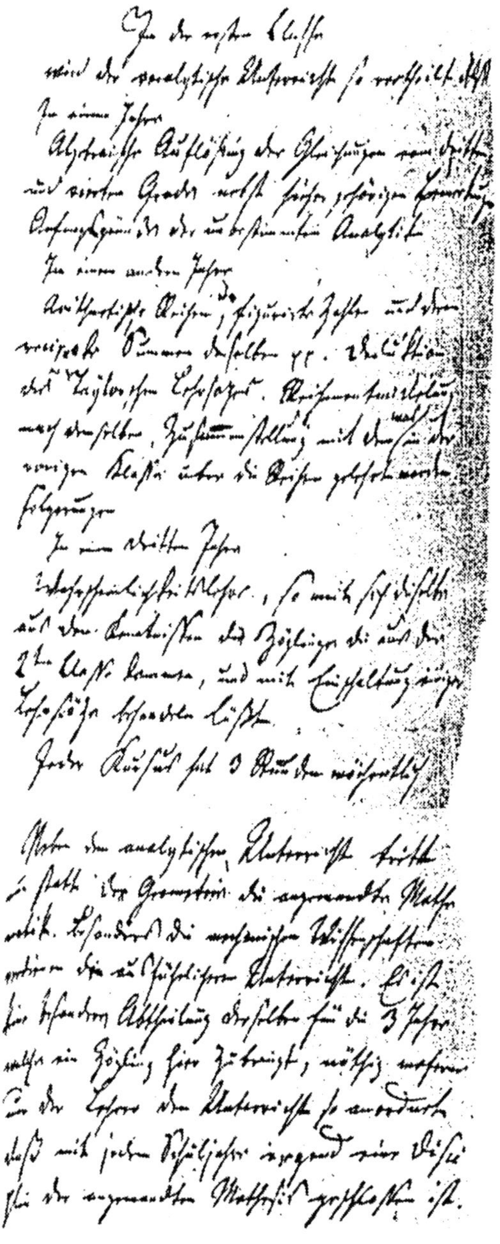

Figure 4.5: Geheimes Staatsarchiv Preussischer Kulturbesitz Berlin, Rep. 76 alt (wissenschaftl. Deputation), Nr. 18, fol. 64v/65r

In the study component developed, the major national focus was given by Prussia, the first German state to be profoundly modernised at the beginning of the nineteenth century. The results of the study were fascinating and entirely different from the traditional type of rather dull listings of administrational decisions about school syllabuses. The emergence of the profession of mathematics teacher in Prussia was shown to be a direct expression of the modernising policy of the Prussian state after 1809: a formerly marginal subject became a major teaching subject and an integral and constitutive component of systematic educational reforms. An example may be seen in figure 4.5, which shows an extract from a mathematics curriculum for the gymnasium, proposed to the Prussian ministry in 1810. The extract shows the novel and ambitious calculus syllabus for the last class (six hours per week for three years), which included Taylor's theorem as well as mechanics. Never entirely realised but an ideal guideline, such a document helps us analyse the historical conditions for implementing curricular change.

In a conscious endeavour, mathematics teacher education was institut-ionalised, together with establishing the profession of mathematics teachers for the new major subject. It is moving to follow the biographies of the first generations of mathematics teachers who, often isolated among their colleagues teaching classical languages and among the local public, struggled desperately to have mathematics acknowledged against all kinds of resistance. The history of mathematics education even in just one country proved to be quite complex, and at the same time provided rewarding structural insights into social and cultural factors of school mathematics and their teachers. Some of these insights are:

- school mathematics in no way constitutes a direct reflection or projection of mathematics as a scholarly body of knowledge—neither in its respective modern state nor in a historically more remote form.

- in the first instance, this is due to the fact (usually systematically neglected in traditional histories of mathematics education) that mathematics is not an isolated subject in school, but has to coexist with many other teaching subjects. The relative status of mathematics as a subject of instruction and examination (and, consequently, the status of its teachers) is shown to be the result of a complex social negotiation process attributing a relative educational weight to each subject of instruction. But not only that, it also emerges that the scope and type of school mathematics emerge as variables which largely depend upon the social functions ascribed to schooling and to the given school structure.

- more particularly, school mathematics is moulded by culturally determined epistemologies characterising the type of mathematics taught in a particular school structure. In fact, one of the most unexpected outcomes of the historical component for the student teachers were concrete visualisations of the continuum of epistemological mouldings ascribed to school mathematics in different social and cultural contexts. These varied from a pure view on mathematics which emphasised formal mental training at the one extreme, and on the other a view of applied mathematics emphasising vocational purposes and usefulness.

- another intriguing dimension is presented by the enormous variability in the relation between school knowledge and scientific knowledge. There are periods where school mathematics constitutes a hermetic body of knowledge, without explicit relations to the academic world, producing its own standards of rigour and its own architecture of mathematics justifying the selection of contents and the chosen hierarchy of concepts. And there are other periods of an 'open' curriculum where school teachers were aiming at following methodological views converging with those of academic mathematics.

The new component for mathematics teacher education has been successfully established at Bielefeld University where courses in history of mathematics education figure in practically all curricula leading up the various teachers' diplomas.

References for §4.3.1.7

Schubring, Gert 1984. 'Essais sur l'histoire de l'enseignement des mathématiques, particulièrement en France et en Prusse', *Recherches en didactique des mathématiques*, **5**, 343-385.

Schubring, Gert 1989a. 'Warum Karl Weierstraß beinahe in der Lehrerprüfung gescheitert wäre', *Der Mathematikunterricht*, **35**: 1, 13-29.

Schubring, Gert 1989b. 'Theoretical categories for investigations in the social history of mathematics education and some characteristic patterns', in: C. Keitel, P. Damerow, A. Bishop, P. Gerdes (eds.), *Mathematics, education and society*, Paris: UNESCO, Science and Technology Education Document Series No. 35, 6-8.

Schubring, Gert 1991. *Die Entstehung des Mathematiklehrerberufs im 19. Jahrhundert. Studien und Materialien zum Prozeß der Professionalisierung in Preußen (1810-1870)* Second, revised edition: Weinheim: Deutscher Studien Verlag.

4.3.2   Current practice in in-service training

4.3.2.1   Denmark: A very short in-service course in the history of mathematics

Torkil Heiede

An in-service course for primary and lower secondary teachers of mathematics, covering the whole history of mathematics in seven three-hour sessions: that is surely impossible!   In earlier years I have given a relatively comprehensive exposition several times in courses consisting of 33 such sessions, stretching from September to May, but that was stopped—not because of too few applicants, but because too few of them had their applications endorsed by the local school authorities, who tended to consider a course devoted entirely to the history of mathematics as a luxury. This took place at the Royal Danish School of Educational Studies, an institution with the purpose of giving further education to teachers in the *folkeskole* (i.e. grades 1-10) in Denmark. We now decided to place a course in the history of mathematics inside a larger course (one six-hour day per week, 33 weeks) in general mathematics.   If it was impossible to cover the whole history of mathematics in seven such sessions, then the solution might be to pick out seven important bits and try to present them in such a way that the participants realised that here was something relevant and interesting, something to return to and to go on with. Also it had to be underlined that hand-outs and other material was meant not to pass unadapted into the participants' own classrooms but to be drawn upon— together with what they could find on their own, helped by a list of references—to colour and maybe improve their mathematics teaching. Here follows a synopsis of what was planned for these seven sessions, with a few commentaries:

John Fauvel, Jan van Maanen (eds.), *History in   mathematics education: the ICMI study*, Dordrecht: Kluwer 2000, pp. 131-134

4.3.2.1.1   Session one: Egyptian mathematics

A base 10 number system which is not a position system. Addition and subtraction algorithms. Multiplication algorithm based on duplication. The similarity to so-called Russian peasant multiplication, and to what goes on inside the calculators of to-day, working in a binary (base 2) system. Calculation with unit fractions. Solving linear equations by trial and error (*regula falsi*). Everything illustrated with original problems and tables from e.g. the Rhind papyrus.

4.3.2.1.2   Session two: Babylonian mathematics

A base 60 system which is also a positional system (as ours), discovered by the participants from a picture of an Old-Babylonian clay tablet containing a multiplication table for 9 (see §8.3.1.1.2). Sexagesimal fractions. Division by table of reciprocals. Solution of quadratic equations (taken directly from pictures of original clay tablets). Deciphering some tablets, e.g. Plimpton 322 tablet; conjectures on its content.

These first two sessions show that number systems and algorithms different from our own can be as valid and efficient as ours. This throws light on our own number system and algorithms.

4.3.2.1.3   Session three: Greek mathematics

Two rather clumsy number systems (also Roman numerals) and the later sexagesimal number system of the Greek astronomers. Pythagorean mathematics. Incommensurability and its consequences. Euclid's *Elements*, in Danish translation. Euclid i, 47-48 (Pythagoras' theorem). Euclid ix, 20: the primes outnumber any number. Proof by exhaustion, especially pyramid, Euclid xii, 7, and cone, Euclid xii, 10. Archimedes, especially the areas of the circle and the surface of a sphere, and the volumes of the cone and the sphere. *Sand-reckoner* and the *Method*. Diophantus, his symbolism and his solutions of equations, especially his treatment of Pythagorean triples. The Greek number concept versus ours.

4.3.2.1.4   Session four: Indian and Chinese mathematics

Number systems and calculations: who invented a symbol for zero? Solution of equations, systems of equations. Indian astronomy and trigonometry. Pythagoras' theorem before Pythagoras; Pascal's triangle before Pascal. The Chinese Suan-Pan and the Japanese Soroban.

4.3.2.1.5   Session five: Arabic and European Mediaeval mathematics

Al-Khwarizmi and his books: the origin of the words *algorithm* and *algebra*. The number system and how it arrived from India. The solution of linear and quadratic equations, possible influence from Babylonia. Omar Khayyam, geometric solution of cubic equations. The translators in Spain and the origin of the word *sinus*. Fibonacci and his books. Maybe also Jordanus and Oresme.

4.3.2.1.6   Session six: European Renaissance and Early Baroque

Figure   4.6:   Mathematics   education c. 1550

Luca Pacioli. The cubic controversy: Del Ferro, Tartaglia, Cardano, Ferrari and the algebraic solutions of cubic and biquadratic equations. A glimpse of Abel and Galois. Trigonometry and navigation. Viète and his notations; his theorem on sums, products etc. of roots in equations.   Stevin and decimal fractions. Descartes and Fermat, the birth of analytical geometry; its importance for the beginning of calculus. Probability theory; Cardano's *Liber   de   ludo   aleae;*   the correspondance between Fermat and Pascal. Fermat's last theorem.

(But next to nothing about the story of the calculus, series, differential equations, real and complex analysis etc. throughout the 18th and 19th centuries, even if it was more or less synonymous with mathematics in this long period. Something has to be left out, and the participants do not themselves teach even the rudiments.)

4.3.2.1.7   Session seven: Non-Euclidean geometry

Its roots in Greek and Arabic mathematics. Saccheri, Lambert, Legendre, Gauss, Bolyai, Lobachevsky, Beltrami, Klein and Poincaré. Most of the participants have never heard about it, and it would come as a shock for them that mathematics has become something separate from physics, in that mathematical statements cannot any more be considered to be true in any straightforward physical sense. Even if the participants never mention non-Euclidean geometry explicitly, their awareness of its existence should influence what they say in class, so that their pupils may get a better impression of the nature of contemporary mathematics.

The course was carried through more or less according to the plan and was repeated, with modifications, in the following years. It was a success with most of the participants, even those who did not know much mathematics and not much general history either, who are victims of our ahistoric times and had no general historical framework on which to hang the history of mathematics. It is obvious that much of the history of mathematics was not even touched upon in this course; it only gave an overview, but with many examples, and with many of them taken as near as possible to original sources.

References for §4.3.2.1

Heiede, Torkil 1992. 'Why teach history of mathematics?', *The mathematical gazette*, **76**, no. 475, 151-157.
Heiede, Torkil 1996a. 'The history of non-euclidean geometry', *Proceedings 2nd UEE Braga 1996*, i, 183-194

4.3.2.2   France: history of mathematics in in-service training for primary and secondary teachers

Hélène Gispert

I would like to give examples of the possible use of the history of mathematics for in-service teacher training. I have given sessions in the history of mathematics to both primary teachers and secondary mathematics teachers. In the case of primary teachers, most of them teach mathematics as just one of many subjects and have no university mathematics education. In spite of the difference in the audience, the aim for these sessions is the same in both cases. The aim is not to train teachers in using history of mathematics in their classrooms. My purpose is, in fact, to use the history of mathematics and the history of mathematics teaching to show the links that have existed in different times between the contents and aims of mathematics as a science on the one hand, and the social, economic and cultural backgrounds in which they were defined, on the other hand.

These historical sessions were organised as part of larger training courses in mathematics which can last from one to four weeks. One or two days are devoted to historical topics chosen in relation to the main topics of the training sessions. These historical sessions consist of both lectures and working groups on original historical texts.

The first example is a four weeks' in-service teacher training for primary teachers which colleagues in mathematics and technology organised and which is called 'geometry with head and hands'. There are two history sessions of three hours each. In the first one, I try to make teachers conscious of the different status that geometry and its teaching had in different societies. I present the cases of Egypt and Mesopotamia, Plato's Academy, China, and several periods of mathematical and intellectual history of Europe, from the Middle Ages to the 19th century. Teachers are then led to question some obvious notions common to their own geometrical experience as secondary pupils (such as rigour, proof, figure, definition) but which seem to contradict their present experience of primary school teaching of geometry. Primary level geometry is actually based on figures (or drawings) and what can be

John Fauvel, Jan van Maanen (eds.), *History in mathematics education: the ICMI study*, Dordrecht: Kluwer 2000, pp. 134-136

seen and handled in the real world. These features often lead teachers to consider that they are not doing 'geometry'. The historical point of view helps them to ask new questions about the relations between geometry at primary level, where proof is not required, and geometry at secondary level where pupils have to gradually learn about proof. It also helps the teachers to work in a new way on the general theme chosen for the session, that is 'geometry with head and hands'.

The second historical session deals with the history of geometry teaching and teaching through the practical tools ('travail manuel') in use for primary schools during the 19th and 20th centuries in France. During this session, teachers work on texts—mostly public documents, official curricular documents and commentaries on the curriculum. The declared purposes of the two parallel educational systems that existed in France until the 1950s can be clearly seen: primary education, intended for the education of the masses, whose curriculum had to be practical, limited, concrete and useful; as opposed to secondary education, reserved for the élite and having only cultural purposes. Geometry, together with its content, teaching, and practical applications is clearly not the same for both systems. A role of the 'travail manuel', widely adopted at the end of the last century, for the primary level, was to provide mathematics object-lessons, including geometry lessons. This gave us topics that brought us back to the main theme of the session and to ask what in geometry teaching relates to the head and what relates to the hands.

The second example was planned for both secondary mathematics teachers and for those working in teacher education. The first group had a one-week session dealing with the link between primary and lower secondary education, now part of a common curriculum programme but formerly quite distinct, as explained above. For the second group, the session was entirely devoted to the history of mathematics and lasted three days. The focus was the history of mathematics teaching during the last two centuries and considered the questions: who were the main actors? and what were the main issues in the mathematics curriculum? As in the first example, I was interested in showing the differences between the history of the two parallel teaching systems, the primary one and the secondary one, whose aims were re-evaluated several times during the last two last centuries in relation to economic and political changes in France. What is interesting is that the position of mathematics, and more generally of science, in both school curricula appeared to be conditional upon these political changes. The study of these periods of change in the curriculum leads mathematics teachers to become aware of the range and variety of the major actors, their interests and the reasons advanced to defend the mathematics curriculum or to argue for change. As well as pedagogic reasons, economic, ideological and scientific reasons are advanced. This should broaden the perceptions of mathematics teachers by making them aware of what they may not have understood before the course: the fact that factors affecting the teaching of mathematics at each period, and therefore also today, depend also on social influences. Delving into history in this way helps to highlight an aspect of the educational situation of today of which secondary teachers seem hardly aware. The first years of the present secondary level for all pupils from 11 to 15 in France (*collège*), including its mathematics content, is in fact the fruit of two quite distinct school traditions which were combined in the 1960s. The primary one, as we have said, was for most of the 11 to 14 or 15 year old children and was concrete and practical. The secondary one, intended for the small minority who pursued their studies up to university level, claimed to be essentially cultural with no practical application in view. Debates about what should be the nature of mathematics teaching in today's 'collège unique' (comprehensive school), catering for all pupils, will be more meaningful after learning about the mathematical content, as well as the pedagogical methods, of these two former distinct, and opposed, systems.

For both examples cited, the teachers have valued the historical detour offered by the sessions. It has allowed them *first* to appreciate the link between mathematics and the history of the societies where it developed and flourished, and *second* to have a better understanding of the main issues of mathematical teaching, past and present. They found history to be a valuable tool to obtain a better understanding of their profession and its practice. Nevertheless, it remains a detour. Most of them consider it quite enough to insert a few historical sessions among a largely non-historical training course and would not have chosen a specific training course in history of mathematics. For this reason, I think these sessions, brief though they were, were effective in that a larger number of teachers gained the benefit of a historical perspective in mathematics and mathematics education than would have been the case if a course only dealing with the history of mathematics had been offered. When it comes to those concerned with training teachers, my view is that they should have more than these limited insights. History deserves to become a tool in the training of teachers, but there is much to do before that goal can be achieved.

References for §4.3.2.2

For the first session, in the first example, standard historical texts can be used. The second example required French historical books on the history of mathematics and science teaching, either written accounts or collections of official texts. The following proved useful:

Belhoste, Bruno 1995. *Les sciences dans l'enseignement secondaire français, textes officiels,* tome 1: 1789-1914, Paris: Economica-INRP.

Belhoste, B., Gispert, H., Hulin, N. 1996. *Un siècle de réformes des mathématiques et de la physique en France et à l'étranger*, Paris: Vuibert-INRP.

Helayel, J. 1998. 'La géométrie à l'école primaire: textes et contextes de son enseignement dans la société française au 19e et 20e siècles', *Copilerem*, 24ème colloque, IREM Université Paris 7, 179-182.

4.3.2.3   Brazil: The concept of function in in-service training

João Pitombeira de Carvalho

The concept of function provides a good example of the value of a historical perspective about the teaching of mathematics. In particular, it can explain and modify one's work with teachers in continuing education projects, such as in-service courses for mathematics teachers of primary and secondary schools.

In Brazil, many of these teachers react strongly when they are told that a function is a correspondence that assigns to every element of a set A, its domain, a well defined element of another set B. Most of them are familiar only with the definition of a function as a particular kind of relation defined on the cartesian set $A \times B$. Some of them even say that this is the only way of defining a function, and that this is perfectly appropriate for the teaching of mathematics in primary or secondary schools.

4.3.2.3.1   Historical Perspective

A historical perspective about the evolution of mathematics teaching in Brazil explains why teachers understand function this way. A secondary school curriculum in Brazil was first established in 1837, with the creation of the *Colégio Pedro II*, a public school which was set up to correct the laxity and disorganisation prevailing at all levels of teaching up to that time. Even though primary school teaching was regulated only in 1946, the *Colégio Pedro II* was fundamental for the organisation and regulation of secondary school teaching in the Brazilian Empire, and later during the republican years, from 1889 on.

The first appearance of the word 'function' in the official curriculum was in 1889. According to the textbooks then in use (e.g. Sonnet 1869), a function was defined in terms of variables:

A variable y is called a function of another variable x, if y varies with x, and if y assumes one or several well defined values when a definite value is attributed to x.

At the same time, just after the monarchy was overthrown, a major curricular reform introduced the study of the differential and integral calculus in the secondary schools. The textbook specified in the official curricular regulation was that by H. Sonnet. This reform lasted only for a few years, and the teaching of calculus in the secondary schools was then abandoned for almost half a century.

The function concept appears on and off in successive curricula for the *Colégio Pedro II* after this period. Examination of some of the textbooks used during this period shows that they adhere to the function definition given above. The mathematics teachers at this model school used to be trained at the *Escola Militar*, the later *Escola Politécnica*, which trained engineers. Functions were introduced

John Fauvel, Jan van Maanen (eds.), *History in mathematics education: the ICMI study*, Dordrecht: Kluwer 2000, pp. 137-140

there as a rule of correspondence (Dynnikov 1999). The perusal of more recent texts used in the *Colégio Pedro II* in the 1920's shows a similar treatment of the function concept. In particular, the texts *Curso de Matemática I* and *Curso de Matemática II*, written by Euclides Roxo, a staunch defender of Klein's ideas on the teaching of mathematics, stress the use of the function concept in the secondary school curriculum. According to him, the function concept should permeate all the curriculum and not be a particular topic of study.

In the early thirties and forties, the mathematical syllabus for secondary schools experienced a round of major modifications and consolidation. These modifications included a unification of the curriculum, a position strongly defended by Euclides Roxo. This syllabus set the trend for all subsequent ones. In the new textbooks written for these curricula, a function is a well defined correspondence between two sets. They state this in terms of independent and dependent variables, and also allow many-valued functions.

Things changed in the late 1950s, with the arrival, in Brazil, of the ideas of the 'modern math' movement. From then on, in almost all textbooks, a function becomes a particular kind of relation in a cartesian product (see for example Pitombeira 1996 and 1998b). In the late 1950s and the 1960s, many books were written for mathematics teachers along the lines of the modern math movement. Also, many in-service programmes were set up to bring teachers up to date. Textbook writers very quickly took up the new ideas, since the official curricula called for a set theoretical approach to the function concept. A study made in 1995 by this author showed that even then some states still specified that the concept of a function should be presented this way. This study dealt specifically with elementary school mathematics. Notwithstanding, the documentation presented by the States' Secretarias de Educação in most cases contemplated secondary school curricula, and so the claim made in this paper is justified (see Pitombeira 1998a). On account of all this, the new presentation of the function concept became widespread in school mathematics. This trend was reinforced by analogous introductions of set-theoretic definitions in some textbooks for universities and teacher colleges.

Because of all these developments, the presentation of the function concept along the lines laid down by the new maths movement became widespread and self-reproducing: the more it was adopted, the more it was included in the official curricula. Now, because of very strong criticism of the modern math movement, there is a new generation of textbooks which has gone back to the definition of a function as a functional dependence. In a certain way, we have completed the circle.

4.3.2.3.2   What to do

The prevailing conception of a function among teachers, that is, as a special kind of relation in a cartesian product, impedes the use of the function concept in most situations. Even if the teacher proceeds from this definition to give examples of 'honest' functions, that is, numerical functions in which a 'variable' y varies with a 'variable' x, we have observed that they do not connect these two notions of a function, and this leads to a very unsatisfactory situation: the student asks himself what really is a function. In the worst case, the teacher restricts his examples of functions to special subsets of a cartesian product, and the student is completely bewildered when he faces the graph of a numerical function and the teacher tells him that the graph represents a function. Also, some of the textbooks which present the cartesian product definition do not stress that in many applications one is really interested in how the dependent variable varies when the independent one varies. Some of them hardly ever present examples of this dependence.

We set ourselves the task of convincing teachers to abandon the cartesian product definition in favour of the correspondence one. Of course, the simple statement that their definition is 'bad' and that the one we propose is 'good' would not change their conception of a function. Instead we set up the following steps, in which the historical perspective on the teaching of mathematics plays a major role.

1 - Review of the evolution of the function concept.

We stressed Euler's contribution. In particular, they were given old and modern textbooks used in Brazil and asked to see how their definitions and examples comply with Euler's point of view, that is, if we still had Euler's definition, would the examples presented in the textbooks be functions? We then studied the evolution of the function concept after Euler's definition, presenting Cauchy's conception of a function, and ending with the Cauchy-Dirichlet-Bourbaki definition. We point out that this part is not a course on the history of mathematics. From the very beginning of our programme the teacher is immersed in school mathematics, via the textbooks.

2 - Discussion of some of the ideas of the modern math movement.

We dealt with its use of set theory and how it attempted to build up the concepts of school mathematics starting from set theory. The teachers received examples of textbooks which followed the ideas of the modern math movement and we asked the teachers to compare the examples they presented with the examples found in the older textbooks. The teachers were also asked to evaluate how the modern math textbooks dealt with the transition from their abstract definition to the presentation of the usual functions of school mathematics.

The teachers were also asked to look for the official curricular instructions issued by the state Secretaria de Educação from the 1950s to 1990s and to discuss their presentation of the function concept.

3 - A presentation of the history of the teaching of mathematics in Brazil

This was along the lines of the historical perspective given above. For this, the teachers were given extracts of textbooks used (from the 1850's to the 1950's) and asked to compare their treatment of the function concept.

4 - A discussion of their own learning of the function concept in their high school and college years.

The teachers were asked to bring their old school and college textbooks and to discuss with the group how these texts present the concept of function and work with it (examples, applications, exercises). A comparison of how some widely used current textbooks present the concept of function and the examples and exercises they give. The teachers analyse current textbooks and write up an essay comparing them, their examples and exercises they give and the coherence between the definitions and the examples presented.

4.3.2.3.3   Conclusion

After trying this approach, which lays heavy stress on the historical aspects of the teaching of mathematics, during several years in programmes of continuing education, we feel that the teachers' ability to present the concept of function meaningfully to their students was much improved. The teachers do not have to live any more with two different function concepts. They know how to translate one into the other, and can emphasise the work with numerical functions, correlating graphs, tables and analytical definitions.

This consideration of the history of mathematics education was essential to show the teachers that the way they see things is a consequence of past ideas, movements, influences. Making them retrace this history will allow them to take account of the different presentations of the function concept, without adopting a hostile position towards some of them.

References to §4.3.2.3

Dynnikov, Circe M. Silva da Silva 1999. *Matematica positivista e sua difusao no Brasil*, Vitoria: EDUFES.

Pitombeira , João Bosco, *et al.* 1996. *Guia de livros didáticos de 1a à 4a séries- Livros recomendados*, Brasília: DF, MEC, CENPEC.

Pitombeira, João Bosco, et alii 1997. *Guia de livros didáticos de 1a à 4a séries*, Brasília: DF, MEC.

Pitombeira, João Bosco, 1998a. 'As propostas curriculares de matemática', in Barreto, Elba Siqueira de Sá (org.) *Os currículos do ensino fundamental para as escolas Brasileiras*, Campinas, SP: Autores Associados- São Paulo: Fundação Carlos Chagas, Coleção Formação do Professor, 91-126.

Pitombeira, João Bosco, et alii,1998b. *Guia de livros didáticos de 5a à 8a séries* Brasília: DF, MEC.

Roxo, Euclides 1937. *A matemática na educação secundaria*, São Paulo: Companhia editora nacional.

Sonnet, H. 1869. *Premiers Eléments de Calcul Infinitésimal, à l'usage des jeunes gens qui se destinent à la carrière d'ingénieur*, Paris: Hachette.

4.4   Issues of Concern

We can summarise the following issues of concern which arise from the evidence presented earlier about the practice and experiences of teacher trainers, both pre-service and in-service, in a number of countries.

1. An evident obstacle for the effective use of mathematics history in classrooms is that mathematics teachers are still rather weakly qualified in their historical

Figure 4.7: the seventeenth century Japanese teacher had to be trained in a variety of subjects.

knowledge and confidence. Efforts are being made to improve the situation by extending and generalising the historical component in teacher training. An analogous problem, however, is the level of qualification of the teacher trainers themselves, i.e. those who have to teach mathematics history to future teachers and are expected to impart a critical use of historical sources and to judge the value of secondary literature. The quality of the training component is dependent on the competence of the teacher trainers in mathematics history. Given that a considerable proportion of these professors are beginning their own teaching courses as autodidacts, the major bottle-neck is the access to reliable secondary literature.

2. It is not feasible for all the necessary general historical background knowledge to be transmitted within the context of training in the history of mathematics; rather, this component can only be effective if the trainee teachers have been provided with a more general knowledge of history: not only of their respective national history and of international history: of political history and of economic history, but also of the history of civilisation. The major part of such general knowledge (and interest in it!) should be assured by the school curriculum but it is necessary to establish a cooperation with historians for the historical component at the university level, too.

3. Due to the close relation of the evolution of mathematical ideas with the development of philosophical and epistemological conceptions, an analogous cooperation with philosophers should be established as well.

4. The new tendency to integrate mathematics history into the training of future primary teachers, meets with other difficulties. These trainee teachers have in

general to study several disciplines, so that there is not much time provided for the mathematical studies proper. In any event, these trainee teachers were usually not high achievers in mathematics in their own schooling. It is therefore highly important to assure sufficient competence in mathematics for them. The results from Cyprus are encouraging and show that the historical component can contribute to improve a positive relationship towards mathematics itself.

5. Up to now, there is only scattered evidence about the effectiveness of the historical training in the later teaching practice. Even at the level of university teaching itself, the evidence of its impact is very meagre, since the lectures are but rarely accompanied by exercises or followed by seminars allowing for a deepening of subjects and self-activity; and assessments via examinations are just as rare. It is desirable both to establish more systematic assessments at the university level on the one hand, and also to foster stronger relations between teacher trainers and their graduates during their later teaching practice in schools.

Such relations would permit an exchange of information which would facilitate reducing unrealistic or exaggerated views about the impact of the use of history in classrooms on the one hand and to transmit recent progress in historiography to practising teachers, thus clarifying older, rather mythical presentations of the evolution of mathematics.

6. To make further progress in integrating a historical component, more staff specialised in teaching history of mathematics are needed, as well as the development of better adapted teaching material and of exemplary modules as models and guides.

Chapter 5

Historical formation and student understanding of mathematics

Luis Radford

with Maria G. Bartolini Bussi, Otto Bekken, Paolo Boero, Jean-Luc Dorier, Victor Katz, Leo Rogers, Anna Sierpinska, Carlos Vasco

Abstract: *The use of history of mathematics in the teaching and learning of mathematics requires didactical reflection. A crucial area to explore and analyse is the relation between how students achieve understanding in mathematics and the historical construction of mathematical thinking.*

5.1   Introduction

Luis Radford

The history of mathematics may be a useful resource for understanding the processes of formation of mathematical thinking, and for exploring the way in which such understanding can be used in the design of classroom activities.

It is in this spirit that in the last decades some mathematics educators have had recourse to the history of mathematics. However, such a task demands that mathematics educators be equipped with a clear and rich theoretical framework accounting for the general formation of mathematical knowledge. In addition to offering a clear epistemological stance, the theoretical framework has to ensure a fruitful articulation of the historical and psychological domains as well as to support a coherent and fecund methodology (see figure 5.1).

John Fauvel, Jan van Maanen (eds.), *History in mathematics education: the ICMI study*, Dordrecht: Kluwer 2000, pp. 143-170

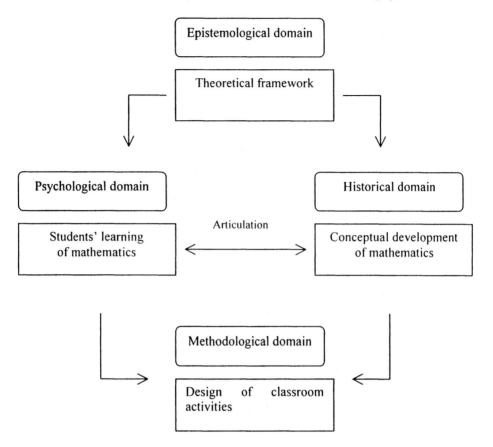

Figure 5.1: Theoretical framework allowing an articulation between the account of students' learning of mathematics and the account of the historical development of mathematics, and supporting a methodology for the design of historically based classroom activities.

The lack of such a suitable framework often leads to oversimplifying views about the way in which mathematical concepts have developed historically (see 'historical domain' in figure 5.1). Indeed, even though new historiographic paradigms have emerged in the past few years (see Gillies 1992, Høyrup 1995, Lizcano 1993, among others), the history of mathematics is all too often read in an *unhistorical* way. That is, narratives are presented which implicitly assume that past mathematicians were *essentially* dealing with our modern concepts, but just did not have our modern notations at their disposal. Reading history like this, in what might be called a *teleological* way, the historian seems to assume, in effect, that there was a course that the historical developments just had to take. In making this assumption, a *normative* dimension is introduced into the account, through which the historian endows other cultures and mathematicians of other epochs with rationalities and conceptualisations that were completely alien to them.

Besides this problem of conveniently framing the historical conceptual development of mathematics, the link between historical developments in mathematical thinking and the students' learning of mathematics (see horizontal arrow in Figure 5.1) has often been done in terms of a naïve psychological version of biological *recapitulationism*. Briefly stated, biological recapitulationism, an idea introduced at the end of the last century, following Darwin's writings on the evolution of species, posits that the development of the individual (*ontogenesis*) *recapitulates* the development of mankind (*phylogenesis*). The German biologist Ernst Haeckel seems to have been the first to transfer this 'biological' law to the psychological domain. He said that "the psychic development of the child is but a brief repetition of the phylogenetic evolution" (quoted by Mengal 1993, 94).

The concept of *genetic development* was partly elaborated in the 1970s, in the work of the psychologists Jean Piaget and Rolando Garcia, as a reaction to this simplistic psychological version of recapitulationism. In their book *Psychogenesis and the history of science* (1989)—a book that has had a significant influence on mathematics educators interested in the use of the history of mathematics—they presented a different perspective. They argued that we should try to understand the problem of knowledge in terms of the intellectual instruments and mechanisms allowing its acquisition.   According to them, the first of those mechanisms is a general process which accounts for the individual's assimilation and integration of what is new on the basis of his or her previous knowledge. (This is a view that runs against the positivist view that knowledge simply accumulates in a straightforward way.)   But then there is an apparent dilemma.   On the one hand, in gaining knowledge the individual is seen as selecting, transforming, adapting and incorporating the elements provided by the external world to his or her own cognitive structures (Piaget and Garcia 1989, 246); while, on the other hand, there can be no assimilation of 'pure' objects divorced from their context, insofar as objects always have a social signification (p. 247).   This paradox led Piaget and Garcia to discuss the influence of the social environment on the evolution of knowledge in the individual.

Pursuing this further led Piaget and Garcia to ask whether two different social environments could lead to two different psychogenetic developments.   Since the works of Bachelard, Kuhn and Feyerabend had stressed the significant role played by social settings in the formation of conceptual systems and theoretical knowledge, Piaget and Garcia's question was hardly inevitable.   The question has become even more urgent nowadays in the light of recent cognitive, anthropological and sociological discussions about the mind.   In an interview given in the mid 1970s, when their book was still in preparation, Piaget clearly stated that one of the problems that led him to write the book was to investigate if there is only one possible line of evolution in the development of knowledge or if there are many, and he replied (Bringuier 1980, 100):

Garcia, who is quite familiar with Chinese science, thinks that they have travelled a route very different from our own. So I decided to see whether it is possible to imagine a psychogenesis different from our own, which would be that of the Chinese child during the greatest period of Chinese science, and I think that it is possible.

However, in their book the problem was dealt with in terms of the difference between the individual's acquisition of knowledge and the 'epistemic paradigm' in which the individual finds him or herself subsumed. By epistemic paradigm they meant "a conception [of science] that has become part of accepted knowledge and is transmitted along with it, as naturally as oral or written language is transmitted from one generation to the next" (Piaget and Garcia 1989, 252). This concept was explicitly presented as an epistemological alternative to Kuhn's concept of paradigm and—in particular—its socially imposed norms. Thus, the 'failure' of Greek and Mediaeval thinkers to conceive the principle of inertia in physics, and the success of the Chinese in conceiving such a principle—which they apparently considered "as obvious as the fact that a cow is not a horse" (p. 253)—was explained in terms of the different epistemic paradigms in which Greek and Chinese science were couched (p. 254). Although the individual was seen as being in dialectical interaction with the object of knowledge, and it was recognised that society provides objects with specific meanings, Piaget and Garcia traced a clear frontier dividing the social and the individual. For them, a distinction must be made between mechanisms to acquire knowledge and the way in which objects are conceived by the subject. In a concise and clear phrase, they said: "Society can modify the latter, but not the former." (p. 267).

In their approach to the relations between ontogenesis and phylogenesis, Piaget and Garcia did not seek for a parallelism of contents between historical and psychogenetical developments but for the mechanisms of passage from one historical period to the following. They tried to show that those mechanisms are analogous to those of the passage from one psychogenetic stage to the next. In addition to the assimilation mechanism previously mentioned, they identified a second mechanism of passage. This was described as a process that leads from the *intra-object*, or analysis of objects, to the *inter-object*, or analysis of the transformations and relations of objects, to the *trans-object*, or construction of structures. The two mechanisms were considered as invariable and omnipresent, not only in time but in space too. That is, we do not have to specify what they are in a certain geographical space at a particular time since it is considered that they do not change from place to place and from time to time.

The Russian psychologist Lev Vygotsky was also concerned with the relationship between ontogenesis and phylogenesis, but—starting from a distinct conception of the mind—took a different approach. Instead of posing the problem in terms of some invariable mechanisms of acquisition of knowledge, he felt that thinking developed as the result of two lines or processes of development: a biological (or natural) process and a historical (or cultural) one. One of his fundamental differences with Piaget and Garcia's approach lies in the epistemological role of culture. For Piaget and Garcia, culture cannot modify the essential instruments of knowledge acquisition, for they saw these instruments as originating in the biological realm of the individual (Piaget and Garcia 1989, 184). In Vygotsky's approach, though, culture not only provides the specific forms of scientific concepts and methods of scientific inquiry but overall modifies the activity of mental functions through the use of tools —of whatever type, be they artefacts used to write as clay tablets in ancient Mesopotamia, or computers in contemporary societies, or intellectual artefacts such as words, language, or inner speech (Vygotsky 1994).

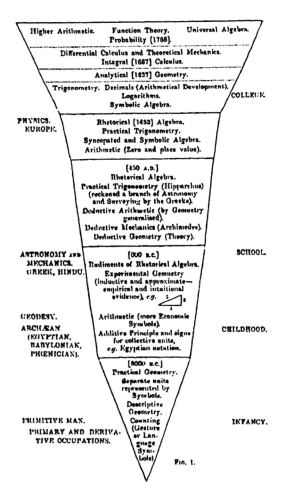

It is this cultural line of development in Vygotsky's account that renders any recapitulationism impossible. For instance, in one of the many passages in which he dealt with this topic, he discusses the development of higher mental functions in history and in the child, and goes on to say that "we do not mean to say that ontogenesis in any form or degree repeats or produces phylogenesis or is its parallel." (Vygotsky 1997, 19). One of the reasons is the variability introduced by the sociohistorical conditions, which are different in each period of the history. In this view, ontogenesis runs, so to speak, underpinned by biological phylogenesis *and* the sociohistorical conditions where ontogenesis takes place (pp. 19-20):

Figure 5.2: Comparison of phylogenesis and ontogenesis have been made since the late 19th century, as seen in this 'Diagram of the development of mathematical experience in the race and in the individual' by Miss Barwell in the Mathematical Gazette *of 1913.*

The growing of the normal child into civilisation usually represents a single merging with the process of his organic maturation. Both planes of development —the natural and the cultural—coincide and merge. Both orders of changes mutually penetrate each other and form in essence a single order of social-biological formation of child personality.

The examples of Piaget and Garcia, and of Vygotsky, uncover the complexity of the problem of the relationship between phylogenesis and ontogenesis and the importance of working towards a clear theoretical framework.

This chapter summarises different ways in which the history of mathematics contributes to a better understanding of the student processes of learning mathematics and the design and analysis of teaching activities. In reference to the different domains mentioned in Figure 5.1, the sections presented in this chapter may be described as follows. In section 5.2, Victor Katz and his colleagues sketch some case studies dealing with the relations between the historical and psychological domains. More specifically, they give some examples from the history of mathematics where we see mathematicians struggling with problems that appear to present difficulties analogous to those faced by our students today, when they tackle the contemporary version of those problems in their school curriculum. They emphasise the importance of teachers having some knowledge of the history of mathematics, as it may help them to help their students overcome some important difficulties which arise in the mathematics classroom.

In section 5.3, Maria Bartolini Bussi and Anna Sierpinska present some sophisticated methodological approaches recently developed by mathematics educators. In these approaches, one of the goals is to study the historical conditions which made possible the emergence of a certain type or domain of mathematical knowledge (historical domain) and to adapt and integrate those conditions into the design of classroom activities (methodological domain) and the analysis of students' forms of mathematical thinking (psychological domain).

In section 5.4, Luis Radford, Paolo Boero and Carlos Vasco focus on the epistemological assumptions (epistemological domain) which underline three current teaching/research approaches using the history of mathematics: Brousseau's epistemological obstacles, Radford's socio-cultural perspective and Boero's Voices and Echoes Games. They make it evident that the interpretation of the conceptual development of mathematics (historical domain), and the investigation of the psychological processes underlying the learning of mathematics (psychological domain), as well as the linking of these phenomena with the design of classroom activities (methodological domain), will all depend upon the chosen framework.

References for §5.1

Bringuier, J-C. 1980. *Conversations with Jean Piaget*, Chicago: Chicago University Press

Gillies, D. (ed.) 1992. *Revolutions in mathematics*, Oxford: Clarendon Press

Høyrup, J. 1995. *In measure, number and weight*, Albany: State University of New York Press

Lizcano, E. 1993. *Imaginario colectivo y creación matemática*, Barcelona: Editorial Gedisa

Mengal, P. 1993. 'Psychologie et loi de récapitulation', in: *Histoire du concept de récapitulation*, P. Mengal (ed.), Paris: Masson, 93-109

Piaget, J. and Garcia, R. 1989. *Psychogenesis and the history of science*, New York: Columbia University Press

Vygotsky, L. S. 1994. 'Tool and symbol in child development', in: René van der Veer and Jann Valsiner, *The Vygotsky reader*, Oxford UK and Cambridge USA: Blackwell

Vygotsky, L. S. 1997. *The history of the development of higher mental functions*, (Coll. wks of L. S. Vygotsky, Vol. 4), R. W. Rieber (ed.), New York: Plenum Press

5.2   The role of historical analysis in predicting and interpreting students' difficulties in mathematics

Victor Katz, Jean-Luc Dorier, Otto Bekken and Anna Sierpinska

As noted in the introduction of this chapter, Piaget and Garcia (1989, 27-28) claim that

the advances made in the course of the history of scientific thought from one period to the next, do not, except in rare instances, follow each other in random fashion, but can be seriated, as in psychogenesis, in the form of sequential 'stages.'... [and] the mechanisms mediating transitions from one historical period to the next are analogous to those mediating the transition from one psychogenetic stage to the next.

Anna Sfard has noted (private communication) that this analogy "is particularly striking at those special junctures where in order to assimilate or create or learn a new concept, the already constructed knowledge has to undergo a complete reorganisation, and the whole epistemological foundation has to be reconstructed as well." The claim of Piaget, which is supported by Sfard, needs of course to be supported by research into students' shifts in understanding mathematical difficulties. This research has been done in several specific cases of student difficulty, where there was a historical reason to believe that such a difficulty might exist. We summarise the results of some of these research studies below.

A first example of this phenomenon of students finding difficulties analogous to those of past mathematicians is familiar to most calculus teachers: the concept of a 'limit' in analysis. Teachers are aware that it is generally difficult to explain the formal notion of limit at the beginning of an elementary calculus class, where it 'logically' belongs. Students certainly 'know' that the limit of $2x+3$ as x approaches 7 is 17, but resist trying to prove such an obvious result using epsilons and deltas. They cannot comprehend why such a proof would be necessary.

To set this in context, historians are aware that the formal idea of a limit was not developed until a century and a half after the basic concepts of the calculus were invented by Newton and Leibniz. During that period, from about 1670 to 1820, many mathematicians used the concept of limit with great understanding —and could calculate limits in many important cases— but they did not have a definition which would enable the statement "the limit of $f(x)$ as x approaches a is L" to be proved with the rigor of classical Greek mathematics. Analysing the historical conditions and reasons why the shift from an intuitive to a formal understanding of limits took mathematicians so long to accomplish gives us valuable information which can help us both predict and interpret our students' difficulties in accomplishing this shift in a few short weeks (see Cornu 1991, Sierpinska 1988, Burn 1993).

John Fauvel, Jan van Maanen (eds.), *History in   mathematics education: the ICMI study*, Dordrecht: Kluwer 2000, pp. 149-154

Besides the difficulty related to the passage from an intuitive to a rigorous understanding and use of the concept of limit, other difficulties arise from this concept in the comprehension of curvilinear area, tangent line and instantaneous flow. An intensive historical search on the development of calculus allowed M. Schneider (1988) to demonstrate that these difficulties surface from the same epistemological obstacle: the absence of separation, in the mind of students, between mathematics and an illusory 'sensible' world of magnitudes. This investigation provided Schneider with a research methodology to render such learning difficulties apparent: for example, the reactions of students in learning about Cavalieri's principles, indivisibles and related paradoxes reveal mental shifts in meaning from the world of magnitudes to their measures.

Jean-Luc Dorier (1998), in his studies of how best to teach the concepts of linear dependence and linear independence in linear algebra, has noted that although students entering university often have certain conceptions of these notions in concrete situations, they have difficulty in understanding the connection of the formal definition with these earlier situations. A historical analysis of the development of these concepts provides help in understanding the students' difficulties.

The twin concepts of linear dependence and independence emerged historically in the context of linear equations and, in particular, in Euler's analysis of Cramer's paradox dealing with the number of intersection points of two algebraic curves. Euler found that the paradox was based on the 'fact' that n linear equations determine exactly n unknown values, but realised that this latter statement is not always true. He discussed several examples in which systems of n equations in n unknowns do not have a single n-fold solution and realised that in certain cases the actual constraints imposed on the unknowns by the equations are fewer than n. That is, Euler stated that certain of the equations are "contained" in the others; this is his notion of what we can call inclusive dependence. After Euler's work, many mathematicians considered this problem of dependence and tried to determine conditions on the determinant of a dependent system which would show the nature of the set of solutions. But it was not until 1875 that Georg Frobenius pointed out the similarity of dependence of a set of equations to dependence of a set of n-tuples. He could then give a formal definition of the concept of 'linear dependence' and show how the notion of 'rank' of a system enabled one to determine the dimension of the set of solutions.

The teaching experiment reported by Dorier, based on a historical analysis of the development of the concept of rank, was designed to help the students understand the power of linear dependence as a formal and unifying concept. Indeed, from their secondary school practice of solving equations, students entering university usually have an Eulerian 'inclusive dependence' idea of equations. But at the university level, it is necessary for the students to move to the stage where they understand the formal concept of dependence in a global context. That is, they need to understand that the equations, and not just n-tuples, must be regarded as objects in their own right and that there needs to be a definition of linear dependence which applies to both of these cases, as well as in even more general contexts. Thus it was necessary to devise a teaching strategy to meet these needs.

On a more elementary level, students often have trouble making the shift from solving concrete problems using words and numbers to the more abstract problem of using letters to designate unknown quantities. Again, we know that, historically, it was a difficult conceptual switch. In order to help students understand the role of letters as representing unknowns, Radford and Grenier (1996a, 1996b) designed a teaching sequence in which students were asked to solve some word problems using manipulatives. These manipulatives were conceived in such a way that the unknown quantity was modelled by a hidden number of candies in a bag or a hidden number of hockey cards in an envelope, and so on. The teaching sequence was structured to allow the students to master two important rules of Islamic algebra, those of *al-muqabala* and *al-jabr*. In the second step of the teaching sequence, instead of using

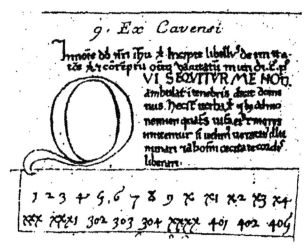

Figure 5.3: Not only 'hidden quantities' are hard to understand. The Hindu-Arabic numerals themselves were difficult for early European users, as this medieval Italian manuscript testifies. The scribe has rendered as "xxx xxx1 302 303 . . ." what we would write as "30 31 32 33". Such a text helps today's teachers to appreciate how difficult it is for pupils to learn positional notation.

manipulatives, the students had to make drawings (e.g. of a bag containing an unknown number of candies) and, in the third step, the students had to use letters instead of drawings. The teaching sequence was inspired by a historical analysis of medieval Italian algebra (Radford 1995, 1997), in particular by an idea of the fourteenth century mathematician Antonio de Mazzinghi, who explained the concept of unknown as a 'hidden' quantity.

Anna Sfard (1995) found furthermore that even if high school students could solve linear equations or systems of linear equations with numerical coefficients, it was still difficult for them to make the jump to solving systems with literal coefficients. She notes that at first she was "quite insensitive to the huge conceptual difference between equations with numerical coefficients and equations with parameters." And it took several weeks of hard work before the students could cope with such equations in a reasonable manner. Sfard found that colleagues had encountered similar difficulties. Again, a historical analysis shows that this difficulty is not surprising. Even though by the late medieval period, letters and other abbreviations were being used in algebra to designate unknowns and their powers, the rules for solving equations were always stated in terms of concrete examples.

Thus one could solve $x^2 + 10x = 39$, but not $x^2 + bx = c$. It was François Viète in the late sixteenth century who first introduced letters to designate known values (parameters) and in this way brought a great conceptual change to algebra. It was Viète's work that enabled formulas to be written to solve quadratic and cubic equations, for example, and that led, in general, to structural manipulations in algebra rather than purely operational ones. The historical difficulties in this shift from numerical to purely symbolic algebra again leads us to believe that teachers must be aware of the conceptual difficulties their students may have in making the same shift.

Lisa Hefendehl-Hebeker (1991) analysed the always difficult task of helping students understand the meaning of a negative number, and the reasons for the rules governing operations with these numbers. Negative numbers have, of course, been used for two millennia in China, but mathematicians in the West have always been suspicious of them, even though the rules for operation on them were known by the sixteenth century. Even as late as the nineteenth century, there were some English mathematicians who tried to reformulate algebra without the use of negative numbers, because they believed that they were nonsensical. The question, in fact, became whether negative numbers were 'quantities' and then what it meant for a 'quantity' to be less than zero. There were, of course, numerous attempts throughout the centuries to justify negative numbers, either by using them to model a particular idea (debt, for example) or by deriving the rules of operation by arguments based on the "principle of permanence of equivalent forms" (Peacock 1830), in particular the distributive and associative laws. Hefendehl-Hebeker shows in her article how modern students' confusions about these laws are mirrored in confusions of such authors as Stendhal and d'Alembert in the 18th century. A teacher would do well to study these 'confusions' to see why his or her own students could be confused. But Hefendehl-Hebeker also notes that Hermann Hankel in the mid-19th century advocated a change in point of view by looking at negatives as an extension of the number system rather than as quantities in their own right. That is, he urged that these numbers be introduced in a purely formal manner, without worrying about what kind of quantity they represent. Again, this history shows how one might try to introduce and justify negative numbers in the classroom.

Another set of numbers which often causes difficulties for students is the complex numbers. At one time in school they are told that negative numbers do not have square roots, and later they are told that in fact they do have square roots. Why have the rules changed? A historical analysis here shows again that there was a long period of development between the first discovery of complex numbers by Cardano and Bombelli in their studies of solutions of cubic equations in the fifteenth century and the general acceptance of these numbers into mathematics in the nineteenth. As in the case of negatives, it took centuries for mathematicians to give up the idea that 'number' must represent the measure of a quantity. The final acceptance of these numbers came only through their geometric interpretation, that is, on their modelling in a well-understood area of mathematics. Again, many textbooks today seem to violate this historical analysis by simply defining the square root of -1 by fiat, without any motivation whatsoever.

Non-Euclidean geometry was developed by three mathematicians early in the nineteenth century. Carl Friedrich Gauss, who developed it first, declined to publish anything on this topic, because he did not want to deal with the controversies he was sure would erupt. But two less famous mathematicians, Janos Bolyai in Hungary and Nikolai Lobachevsky in Russia, both published their studies in this field around 1830. Nevertheless, it proved very difficult for mathematicians to give up the very strong conviction that geometry describes a unique reality and, as such, can not admit a plurality of axiom systems. It was not until several mathematicians showed how non-Euclidean geometry could be modelled in Euclidean geometry that the mathematical community began to accept the validity of non-Euclidean geometry. So again, we should not be surprised when there is difficulty for students to understand that Euclidean geometry may not in fact be the 'best' geometry to describe the space in which we live.

A final common student difficulty involves the transition to abstraction. As a typical example, many instances of what today are called groups were known in the first eight decades of the nineteenth century—and some were known even earlier. Yet it was not until 1882 that the first complete formal definition of this abstract concept was given. Nevertheless, many current textbooks in abstract algebra begin by giving a formal definition of a group before the student has experienced many of these examples. It is not surprising that students have difficulties making the leap to abstraction; too little attention has been paid to the necessary steps that historically preceded this leap.

As these examples demonstrate—and there are numerous others—a teacher who is knowledgeable in the history of mathematics will anticipate student difficulties in areas where, historically, much work was needed to overcome significant difficulties. Thus the teacher can be prepared with appropriate teaching strategies for these situations, ones which may well be in accord with the historical developments and which will help the students overcome these obstacles to understanding. And as some of the research results in this area demonstrate, these strategies may well be effective. Yet the knowledge of history of mathematics is not sufficient to develop teaching strategies; if the analysis of historical conditions of the emergence of a concept is an important source of information to predict and analyse students' difficulties, teachers still must take into account the reality of teaching at a certain level with a certain type of student. There is no automatic transfer from history to teaching. First, the knowledge of history must be as complete as possible, involving primary sources whenever feasible.   Second, there must exist a preliminary didactical investigation about students' difficulties.   Finally, the confrontation of the historical and didactical situations must be made with great care, taking into account the conditions and constraints of the two different environments, the historical and the classroom.

Such work needs competence both in history and in mathematics education research and shows interesting possible interactions between these two fields for the future.

References for §5.2

Burn, R. P. 1993. 'Individual development and historical development: a study of calculus', *International journal of mathematics education, science and technology*, **24**, 429-433

Cornu, B. 1991. 'Limits', in D. Tall (ed.), *Advanced mathematical thinking*, Dordrecht: Kluwer, 153-166

Dorier, J.-L. 1998. 'The role of formalism in the teaching of the theory of vector spaces', *Linear algebra and its applications*, **275**, 1-4, 141-160

Hefendehl-Hebeker, L. 1991. 'Negative numbers: obstacles in their evolution from intuitive to intellectual constructs', *For the learning of mathematics*, **11** (1), 26-32

Peacock, George 1830. *A treatise on algebra*, Cambridge

Piaget, J. and Garcia, R. 1989. *Psychogenesis and the history of science* (trans. by Helga Feider), New York: Columbia University Press

Radford, L. 1995. 'Before the other unknowns were invented: didactic inquiries on the methods and problems of medieval Italian algebra', *For the learning of mathematics*, **15** (3), 28-38

Radford, L. and Grenier, M. 1996a. 'Entre les choses, les symboles et les idees. . . une sequence d'enseignement d'introduction à l'algèbre', *Revue des sciences de l'éducation* **22**, 253-276

Radford, L. and Grenier, M. 1996b. 'On the dialectical relationships between symbols and algebraic ideas', in: L. Puig and A. Gutierrez (eds.), *Proceedings of the 20th International Conference for the Psychology of Mathematics Education*, Valencia: Universidad de Valencia, vol. **4**, 179-186

Radford, L. 1997. 'L'invention d'une idée mathématique: la deuxieme inconnue en algebra', *Repères, Revue des IREMs* **28** (July), 81-96

Schneider M. 1988. *Des objets mentaux 'aire' et 'volume' au calcul des primitives*, Thèse de doctorat, Louvain-la-Neuve

Sfard, Anna 1995. 'The development of algebra: confronting historical and psychological perspectives', *Journal of mathematical behavior* **14**, 15-39

Sierpinska, Anna 1988. 'Sur un programme de recherche lié a la notion d'obstacle epistemologique', in N. Bednarz and C. Garnier (eds.). *Construction des savoirs: obstacles et conflits*, Ottawa: Agence d'Arc Inc., 130-148

5.3   The relevance of historical studies in designing and analysing classroom activities

Maria G. Bartolini Bussi and Anna Sierpinska

With contributions by Paolo Boero, Jean Luc Dorier, Ernesto Rottoli, Maggy Schneider, and Carlos Vasco

When a mathematics educator draws on the history of the domain in designing activities for the students he or she may be looking for facts: Who were the authors of that particular piece of mathematics? When did they live? What were their lives?

John Fauvel, Jan van Maanen (eds.), *History in mathematics education: the ICMI study*, Dordrecht: Kluwer 2000, pp. 154-161

By introducing historical anecdotes in his or her classes he or she may increase the students' motivation to learn mathematics. But a historical study may have other goals as well: looking for geneses of mathematical ideas or contexts of emergence of mathematical thinking, in the aim of defining conditions which have to be satisfied in order for the students to develop these ideas and thinking in their own minds.

5.3.1   Bringing historical texts into the classroom: the 'voices and echoes' games

For example, Boero *et al.* (1997, 1998) concerned themselves with the conditions of emergence of theoretical knowledge. Mathematical thinking is theoretical par excellence, and without developing this special attitude of mind in the students there is less opportunity for deepening their understanding of mathematics. A historico-epistemological analysis was, for these authors, a basis for an analytical definition of theoretical knowledge which included parameters such as organisation, coherence and systematic character, the role played by definitions and proofs, the speech genre characteristic of theoretical discourse, and the ways of viewing the objects of the theory. This definition became subsequently a basis for a didactic theory: indeed, Boero *et al.* have designed and implemented an innovative educational methodology in the classroom called the 'voices and echoes game', which draws on the Vygotskian distinction between everyday and scientific concepts and the Bakhtinian construct of 'voice'.

The main hypothesis of this methodology is the introduction, into the classroom, of 'voices' from the history of mathematics (in the form of selected primary sources, with commentaries). This might, by means of well chosen tasks, develop into a 'voices and echoes game' suitable for the mediation of some important elements of theoretical knowledge. The chosen examples of theoretical knowledge are conceptual leaps in the cultural history of mankind: the theory of falling bodies of Galileo and Newton, Mendel's probabilistic model of the transmission of hereditary traits, mathematical proof and algebraic language. All these feature aspects of a counterintuitive character. The authors claim that the 'new' manners of viewing and the methodological requirements are expressed by the 'voices' of the protagonists themselves in the speech genre that belongs to their cultural tradition. Such voices act as voices belonging to real people with whom an imaginary dialogue can be conducted beyond space and time. The voices are continuously regenerated in response to changing situations: They are not passively listened to but actively appropriated through an effort of interpretation. The authors describe a number of teaching experiments whereby they introduce some analytical tools (i.e. different types of echoes) which, on the one hand, are used to interpret classroom processes and, on the other, are used to design classroom activity. For instance, a *'mechanical echo'* consists in a precise paraphrasing of a verbal voice, whilst an *'assimilation echo'* refers to the transfer of the content/method conveyed by a voice to other problem situations. A *'resonance'* is a student's appropriation of a voice as a way of reconsidering and representing his or her experience. The most delicate issue in this methodology is, certainly, the selection of historical sources capable of conveying the crucial ideas of a scientific revolution in a concise manner, so as to comply with

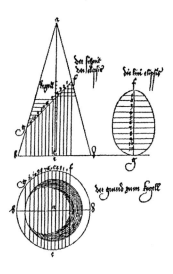

Figure 5.4: Whether a section of a cone is the same as a section of a cylinder, and whether either is egg-shaped, has long been debated. Here Dürer's discussion of the ellipse (Underweysung der Messung, 1525)

the space and time constraints of institutionalised teaching. Boero's published experiments concern mainly grade 8 secondary school students, but studies presently in progress (with voices taken from Plato's dialogues) have given evidence that similar processes can be implemented also in primary schools and with pupils from a range of socio-cultural backgrounds (Garuti *et al* 1999). We analyse below (§5.4) the epistemological assumptions of this methodology.

This approach is consistent with the approach of Bartolini Bussi *et al.* (1996, 1999) who also introduced a guided reading of historical sources in primary school, in two long-term teaching experiments concerning perspective drawing and gears. Even if no explicit voices and echoes game was introduced in the classroom, the guided reading and interpretation of well selected historical sources had been used to institutionalise the pieces of knowledge built in the classroom by shifting them to a theoretical level. In both experiments the appropriation of the theoretical dimension of mathematical knowledge had led the pupils to produce theorems, i. e. statements with proofs inside a reference theory (Mariotti *et al.* 1997). The above experiments concern early grades of school (4-8).

Other experiments have been carried out successfully in the 11th grade (Ernesto Rottoli, personal communication), using original texts of Greek authors and excerpts from historical studies, in order to integrate the knowledge acquired during philosophy lessons and the knowledge acquired during mathematics lessons. The aim was to organise a deeper level of knowledge. The design was based on the awareness that in ancient times mathematics and philosophy were strictly linked to each other and some traces of this link are still present in highly organised and culturally rooted linguistic patterns.

5.3.2   Indirect use of historical and epistemological studies in the design of activities for students

In the research projects described above, elements of the history of science (fragments of original texts) were used in an explicit manner in the teaching sequences, and historico-epistemological studies were directly linked to the contents of teaching. The links between the historical studies and the teaching design can be much more implicit and indirect, and the relevance of these studies for the didactic activity somewhat less obvious.

5.3.3   The example of linear algebra

This is certainly the case of the research projects on the teaching and learning of linear algebra conducted, independently, by Dorier and Sierpinska.

The motivation of these research projects has been students' commonly stated difficulty with the axiomatic approach used in undergraduate linear algebra courses. This difficulty is often hard for mathematicians to understand, for whom the axiomatic approach is indeed the royal road to linear algebra, at last allowing the subject to be presented in a simple, neat and coherent way. The questions that naturally arise in this situation are: why is it difficult to understand a simple axiomatic theory? What are the conditions of coming to construct or understand this or that particular concept of this theory? What can be done to facilitate the understanding of this theory by the students? Some answers to these questions make no reference to history. For example, one may say that the axiomatic theories that constitute linear algebra are simple only in appearance. A slightly deeper mathematical analysis of the basic concepts of linear algebra shows their inner complexity (see, e.g. Sierpinska, Dreyfus, Hillel, 1999). This complexity may not be accessible to an undergraduate student, and therefore, he or she will have to accept the teacher's word that, for example, it makes sense to accept this definition rather than a different one. This happens so often in a linear algebra course, that many students end up developing what is called 'the obstacle of formalism' (Dorier *et al* 1997). It may not have been necessary to refer to history to answer these questions. But it proved useful and inspiring, both in explaining students' difficulties and in designing activities for them.

For example, a look at the history of linear algebra from a very broad perspective of currents of thought allowed the identification of three interacting modes of reasoning, labelled 'synthetic-geometric', 'analytic-arithmetic', and 'analytic-structural' (Sierpinska *et al.* 1997). These modes of reasoning are linked to different theoretical perspectives and imply different meanings of concepts. They are not equally accessible to beginning linear algebra students, and the students tend to be inflexible in using them in different contexts. An awareness of these modes of reasoning and their role in linear algebra helps in both designing activities for students and reacting to the students' responses to them in a teaching situation.

A more fact-focused look at the history of linear algebra allowed the identification of the contexts in which the basic linear algebra concepts emerged: analytic geometry, vector algebra, vector analysis and applications in physics; linear equations and determinants, linear differential and functional equations, abstraction of vector structures in functional analysis (Dorier 1995a, 1997). Specific contexts have been used in the design of history-inspired classroom activities. For example, instead of simply giving the definition of a linearly independent set of vectors and following it by a series of exercises, Dorier (1998a, 1998b) proposed to anchor the students' understanding in their experience of the Gaussian elimination method for solving systems of equations, which is introduced in secondary schools in France. The task for the students was to discuss and analyse this method. In this research, history was a source of inspiration and a means of control in the building of the didactic experiment, but the experiment did not aim at a reconstruction in the classroom of the historical development or even at commenting on historical texts.

Case study: Fermat as an inspiration for work with Cabri

The reference to history is also implicit in recent research by Sierpinska, Hillel & Dreyfus (submitted), which focuses on the students' understanding of the notion of vector and its coordinates in a basis. This research involved designing and evaluating a teaching sequence in the Cabri dynamic geometry environment. What emerged was the striking difference between the way in which Fermat approached the problem of finding a canonical equation of a conic in his *Ad locos planos et solidos isagoge* (c.1635) and the algorithmic procedure which is normally used in present day linear algebra courses. This triggered an understanding of the difference between geometric and arithmetic spaces, and a coherent explanation in these terms of the students' difficulties and conceptions. A brief outline of this explanation follows.

Elements of an *n*-dimensional arithmetic space are *n*-tuples of real numbers. By defining operations of addition and scalar multiplication on the *n*-tuples in a coordinate-wise fashion one obtains a vector space structure usually denoted by $\mathbb{R}^n$. There is a long-standing tradition of referring to the elements of the arithmetic spaces as 'points', and of using the language of Euclidean geometry to refer to their subsets such as straight lines and planes. This is what we do in linear algebra classes, without, however, discussing with the students the status, in the theory, of the geometric objects thus evoked. There are important differences between the 'arithmetic spaces' underlying vector spaces $\mathbb{R}^n$ and the 'geometric spaces' of Euclidean geometry. The objects of the arithmetic spaces are sets of *n*-tuples of real numbers defined by conditions (in the form of equations, inequalities, etc.) on the terms of the *n*-tuples belonging to the sets. These objects can be represented by geometric figures like lines or surfaces. The representations will depend on the choice of a coordinate system.

A set $\{(x, y) : x^2 + y^2 = 1\}$, for example, will be represented by a geometric circle in an orthonormal coordinate system, and by a geometric ellipse in a non-orthonormal coordinate system. (Here *geometric circle* means the locus of points equidistant from a given point.) In geometric spaces, the roles of objects and representations are reversed. Objects, given by relations between their parts, can be represented by sets of *n*-tuples defined by conditions on their terms, e.g. by equations. These equations will be different depending on the choice of the coordinate system.

Fermat and Descartes worked with geometric spaces, and for them, equations were representations of geometric objects: they were introducing a system of coordinates into a pre-existing geometric space. But, in a process which started by the end of the 17th century with the work of Newton and other creators of calculus, representations started to play the role of objects: "Before Descartes, the solution of an algebraic equation was nothing but a tool to solve other problems. After Descartes and particularly at the end of the 17th century, to give an equation or a symbolic expression was just to give a curve, and to give an integral was just to give an area, even if the curve and the area are geometric objects that we can perfectly characterise without mentioning any equation or integral." (Panza 1996, 245). This process led to the replacement of the geometric space with, as it were, a system of coordinates without an underlying geometric space.

The geometric language and drawings of lines and planes in today's linear algebra textbooks are used as mere didactic aids in the introduction of the $\mathbb{R}^n$ spaces, illustrations which play no role in the building of the theory. But thinking of vectors as *n*-tuples leads, notoriously, to students' difficulties with the notions of 'change of basis' and 'coordinates of a vector in a basis', especially when these notions are introduced in the context of $\mathbb{R}^n$ spaces (Hillel & Sierpinska 1994). Indeed, for a student who is thinking in terms of arithmetic spaces, the notion of change of coordinates may not make sense. Insofar as an arithmetic space is nothing but a system of coordinates, changing the system means changing the space, so one should maybe speak of transformations of the space. The very notion of coordinates of a vector does not seem to make sense in the arithmetic frame of mind, where a vector is nothing but coordinates. In our courses we often try to give some meaning to the notion of change of basis by introducing the topic of canonical equations of conics. But in doing this, without warning the student, we revert to thinking in terms of geometric spaces: conics are again geometric objects which can be represented by different equations depending on the choice of the coordinate system. This only adds to the confusion in the students' minds. The notions of coordinates of a vector in a basis and change of basis make more sense for the students when they start working with vector spaces other than $\mathbb{R}^n$ (especially with function spaces) but, at an early stage in the teaching of linear algebra, it seems useful to restore the geometric genesis of the $\mathbb{R}^n$ spaces. This was the guiding idea of the teaching design and an important part of the rationale behind the choice of the computer environment, namely the preference of a Dynamic Geometry Software over a Computer Algebra System.

A posteriori, it is clear that it was not necessary to study Fermat's *Isagoge* to come to this understanding of the students' difficulties. But it helped a lot in clarifying ideas and making distinctions between blurred concepts. The simple reason for this can be that understanding ideas gains much from analyzing contrasting ways of thinking, from having access to their articulated exposition, and from following their evolution over long periods of time. All this is made possible in a historical study.

5.3.4   The example of calculus

Another example of the use of historical studies in understanding students' difficulties and designing activities for them is found in a research project conducted by Schneider (details in §8.2.2). This is a project concerned with calculus, which takes into account the order and choice of historical contexts, the historical forms of the central concepts, and the analysis of the evolution of these concepts in terms of epistemological obstacles (Schneider 1988). Activities for the students are designed with the intention of allowing the students to put to test, individually and collectively, their previous beliefs and to become aware of the limitations of these. The problem situations generated in these activities are expected to give rise to cognitive and socio-cognitive conflicts and to create favourable conditions for students to reach a better understanding.

Although the project is framed by a constructivist view, it is not assumed that the students construct theoretical knowledge only as described by the constructivist model. Indeed, in this project, students' understanding is seen as dependent, to a certain extent, on the didactic mediation of the teacher. For example, a game of 'voices and echoes' (in Boero's sense, see above) between Berkeley's text and the students about instantaneous velocity, with a meta-level type of intervention of the teacher (see Dorier 1995b), makes the students better aware of their own perception of mathematics and of the connections of this discipline with the perceptible phenomena of the physical world. In this project, the theory of epistemological obstacles and the constructivist approach are conceived of as hypotheses whose efficiency should be tested case by case, taking into account the specificity of the mathematical contents, the socio-cultural origin of students, the problem situations as described by some precise didactic variables, each situation having to be studied didactically (for an example of a didactic study of a situation related to instantaneous flow see Schneider 1992).

5.3.5   Research on the methodology of history-based design of activities for students

In neither of the examples of research given in this section was the methodology of history-based design and analysis of student activities an object of explicit discussion. Other research in mathematics education is concerned with this particular question, especially in the context of the theory of epistemological obstacles (e.g. Schneider 1988, 15-16; Sierpinska 1994, 120-125). Here, let us mention in more detail only a methodology proposed by Vasco (1995), which is not related to the framework of epistemological obstacles. The heuristics proposed in this work, called 'forward and backward heuristics', are aimed at helping to find hypotheses for potentially optimal sequencing of mathematics curricula. The 'forward heuristics' are meant to propose efficient ways of reviewing the phylogenesis of the particular mathematical subject, in order to optimise the ontogenetic mastery of that conceptual field. The 'backward heuristics' propose ways to trim, compress, and even alter the sequences found through the forward heuristics. Forward heuristics lay out the rough draft of the roads on the mathematical map; backward heuristics do the redesigning, the short-cutting, and the road signalling (Vasco 1995, 62).

References for §5.3

Bartolini Bussi M. 1996. 'Mathematical discussion and perspective drawing in primary school', *Educational studies in mathematics* **31**, 11-41

Bartolini Bussi M., Boni M., Ferri F. and Garuti R. 1999. 'Early approach to theoretical thinking: gears in primary school', *Educational studies in mathematics* **39**, 67-87

Bartolini Bussi, M., and M. A. Mariotti, 1999. 'Semiotic mediation: from history to the mathematics classroom', *For the learning of mathematics* **19** (2), 27-35

Boero P., Pedemonte B. and Robotti E. 1997. 'Approaching theoretical knowledge through voices and echoes: a Vygotskian perspective', *Proceedings of the 21st International Conference on the Psychology of Mathematics Education*, Lahti, Finland, vol. **2**, 81-88

Boero P., Pedemonte B., Robotti E. and Chiappini G. 1998. 'The 'voices and echoes game' and the interiorization of crucial aspects of theoretical knowledge in a Vygotskian perspective: ongoing research', *Proceedings of the 22nd International Conference on the Psychology of Mathematics Education*, Stellenbosch, South Africa, vol. **2**, 120-127

Dorier, J.-L., Robert, A., Robinet, J. A. and Rogalski, M. 1997. 'L'algèbre linéaire: l'obstacle du formalisme à travers diverses recherches de 1987 à 1995', in: J.-L. Dorier (ed.), *L'enseignement de l'algèbre linéaire en question*, Grenoble: La Pensée Sauvage Èditions, 105-147

Dorier, J.-L. 1995a. 'A general outline of the genesis of vector space theory', *Historia mathematica* **22**, 227-261

Dorier, J.-L. 1995b. 'Meta level in the teaching of generalizing concepts in mathematics', *Educational studies in mathematics* **29**, 175-197

Dorier, J.-L. 1997. 'Une lecture épistémologique de la genèse de la théorie des espaces vectoriels', in: J.-L. Dorier ed., *L'enseignement de l'algèbre linéaire en question*, Grenoble: La Pensée Sauvage Éditions, 27-105

Dorier, J.-L. 1998a. 'The role of formalism in the teaching of the theory of vector spaces' *Linear algebra and its applications* **275**, 141-160

Dorier, J.-L. 1998b. 'État de l'art de la recherche en didactique à propos de l'enseignement de l'algèbre linéaire', *Recherches en didactique des mathématiques* **18** (2), 191-230

Garuti, R, P Boero & G Chiappini, 1999. 'Bringing the voice of Plato in the classroom to detect and overcome conceptual mistakes', *Proceedings of the 23rd PME Conference, Haifa, Israel* iii, 9-16

Hillel, J. & Sierpinska, A. 1994. 'On one persistent mistake in linear algebra', *Proceedings of the 18th International Conference on the Psychology of Mathematics Education*, Lisbon, Portugal, vol. **3**, 65-72

Mariotti M. A., Bartolini Bussi M., Boero P., Ferri F. and Garuti R. 1997. 'Approaching geometry theorems in contexts: from history and epistemology to cognition', *Proceedings of the 21st International Conference on the Psychology of Mathematics Education*, Lahti, Finland, vol. **1**, 180-195

Panza, M. 1996. 'Concept of function, between quantity and form, in the 18th century', in: H.N. Jahnke, N. Knoche and M. Otte (eds.), *History of mathematics and education: ideas and experiences*, Göttingen: Vandenhoeck and Ruprecht, 241-269

Schneider M. 1988. *Des objets mentaux 'aire' et 'volume' au calcul des primitives*, Thèse de doctorat, Louvain-la-Neuve

Schneider M. 1992. 'A propos de l'apprentissage du taux de variation instantané', *Educational studies in mathematics* **23**, 317-350

Sierpinska, A. 1994. *Understanding in mathematics*, London: Falmer Press

Sierpinska, A., Defence, A., Khatcherian, T., Saldanha, L. 1997. 'A propos de trois modes de raisonnement en algèbre linéaire', in: J.-L. Dorier (ed.), *L'enseignement de l'algèbre linéaire en question*, Grenoble: La Pensée Sauvage Éditions, 249-268

Sierpinska, A., Dreyfus, T., Hillel, J. 1999. 'Evaluation of a teaching design in linear algebra: the case of linear transformations' *Recherches en didactique des mathématiques* **19**, 7-40

Vasco, C. E. 1995. 'History of mathematics as a tool for teaching mathematics for understanding', in: D. N. Perkins, J. L. Schwartz, M. M. West, and M. S. Wiske, *Software goes to school: teaching for understanding with new technologies*, New York: Oxford University Press, 56-69

5.4   Epistemological assumptions framing interpretations of students understanding of mathematics

Luis Radford, Paolo Boero and Carlos Vasco

Two different phenomena need to be linked, in using the history of mathematics to understand better the student processes of learning mathematics and the way in which such an understanding can be used in the design of classroom activities. On the one hand, the learning processes of contemporary students; on the other hand, the historical construction of mathematical knowledge. These phenomena belong to two different theoretical realms: the former to the psychology of mathematics, the latter to an opaque field where epistemology and history (to mention only two disciplines) encounter each other.

The linking of psychological and historico-epistemological phenomena requires a clear epistemological approach.   Within the field of mathematics education, different approaches have been used.   They differ in their epistemological assumptions and, as a result of this, they provide different explanations of the history of mathematics. They also offer different interpretations of students' understanding of mathematics and suggest different methodological lines of pedagogical action.   The aim of this section is to provide an overview of some approaches and their corresponding epistemological frameworks.

5.4.1   The 'epistemological obstacles' perspective

This approach is based on the idea of epistemological obstacles developed by G. Bachelard and later introduced into the didactics of mathematics by G. Brousseau in the 1970s. Brousseau's approach is based on the assumption that knowledge exists and makes sense only because it represents an optimal solution in a system of constraints.   For him, historical studies can be inspiring in finding systems of constraints yielding this or that particular mathematical knowledge: these systems of constraints are then called '*situations fondamentales*'.   In Brousseau's view, knowledge is not a state of mind; it is a solution to a problem, independent of the solving subject.   Within this context, an epistemological obstacle appears as the source of a recurrent non-random mistake that individuals produce when they are trying to solve a problem.

A clear assumption underlying this approach is that an epistemological obstacle is something wholly pertaining to the sphere of the knowledge—a sphere that Brousseau conceives as separated from other spheres.   Thus he distinguishes the *epistemological obstacles* from other obstacles, e.g. those related to the students' own cognitive capacities according to their mental development (*ontogenetic obstacles*), those which result from the teaching choices (*didactic obstacles*)

John Fauvel, Jan van Maanen (eds.), *History in   mathematics education: the ICMI study*, Dordrecht: Kluwer 2000, pp. 162-167

(Brousseau 1983, 177; Brousseau 1997, 85-7) and those whose origin is related to cultural factors (*cultural obstacles*) (Brousseau 1989; Brousseau 1997, 98-114). Of course, the clear-cut division of obstacles into ontogenetic, didactic, cultural and epistemological categories is in itself an epistemological assumption.

The link between the psychological and the historical phenomena to which we referred previously is ensured by another epistemological assumption: in Brousseau's account, an epistemological obstacle is precisely characterised by its reappearance in both the history of mathematics and in contemporary individuals learning mathematics. He says (translation from Brousseau 1983, 178; Brousseau 1997, 87-8): "The obstacles that are intrinsically epistemological are those that cannot and should not be avoided, precisely because of their constitutive role in the knowledge aimed at. One can recognise them in the history of the concepts themselves."

A third epistemological assumption is to be found in the articulation 'student/milieu'. According to Brousseau, the teacher sets the situation, but the knowledge which will result is due to the student's appropriation of the problem. Thus, the motivation is an exclusive relationship between the problem-situation and the student. In doing this, Brousseau supposes that a kind of isolation between the teacher and the student takes place during the process of solving the given problem.

The interpretation of the student's understanding of mathematics is framed here by the idea that the development of knowledge is a sequence of conceptions and obstacles to overcome (Brousseau 1983, 178). Consequently, the pedagogical action is focused on the elaboration and organization of teaching situations built on carefully chosen problems that will challenge the previous students' conceptions and make it possible to overcome the epistemological obstacles, opening new avenues for richer conceptualisations (for an example, see the way Schneider organised her calculus teaching, §5.3.4).

Sierpinska has stressed that, although the new conceptualisations may be seen as more complex than the previous ones, these do not have to be necessarily related to steps in the development or progress of knowledge: "Epistemological obstacles are not obstacles to the 'right' or 'correct' understanding: they are obstacles to some change in the frame of mind." (Sierpinska 1994, 121).

5.4.2   A socio-cultural perspective

Some Vygotskian perspectives in mathematics education choose, from the outset, a different set of epistemological assumptions.   Thus, in Radford's socio-cultural perspective, knowledge is not restricted to the technical character which results when knowledge is seen as essentially related to the actions required to solve problems. Following a socio-historical approach (see eg Mikhailov 1980, Ilyenkov 1977) and a cultural tradition (see eg Wartofsky 1979), knowledge is conceived as a culturally mediated cognitive praxis resulting from the activities in which people engage. Furthermore, the specific content with which knowledge is provided is seen as framed by the rationality of the culture under consideration. It is the mode of that rationality which will delimit the borders of what can be considered as a scientific problem and what shapes the norms of scientific inquiry-for instance, what is an accepted scientific discourse and what is not, what is accepted as evidence and what is not. The mode of the rationality relates directly to the social, historical, material and symbolic characteristics underpinning the activities of the individuals (Radford, submitted). Hence, from a sociocultural epistemological viewpoint, knowledge can only be understood in reference to the rationality from which it arises and the way the activities of the individuals are imbricated in their social, historical, material and symbolic dimensions.

In this line of thought, a problem is never an object on its own, but is always posed, studied and solved within the canons of rationality of the culture to which it belongs (Radford 1997a). For example, the supposed numerical patterned cosmo-logical nature of the universe was an important belief in the culture of the Neoplatonists (as it was in the early pythagorean schools). Another belief from that early Greek period was that "the paradigmatic relation between the world and numbers is such that what is true of numbers and their properties is also true of the structure and processes of the world" (O'Meara 1989, 18). The problems that they posed, resulting from the aforementioned assumed numerical structure of the world and the investigation of this structure through non-deductive methods (Radford 1995), were seen as being completely genuine and valid within their rationality and beliefs.

In Radford's socio-cultural approach, the student/milieu relation is sustained by the epistemological assumption according to which knowledge is socially constructed. Instead of seeing such a construction as a diachronic move between the teacher and the student, as is often the case in socio-constructivist accounts, the student is seen as fully submerged in his cultural milieu, acting and thinking through the arsenal of concepts, meanings and tools of the culture. The way in which an individual appropriates the cultural knowledge of his or her culture is often referred to in Vygotskian perspectives as *interiorisation*. Different accounts of interiorisation can be provided. In the socio-cultural approach under consideration, a semiotic, sign-mediated, discursive account sees interiorisation not as a passive process but an active one, in which the individual (through the use of signs and discourse) re-creates concepts and meanings and co-creates new ones (Radford 1998). An experimental historically-based classroom study concerning the re-creation of concepts can be found in Radford and Guérette (1996). A historical case study about the co-creation of new mathematical objects is provided by the invention of the second unknown in algebra by Antonio de Mazzinghi in the 14th century (see Radford 1997b).

In this socio-cultural perspective, the classroom is considered as a micro-space of the general space of culture, and the understanding that a student may have of mathematics is seen as a process of cultural intellectual appropriation of meanings and concepts along the lines of student and teacher activities. Understanding is not seen merely as a unidirectional stage reached by a fortunate student resulting from the sudden awareness of something becoming clear. As Voloshinov (1973, 102) put the matter, "Any true understanding is dialogical in nature", meaning that at the very core of understanding resides a hybrid semiotic matching of different views. Since such a semiotic matching is contextually situated and culturally sustained, there is no question, in this approach, of reading the history of mathematics through recapitulationistic lenses (whether of contents or mechanisms). The history of mathematics is a rather marvellous locus in which to reconstruct and interpret the past, in order to open new possibilities for designing activities for our students. Although cultures are different they are not incommensurable; as explored in Voloshinov's concept of understanding, cultures can learn from each other. Their sources of knowledge (e.g. activities and tools) and their meanings and concepts are historically and panculturally constituted. This is made clear by the fact that most of our current concepts are mutations, adaptations or transformations of past concepts elaborated by previous generations of mathematicians in their own specific contexts.

5.4.3   The 'voices and echoes' perspective

Let us now turn to the epistemological assumptions underlying Boero's 'voices and echoes' perspective (see §5.3.1). His point of departure is the fact that some verbal and non-verbal expressions (especially those produced by scientists of the past) represent in a dense way important leaps in the evolution of mathematics and science. Each of these expressions conveys a content, an organisation of the discourse and the cultural horizon of the historical leap. Referring to Bachtin (1968) and Wertsch (1991), Boero & al (1997) called these expressions *voices.* Performing suitable tasks proposed by the teacher, the student may try to make connections between the voice and his/her own interpretations, conceptions, experiences and *personal senses* (Leont'ev 1978), and produce an *echo,* a link with the voice made explicit through a discourse. What the authors have called the *Voices and echoes game* (VEG) is a particular educational situation aimed at activating students to produce echoes through specific tasks: *"How might X have interpreted the fact that Y?";* or *"Through what experiences might Z have supported his hypothesis?";* or: *"What analogies and differences can you find between what your classmate said and what you read about W?".*

The epistemological assumptions underlying the VEG, partly presented in Boero & al (1998), concern both the nature of 'theoretical knowledge' (the content to be mediated through the VEG), and the cognitive and educational justifications of the VEG. As regards the nature of theoretical knowledge, in mathematics and elsewhere, some characteristics were highlighted drawing on the seminal work of Vygotsky about scientific concepts (see Vygotsky 1990, chapter 6). In particular, theoretical knowledge is systematic and coherent; validation of many statements depends on logico-linguistic developments related to basic assumptions (axioms in mathematics, principles in physics, etc.).

In relationship to the problem of transmitting mathematical theoretical knowledge in school, the preceding description was refined by taking into account Wittgenstein's philosophy of language as well as recent developments in the field of mathematics education by Sfard. The following aspects of theoretical knowledge in mathematics were considered as crucial, concerning both the *processes of theory production* (especially as regards the role of language) and the *peculiarities of the produced theories:*

– theoretical knowledge is organised according to explicit *methodological requirements* (like coherence, systematicity, etc.), which offer important (although not exhaustive) guidelines for constructing and evaluating theories;

– definitions and proofs are key steps in the progressive extensions of a theory. They are produced through *thinking strategies* (general, like proving by contradiction; or particular, like 'epsilon-delta reasoning' in mathematical analysis) which exploit the potentialities of language and belong to cultural tradition;

– the *speech genre* of the language used to build up and communicate theoretical knowledge has specific language keys for a theory or a set of coordinated theories—for instance, the theory of limits and the theory of integration, in mathematical analysis. The speech genre belongs to a cultural tradition;

– as a coherent and systematic organisation of experience, theoretical knowledge vehiculates specific *'manners of viewing'* the objects of a theory (in the field of mathematical modelling, we may consider deterministic or probabilistic modelling; in the field of geometry, the synthetic or analytic points of view; etc.).

In Boero *et al.* (1998), the authors claim that the approach to theoretical knowledge in a given mathematics domain must take these elements into account, with the aim of mediating them in suitable ways. Concerning the problem of 'mediation', the assumption is made that, depending on its very nature, *each of the listed peculiarities is beyond the reach of a purely constructivistic approach.*

The authors' working hypothesis is that the VEG can function as a learning environment where the elements listed above can be mediated through suitable tasks, needing 'active imitation' in the student's 'zone of proximal development'. The first teaching experiments, reported in Boero *et al* 1997, Boero *et al.* 1998, Garuti 1997, Lladò & Boero 1997, Tizzani & Boero 1997, were intended to provide experimental evidence for this hypothesis.

The three perspectives mentioned in this section have shown a variety of ways of conceiving the production of knowledge. Each of them relies on different epistemological assumptions. It is evident from this that different epistemological assumptions lead to different interpretations of the history of mathematics, as well as different ways of linking historical conceptual developments to the conceptual developments of contemporary students.

References for §5.4

Bachtin, M. 1968. *Dostoevskij, poetica e stilistica*, Torino: Einaudi

Boero P., Pedemonte B. and Robotti E. 1997. 'Approaching theoretical knowledge through voices and echoes: a Vygotskian perspective', *Proceedings of the 21st International Conference on the Psychology of Mathematics Education*, Lahti, Finland, vol. 2, 81-88

Boero P., Pedemonte B., Robotti E. and Chiappini G. 1998. 'The 'voices and echoes game' and the interiorization of crucial aspects of theoretical knowledge in a Vygotskian perspective: ongoing research', *Proceedings of the 22nd International Conference on the Psychology of Mathematics Education*, Stellenbosch (South Africa), **2**, 120-127

Brousseau, G. 1983. 'Les obstacles épistémologiques et les problèmes en mathématiques', *Recherches en Didactique des Mathématiques*, **4** (2), 165-198

Brousseau, G. 1989. 'Les obstacles épistémologiques et la didactique des mathématiques', in N. Bednarz et C. Garnier (eds), *Construction des savoirs, obstacles et conflits*, Montréal: Agence d'Arc, 41-64

Brousseau, G 1997. *Theory of didactical situations in mathematics*, ed & tr N Balacheff *et al*, Dordrecht: Kluwer

Garuti, R. 1997. 'A classroom discussion and a historical dialogue: a case study ', *Proceedings of the 21st International Conference on the Psychology of Mathematics Education*, Lahti, Finland, **2**, 297-304.

Ilyenkov, E. V. 1977. *Dialectical logic*, Moscow: Progress Publishers

Lladò, C. & Boero, P. 1997. 'Les interactions sociales dans la classe et le role mediateur de l'enseignant', *Actes de la CIEAEM-49*, Setubal, 171-179

Leont'ev, A. N. 1978. *Activity, consciousness and personality*, Englewod Cliffs: Prentice-Hall

Mikhailov, F. T. 1980. *The riddle of the self*, Moscow: Progress Publishers

O'Meara, D. J. 1989. *Pythagoras revived*, Oxford: Clarendon Press.

Radford, L. 1995. 'La transformación de una teoría matemática: el caso de los números poligonales', *Mathesis* **11** (3), 217-250.

Radford, L. and Guérette, G. 1996 'Quadratic equations: re-inventing the formula: a teaching sequence based on the historical development of algebra', in *Proc. HEM Braga* ii, 301-308.

Radford, L. 1997a. 'On psychology, historical epistemology and the teaching of mathematics: towards a socio-cultural history of mathematics', *For the learning of mathematics* **17** (1), 26-33.

Radford, L. 1997b. 'L'invention d'une idée mathématique : la deuxième inconnue en algèbre', *Repères* (Revue des instituts de Recherche sur l'enseignement des Mathématiques), juillet, **28**, 81-96.

Radford, L. 1998. 'On signs and representations: a cultural account', *Scientia paedagogica experimentalis*, **35**, 277-302

Radford, L. (submitted). 'On mind and culture: A post-Vygotskian semiotic perspective, with an example from Greek mathematical thought.'

Sfard, A. 1997. 'Framing in mathematical discourse', *Proceedings of the 21st International Conference on the Psychology of Mathematics Education*, Lahti, Finland, **4**, 144-151

Sierpinska, A. 1994. *Understanding in mathematics*, London: Falmer Press

Tizzani, P. & Boero, P. 1997. 'La chute des corps de Aristote à Galilée: voix de l'histoire et echos dans la classe', *Actes de la CIEAEM-49*, Setubal, 369-376

Voloshinov, V. N. 1973. *Marxism and the philosophy of language*, Cambridge, Mass: Harvard University Press

Vygotsky, L. S. 1990. *Pensiero e linguaggio*, edizione critica a cura di L. Mecacci, Bari

Wartofsky, M .1979. *Models, representations and the scientific understanding*, Dordrecht: Reidel

Wertsch, J. V. 1991. *Voices of the mind*, Wheatsheaf, Harvester

Wittgenstein, L. 1969. *On certainty*, Oxford: Basil Blackwell

5.5   Conclusions: guidelines and suggestions for future research

Jean-Luc Dorier and Leo Rogers

The various issues addressed in this chapter, and the related teaching experiments and didactical analyses briefly described, show clearly that while 'naive recapitulationism' has persisted in many forms, the relation between ontogenesis and phylogenesis is now recognised to be much more complex than was originally believed. The relations between history of mathematics and learning and teaching of mathematics can be extremely varied.   Some teaching experiments may use historical texts as essential material for the class, while on the other hand some didactical analyses may integrate historical data in the teaching strategy, and epistemological reflections about it, in such a way that history is not visible in the actual teaching or learning experience.

While some knowledge of history of mathematics may help in understanding or perhaps even anticipating some of our students' misunderstandings, a careful didactical analysis using history of mathematics is necessary in order to try to overcome students' difficulties.   History may be a guide for designing teaching experiments but it is only one of many approaches, more or less essential, more or less visible, of the whole didactical setting.   Therefore, one of the necessary conclusions of this chapter would be that any use of history in the teaching of mathematics needs an accompanying didactical reflection.

This way of putting things creates an asymmetry between history and didactics which may not reflect their actual relationship. Indeed any attempt to put in relation the history of mathematics and the teaching or learning of mathematics necessarily induces an epistemological questioning both of individual cognitive development and of the interpretations of the historical development of mathematics.   What happened in the past and what may be likely to happen in the classroom are obviously different phenomena because they are based in very different cultural, sociological, psychological and didactical environments and because contemporary didactical contexts and historical periods conform to very different constraints.

Beyond these differences, the act of teaching is legitimated by the belief that what is taught in the classroom bears some similarity with professional mathematics. However, the knowledge to be taught (*savoir enseigné*) is a transformation of the knowledge of 'professional' mathematicians (*savoir savant*) even if it uses the same vocabulary, notions, and so on, and it is rare that historical processes are taken into account explicitly while writing curricula. Historians of mathematics may object that this is a nonsense. On the other hand, it would also be a nonsense to try to impose a reconstruction of history in the teaching process, in a very strict

John Fauvel, Jan van Maanen (eds.), *History in   mathematics education: the ICMI study*, Dordrecht: Kluwer 2000, pp. 168-170

recapitulationist paradigm. As Chevallard says (translated from Chevallard 1991, 48):

Another direction for research consists in being aware that the planned didactical construction of knowledge is a specific project within the teaching process, bearing an *a priori* heterogeneity with the scientific practices of knowledge, and not immediately reducible to the corresponding socio-historical geneses of knowledge.

Nevertheless, teaching is still organised in such a way that there is a social demand that the knowledge to be taught must appear as close as possible to the official knowledge of mathematicians. In this sense, an epistemological reflection on the development of ideas in the history of mathematics can enrich didactical analysis by providing essential clues which may specify the nature of the knowledge to be taught, and explore different ways of access to that knowledge. Nevertheless what appears to have happened in history does not cover all the possibilities.

Figure 5.5: Nicolaus Copernicus, in front of the Polish Academy of Sciences in Warsaw, seen through the interpretative lens first of Polish history, then of the Danish sculptor Bertel Thorwaldsen, then of a British photographer in the 1990s. Now an inspiration to Polish students, in the 19th and 20th centuries many who had only vague understanding of his achievements were nevertheless agitated about whether Copernicus was Polish or German. The sphere and the compasses have long been symbols to represent a mathematician to the gaze of passers by.

We cannot reconstruct the past with any certainty. Not only are we missing essential data (for example, lost texts, ephemera, unpublished material or oral exchanges) but also a historical fact or event is never pristine. A fact or event is always seen through interpretative lenses and hence will only be partial and subjective. We face essentially similar difficulties when analysing didactical events.

To this extent history and mathematical pedagogy share common theoretical issues with regard to the necessity for epistemological reflection. We need not only to look through history in order to try to improve the teaching of mathematics but also to elaborate common ('echoing') ways of exploring historical and didactical situations. This could be a very challenging issue for future research which could be approached from different viewpoints. It could be a new way of raising the issue of cultural influences in the development of mathematics.

We have said above that what happened in history does not cover all the possible ways of access to one specific element of knowledge. Yet, when setting up a teaching programme, one should try to analyse as many ways of access to the knowledge as possible. This is an important part of any didactical analysis where the use of history can be informative. However, this work is usually confined within the limits of an official curriculum. Indeed, traditions in curricula are sometimes so strong that our views, even as researchers in mathematics education, on the organisation of knowledge are limited because of the strong cultural influences that unconsciously guide our thoughts about the different possible organisations of a curriculum. Because history is temporally and culturally distant from the mathematics taught in our usual curricula, it may provide us with some unusual ways of access to knowledge that could be of considerable didactical value. Of course, this can be possible only if one does not look at history through the lens of 'modern mathematics'. In this sense, another line of development for future research would be a reflection on certain parts of the curriculum in relation to an epistemological reflection on its historical developments.

It may be added that, among the areas for further research, it seems important that mathematics educators and teachers should become more closely involved in co-operative efforts to develop and implement lessons and modules using the history of mathematics as we have shown here. In a similar manner, collaborative work between historians of mathematics and mathematics educators can contribute to better elucidation of the problem of the link between the epistemological and psychological aspects of the conceptual development of mathematical thinking.

Reference for §5.5

Chevallard, Y. 1991 *La transposition didactique*, Grenoble: La Pensée Sauvage.

Chapter 6

History in support of diverse educational requirements—opportunities for change

Karen Dee Michalowicz

with Coralie Daniel, Gail FitzSimons, María Victoria Ponza, Wendy Troy

Abstract: *The needs of students of diverse educational backgrounds for mathematical learning are increasingly being appreciated. Using historical resources, teachers are better able to support the learning of students in such diverse situations as those returning to education, in under-resourced schools and communities, those with educational challenges, and mathematically gifted students.*

6.1 Introduction

The scholarly study of mathematics history has, for the most part, taken place within the realm of the universities. Within the universities, one can find the research community of mathematicians interested in mathematics history; within the universities one can find the authors of mathematics history books and texts. At a growing number of universities and colleges, the study of the history of mathematics has become part of the curriculum for mathematics undergraduate and graduate students. Indeed, in recent times many universities and colleges have started to provide courses in history of mathematics for prospective secondary school mathematics teachers.

Nevertheless, there has been minimal interest from the mathematics community in introducing the history of mathematics to pre-college students, or to students who choose alternative directions for their post-secondary education. One can speculate on the reasons for this. Those who teach mathematics history in the university are not the teachers of the primary students, secondary students, or students seeking alternative education at whatever level. Nor are they the teachers of the gifted pre-college students. When one finds a primary or secondary teacher using mathematics history in a pedagogical way, it is usually (although in some countries this is changing) because the teacher is an amateur mathematics historian, not because the teacher had been trained in the area.

John Fauvel, Jan van Maanen (eds.), *History in mathematics education: the ICMI study*, Dordrecht: Kluwer 2000, pp. 171-200

The authors of this chapter, university faculty in mathematics and mathematics education, teachers of secondary and elementary students, teachers of the gifted, and teachers of students seeking or needing alternative education practices, have found that mathematics history has greatly influenced their success in the classroom. Their anecdotal evidence is voluminous.

This chapter has been written to highlight some of the ideas and practices of these teachers. Their experiences circle the globe. They have worked with children and young adults from many economic, cultural, social and educational backgrounds. They have located resources or created and produced their own. From different backgrounds and different countries, they have brought their common love of mathematics history to the students that they teach. They all have seen how inspiring mathematics history can be to their students regardless of their diverse backgrounds.

Although some of the following essays are specific to individual countries, the heart of the issue applies globally. In most countries, similar circumstances can be found. For example, the educational inequity witnessed in some areas of Argentina can be likened to that of impoverished regions in such countries as the United States. Teacher training is another global issue. Curriculum is a volatile issue, especially in countries without a national curriculum.

6.2   Educational, cultural, social and economic diversity in primary, secondary and tertiary settings

6.2.1   Primary education and the use of mathematics history in the classroom

Except within the most impoverished areas of the world, most children receive at least a primary education, including arithmetic. While most students will have some secondary education, students in some areas of the globe may go no further than primary school. In whatever country, it behooves society to provide the best and most solid primary education possible. Unfortunately, mathematics is that part of the primary curriculum many teachers are less than eager to teach. Undoubtedly students realise when their teachers' attitude towards mathematics is one of anxiety. Students will tend to follow the lead of their teacher, their role model, in this.It seems imperative to provide teachers with tools and resources that will reduce their anxiety and that of their students. Part of this strategy lies in the ways teachers and students see the value of mathematics, as something useful and interesting beyond the needs of basic computation. No one will deny that students should know how to compute. Whole number operations, fractions and decimals are a necessary part of primary education. They are skills that most need throughout their lives. However, students and teachers need to know that mathematics is much more than computation.

The history of mathematics is an instrument to enhance the value of mathematics in the classroom and to enlighten students to the breadth of mathematics. When primary teachers are given the opportunity to see how mathematics can be connected to their social studies curriculum (geography, history, etc.) and even to their literature curriculum, arithmetic can begin to take on a more meaningful role in the classroom. While there is little if any research to verify this position, there is a myriad of anecdotal reports given by primary teachers who have found success in the practice of connecting the history, geography, and cultural times of mathematics to the study of primary arithmetic.

Figure 6.1: The Egyptian Rhind papyrus (written by Ahmes the scribe in 1650 BC in hieratic, top image, with its hieroglyphic transcription below), now in the British Museum, is accessible to primary school pupils in deciphering and calculating, as well as in problem solving. Primary school teachers may use it to link mathematics with history in their classes. From The Rhind Mathematical Papyrus *(A.B. Chace et al. eds.),* **ii,** *Oberlin Ohio 1929, pl. 73*

Given that the use of the history of mathematics in the primary classroom is an idea with merit, the question is how can it be accomplished. It appears that the need lies in two areas. One, teachers need to receive the necessary education to be able to understand about the history of mathematics and how it connects to the arithmetic in the classroom. Second, teachers must have access to materials, or at least need guidance on where to look for materials or how to create their own materials for the classroom.

In many countries, mathematics education for primary teachers is minimal. Many of these teachers would not be comfortable with the secondary school mathematics content. Thus, even if it is available, a university course in mathematics history would not be something the pre-service primary teacher would attempt. The mathematics is too sophisticated. The type of course in mathematics history that primary teachers need, which would connect with their prospective curriculum, content and pedagogical concerns is just not available. Even inservice education for primary teachers in the use of mathematics history is seldom found.

Mathematics history resources for the primary classroom do exist in a small number. However, these resources are not available globally and are costly. Is there a solution for providing the primary teacher with the instruction and the materials needed to use mathematics history in the classroom? Probably not at this time in a conventional way. However, there are tiny alternative steps that can be taken to work toward providing for the teachers needs. First and foremost, teacher educators internationally need to make a commitment to providing opportunities for teachers to learn about using the history of mathematics and how to develop materials. They need to look to traditional ways of instruction and to non-traditional, alternative ways of instruction. With the latter idea in mind, it is suggested that the Internet could play a very important role in both education and resources. There are a number of outstanding Web sites that contain mathematics history resources some of which are excellent for the primary teacher (see the section on Internet use, §10.3.2). At some such sites there could be posted a 'Primary Teacher Education Centre' which would include sources for historical readings for the teacher, ways the curriculum could be connected, timelines, maps, and other primary materials. Names of teachers using mathematics history who could mentor other teachers could be listed, too.

The Internet, although still not accessible universally, is becoming more and more available even to remote areas. Globally educators are beginning to realise that Internet access, at a reasonable cost, provides the unlimited information that a school or even town library in an impoverished area could never provide. In many countries, the Internet is already providing for adult education in many fields of instruction. Obviously, providing inservice in mathematics history on-line is a capability that already exists and needs only to be put in place.

Attitudes toward mathematics are developed early in children. We know how well a positive attitude influences learning. Therefore, knowing how the use of mathematics history can provide for affective student needs, it appears that its introduction into the primary grades is very important. It would be desirable if all schools of education would provide teacher training in the use of the history of mathematics for primary school. This ideal appears to be a long term process. However, with a little creative thinking, the Internet could provide information and resources that the primary teacher could use immediately. The challenge is there.

6.2.2   Under-served (limited resources) students

María Victoria Ponza

Among the resources assigned to education in the world, economic funds are essential to our purposes, since pedagogic resources depend on them. There are few qualified teachers, no improvement, renovation or 'up-dating' for such teachers as there are, or for the necessary materials such as buildings, chairs, books, and paper. It is a fact that all over the world the economic resources allotted to education are insufficient. But in some countries they are excessively scarce, and such funds as are available are improperly distributed. This section is a case study in how countries in this situation can use the history of mathematics in the light of these

John Fauvel, Jan van Maanen (eds.), *History in   mathematics education: the ICMI study*, Dordrecht: Kluwer 2000, pp. 174-179

constraints, from the perspective of one such region, the province of Cordoba in Argentina.

In Argentina and many other countries, there are two very different types of schools: government schools and private schools, differing mainly in the economic resources at the schools' disposal. In government schools most of the funds go into paying salaries (at rather a low level), while private schools are able to invest more and more money every year in providing all the necessary resources for teaching already mentioned (teachers, materials, training, buildings, etc.). As a result, the social gap between rich and poor in that country becomes widened.

Frequently, people are heard to refer to the existence of two Argentinas as a consequence of an economic policy that has a direct influence upon educational policy. Families naturally belong to one or the other Argentina, most of them to the one where the lower resources are found. In consequence, government schools are always overcrowded. (This situation is not unique to Argentina, of course; but Argentina is the subject of this case study.)

A difference between some countries and Argentina lies in the fact that many of the Argentine pupils are aware of hardship and realise how fortunate they are to be able to attend high school. These pupils at least begin with a wish to study and to take the best advantage of their opportunities. Similarly, there are for historical reasons a cadre of capable teachers in government schools who take their work seriously. Nonetheless, there are broader social factors working in the opposite direction. One of the main problems of a country such as Argentina today is that, partly in response to a spreading global ethos, people are drawing nearer to short-term individual action and further away from investing in training and excellence. The trend towards immediate gratification is unlikely to bring about important achievements in education, by diverting attention and resources from the long-term competency and skills which are needed for sustained success in the world today.

Mariano Moreno School in Rio Ceballos, Cordoba Province, Argentina, belongs to the group of government schools of Argentina where economic resources are minimal. Any project away from the conventional depends exclusively on the will, drive and creativity of the pupils and teaching personnel without expectations of funding. In 1994, I coordinated and took part in an interdisciplinary project lasting an academic year with 13-year-old pupils and encompassing seven subjects. The project was approached from the context of history in general. Thus in order to obtain a coherent participation of the area of mathematics, it was the history of mathematics that was explored. As a result of the project, many changes were observed in the attitude of pupils towards mathematics: their rejection of the subject decreased and they experienced a surge of interest. Historical investigations of important figures in the mathematical past, their lives and discoveries, enabled pupils to see human aspects of mathematics that they had never previously imagined. From that moment on, I started using the history of mathematics as a resource for teaching pupils of different ages.

In recent years, the province of Cordoba has experienced untimely reforms in the educational system, practically without notice and without providing training about the reforms for the teachers. This has put the teaching staff in a difficult position with regard to curricular contents, and also led to problems in the conduct of pupils.

Thus it was that in 1996, we found ourselves teaching, for the first time, pupils a year younger who had been moved from their previous school where they originally expected to complete their seventh grade. These students were relocated to a school with very bad classroom conditions. As a consequence, the pupils had difficulty in adapting to their new environment and discipline problems were rife. This particular situation forced us ask ourselves what could be done in the light of the reality that we were confronted with. The answer was prompt: make use of other resources; in this case, resort to the history of mathematics.

In the first year course, under my charge, the history of mathematics exercised what can only be described as a magical effect. In moments when disorder impeded hearing any possible explanation of usual mathematical curriculum, I found that telling, by way of story, the history of the symbols + (plus), of – (minus), of mathematicians such as Euclid or Galois, etc., succeeded in calming everyone and aided in the progress of the lesson. In 1997 those same pupils, now in the second year, agreed to work with me on curricular contents from the history of mathematics.

For several years my pupils have been performing mathematical dramatisations, thus establishing a relationship between mathematics and other subjects. The 1997 proposal put forward by me offered a good opportunity to perform drama connected to the history of mathematics. The project aimed at providing pupils with a lively experience of historical facts regarding mathematics by experiencing the life of some famous figures so as to humanise our subject. Also, a comparative study of the social and political contexts at different ages down to the present was undertaken. These ideas crystallised into a practical plan when my pupils suggested writing and performing a play on the life of Evariste Galois (for more details see Ch. 10). The drama production and related interdisciplinary activities provided particular reflections upon the following general questions:

– What role can the history of mathematics play in response to special educational needs?

– What relationship is there between the role or roles we attribute to history and the ways of introducing or using it for educational purposes?

– What consequences will it produce for organisation and practice in the classroom?

Working out the curricula of the whole academic year using the history of mathematics and drama gave us a chance to become aware of several beneficial aspects.

1. In a school with an excessive number of pupils on every course, and where economic scarcity influences the possibility of having texts as teaching aids, the history of mathematics acted as a mobilising element for bringing together resources which on their own would have been wanting.

2. It provided an incentive for reading and encouraging the use of the library.

3. It stimulated the development of expression in language and mathematics, since a large amount of new mathematical terminology was discovered, used and understood by students through reading the history and interpreting the text.

4. The introduction of historical anecdotes served to humanise the mathematics and to lessen students' rejection of it.

5. The dramatisation and the work on generating the text revealed hidden aspects of the personality of the pupils, as well as sensitising them to the realities of the past and the actuality of the present. Pupils identified aspects of Galois' life with the unfair conditions they suffer in their own lives. The public recognition eventually granted to Galois' work, although belated, awakened some hope in them and encouraged them to continue studying and participating. As a teacher, this gave me a chance to become acquainted with other aspects of my pupils' lives and understand them better.
6. The introduction of mathematics history into the course curriculum attracted the attention of pupils, since it was the first time that they had seen such a thing.
7. With regard to the mathematical concepts arising from Galois, the pupils could not, of course, tackle them in depth owing to their rudimentary knowledge; they are 14-year-old pupils and in their 2nd. course of the basic cycle.

The political realities of finding a role for history in the classroom.

At present the history of mathematics is not included in the course of study for pupils. Nor is it present in the curriculum of institutions in charge of preparing nationally qualified mathematics teachers in Argentina (and most other countries for that matter). Those who already teach mathematics are minimally interested, partly through being overworked already, and have had little opportunity to see the history of mathematics as other than a gratuitous accessory. Such teachers do not pursue further information and do not refer to mathematics history during classroom teaching. If, however, consideration of the history of mathematics even in an isolated manner (by only an individual teacher here or there) brings about such important consequences for the teaching and learning of mathematics, as has been seen in this case study, it becomes imperative to work towards the systematisation of these advantages. What is needed is the wider availability of historical texts in the country, which up to now are in the hands of only a very few teaching personnel, and owing in any case only to personal efforts. The public libraries lack this material, and it is essentially completely absent from schools and institutions of teacher training. (The bibliographies elsewhere in chapters 9 and 11 of this book give an indication of what is available.)

At this moment, year 1998, the pupils I worked with on the Galois project are attending the 3rd school year. Since I am still their teacher, I have the opportunity to see the effects of their dramatic work in the previous year. It appears that they have a stronger basis in their mathematical curricula. And they have learned much about being patient when searching for information, and feeling pleasure when discovering. The project continues largely thanks to the pupils' own initiative.

A good example of this is the method we applied to arrive at the concept of irrational number and the enlargement of the numerical field. To begin, I planned for more thorough pupil research about Pythagoras than was done the previous year. My aim was to lead the pupils to discover the square root of 2, starting from the Pythagorean theorem they already knew. One of the self-appointed teams was made up of three students, two of them attracted to mathematics after studying its history. This team contributed a great amount using information found in the town library. They discussed it and pointed out the main details. They then synthesised their findings. When all the pupils had read this synthesis, in a subsequent lesson I had them examine the sociopolitical conditions found during the 6th century BC, starting with the Pythagoreans. The pupils concluded that from the social point of view the Pythagoreans formed a closed community with very specific ground rules for social behavior. Their behaviour seemed to be politically inconsistent: in external affairs they were fighters against the tyranny of Polycrates, whereas in home affairs they were tyrants against one another, living under a very strict regime characterised by the secrecy of its acts, the disclosure of which meant a threat to their lives.

We discussed whether there was any parallelism between conditions in those ages and the present time. They concluded that there is great similarity, since at present, in the pupils' opinion, there are closed groups or lobbies both within governments and outside them, such as in the news media. They investigate each other and find out facts which may have serious consequences upon the population. Yet, important information is kept a secret. These groups look critically towards everything that lies outside them. None the less, they keep secret what does not further their own interests.

We examined the mathematical discoveries of the Pythagoreans in relationship to their intrinsic and social value. The pupils concluded that these discoveries were numerous and very influential, such as philosophy of life based on numbers, musical notes, the notion of one of the first non-geocentric planetary systems and the celebrated theorem (which seems to have already been used by Babylonians and well-known in other cultures). We examined the mathematical consequences of the theorem and its close relationship to the socio-political behavioural ground rules of the Pythagoreans. The pupils discovered that the right triangle with legs 1 and 1 led them to the square root of 2 which the Pythagoreans kept secret. I explained to the pupils that this was an *incommensurable*, as well as what this term meant, and gave reasons for our interest: that it was one of the most famous of non-rationals in history, and explored with them its connection to geometry. Thus I introduced the irrational numbers which enlarged the numerical field and completed the straight line.

Pupils have little notion that they are themselves the subjects of history, that they are the makers of present history. As such, it is imperative that they should learn to be broad minded, working out and holding their own point of view, respecting others ideas. Here interdisciplinary activities play an key role, allowing pupils to establish connections, participate, strengthen and convey ideas, avoiding a mere repetition of other people's concepts.

It appears that the educational reform begun in Cordoba in 1996 may never achieve its goals. One of the causes for this debacle is not having the trained teachers needed to carry out interdisciplinary tasks, and not having restructured the system to implement this. Perhaps my pupils will in due course take their place in helping encourage the use of interdisciplinary tasks and build a better Argentina. My main objective as regards the 1998 project can be synthesised in one question: how might pupils who have learned the importance of knowledge and hard work contribute to change our society being, as they are, subjects of history?

Note

My special thanks to the pupils of Mariano Moreno School, 3rd. year course, 1st. section, year 1998, and to Dr Maria Luiza Cestari.

References for §6.2.2

Boero P., Pedemonte B., Robotti E. 1997. 'Approaching theoretical knowledge through voices and echoes: a Vygotskian perspective', *Proceedings of the 21st International Conference on the Psychology of Mathematics Education*, Lahti, ii, 81-88

Durán, Antonio Jose 1996. *Historia, con personajes, de los conceptos del cálculo*, Madrid: Ed. Alianza Universal, 17-22

Hitchcock, Gavin 1997. 'Teaching the negatives, 1870-1970: a medley of models', *For the learning of mathematics* 17 (1), 17-25

Muñoz Santoja, José, Carmen Castro, María Victoria Ponza 1996. 'Pueden las matemáticas rimar?', *Suma* 22, (Federación Española de Sociedades de Profesores de Matemáticas, Zaragoza), junio, 97-102

Panza Doliani, O, Ponzano, P. 1994. *El saber, sí ocupa lugar*, Córdoba, Argentina: Ciencia Nueva , 13-24

Poincaré, Henri 1995. 'La creación matemática', in *Investigación y Ciencia: Grandes Matemáticos*, Barcelona: Ed. Prensa Científica SA, 2-4

Ponza, María Victoria 1996. 'La experiencia interdisciplinaria en la realidad educativa de hoy', *Suma* 21 (Federación Española de Sociedades de Profesores de Matemáticas, Zaragoza), febrero, 97-101

Ruiz Ruano, Paula; Perez, Pilar 1996. 'Hipatia en el país de las empatías', Jaén: Consejería de Educación y Ciencia (*Centro de profesores de Linares)*, 9-18.

Savater, Fernando 1997. *El valor de educar*, Barcelona: Ed. Ariel SA, 47-54, 92-100, 110-111, 116-142

6.2.3 Alternative educational pathways: adult learners returning to mathematics education, vocational education and training

Gail FitzSimons

In an era of economic rationalism the education of adult learners is assuming increasing importance, both in general return-to-study classes and in specifically oriented vocational classes. For the purposes of this discussion, adult learners are taken to be people who have been out of formal education systems for some length of time, or participating for the first time; vocational students include those who are returning to, or continuing with, post-compulsory education. (In some countries specialised vocational education begins during the secondary years of schooling.) These educational settings include both formal institutions as well as informal, community, and workplace sites. It is not possible to make universal statements on the provision of adult and vocational education—they each vary in the degree of emphasis placed on general versus specific vocational content and in the importance placed on credentials and pathways to further study. In different countries the

John Fauvel, Jan van Maanen (eds.), *History in mathematics education: the ICMI study*, Dordrecht: Kluwer 2000, pp. 179-184

responsibility for the costs of such education is distributed variously among governments, industries, and individual students, but the overall intention would appear to be the improvement of the economic and/or the social well-being of the individual and the community at large.

Mathematics is seen as having a crucial role in respect of social end economic development. However, mathematics education for adults has generally followed the form of selections from the entire range of school and sometimes undergraduate curricula, with adult life skills or vocational examples inserted as deemed appropriate. In other words, many (but not all) courses offered are likely to be fairly traditional and based on a utilitarian framework. By and large they are also been premised on a deficit model of the learner, seeking to remedy perceived gaps in their mathematical knowledge when compared to official checklists of so-called essential skills. For a more extensive comparison of the international situation regarding adult learners of mathematics see FitzSimons (1997).

This section concerns the teaching of mathematics to adult learners from the perspective of the use of the history of mathematics and mathematics education. I argue that its adoption will ultimately be more effective and empowering for the individual, the wider community, and national and even global interests. Thus the opportunities for and constraints on the use of history of mathematics in adult and vocational education will be discussed.

In vocational education there appears to be little or no place for the history of mathematics: should it make an appearance in texts, it is generally trivial, sometimes inaccurate, and not integrated with the main thrust of the lesson (Maass & Schloeglmann 1996). Adult education may or may not have such an instrumental focus: inclusion of the history of mathematics is a matter of chance in terms of quantity and quality of effort. Contributing factors are the rigidity of the curriculum, and the background of the teacher in terms of philosophical beliefs about mathematics and pedagogical content knowledge and reasoning (Brown & Borko 1992), as well as knowledge of the history of mathematics itself. Clearly there is a need for quality resource material and appropriate professional development.

The diversity of social, cultural, and economic backgrounds in society at large may be somewhat reduced in any particular study group, although the life experiences of each person will be unique. In addition there will be variations in educational background and of expectations in the cognitive and affective domains. Each person may, in different ways, be likely to seek to increase their economic, social, cultural, and/or symbolic capital (Bourdieu 1991). Use of the history of mathematics in teaching provides an opportunity for the learner to appreciate the struggles of people throughout history to overcome difficulties similar to those they are facing. Viewing the study of mathematics through the lens of a study of humanity, rather than as cold, hard science, can play an important role in the overcoming of mathematics anxiety (FitzSimons 1995). The study of ethnomathematics in particular is a powerful means of valuing experiences and cultures of members of minority groups while expanding the horizons of all participants. However, Knijnik (1993) warns against placing too high a value on the popular knowledge of subordinate groups, and recommends that students have the opportunity to become aware of the possible limitations, which may be transcended through a process of cultural synthesis. The term 'ethnomathematics' may also be applied to the mathematics practised by people in the workplace, its form evolving from the adaptation of strategies for solving problems that arise within this particular culture which has, of course, its own discourse and literacies (O'Connor 1994).

There are constraints within formal educational institutions providing mathematics education for adult and vocational students. For example, the use of the history of mathematics may be constrained by the inertia generated by structural rigidities of curriculum, particularly in the case of competency-based education and training (FitzSimons 1996). Educational systems which do not value professional development specific to mathematics teaching and learning exacerbate the problem, as does the trend (in Australia at least) towards the removal of any requirement for educational qualifications in teachers in a profession that is becoming increasingly deregulated and casualised. On the other hand, systems which allow flexibility in curriculum and assessment enable creativity on the part of informed teachers and their students in the pursuit of knowledge for its own sake, and it is here that the use of history of mathematics has the possibility of flourishing.

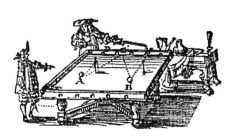

Figure 6.2: Flexibility in the curriculum for adult learners (from the algebra text by A. de Graaf, Amsterdam 1672)

A further constraint on the use of history of mathematics is the power of industry to determine narrowly focused curricula for vocational students. Even the more general adult education is frequently subjected to being framed within notions of 'usefulness'. Given the underlying expectation of economic efficiency, especially in government- or industry-subsidised education and training, excursions into the history of mathematics are likely to be seen as a waste of time and money. This is a particularly short-sighted approach in view of the attention paid to the development of so-called key competencies (Mayer 1992) described as being essential for effective participation work and in other social settings, for example:

a) collecting, analysing and organising information,
b) communicating ideas and information,
c) planning and organising activities,
d) working with others and in teams,
e) using mathematical ideas and techniques,
f) solving problems, and
g) using technology.

The development of each of these would be enhanced through the use of the history of mathematics, since the skills developed through historical activity are precisely those leading to these competencies. It would also seem beneficial to increase the understanding of vocational students by including something of the history of ideas that have led to the present situation of an increasingly technological environment.   It is through such understanding, as well as their developing competencies, that the possibilities for progress by workers and students might be realised.

In considering how the educational benefits might be brought about it is, of course, particularly essential to convince policy makers of the appropriateness of including history of mathematics, as well as to provide encouragement and support for teachers.

Justification for the use of history of mathematics can be made on grounds of enhancing individual development, especially in the possibilities it offers for overcoming mathematics anxiety, in broadening the socio-cultural perspectives of learners, and also in stimulating further mathematical and scientific enquiry. This leads to social benefits for the individual in increased self-confidence and respect from others, and for the community in terms of that person's participation in decision-making processes required in a social democracy as well as in the workplace. In addition, adult learners are able to share their legitimate knowledge with others such as family and friends for whom they can act as role model, mentor, or even collaborator. There may be economic benefits at the personal level, and ultimately national, even international economic benefits flowing from enhanced participation in mathematics education. It is recognised that there is considerable debate about the paradox of an increasingly technological society, formatted by hidden mathematics (Skovsmose 1994), apparently needing to know less mathematics. However, Noss (1997) has argued that there is now a greater need for people to be able to use mathematics in a constructive, interpretive way in situations of conflicting information and to be able to find practical solutions in the inevitable situation of technological breakdown. Learners at all levels need to have the self-confidence to persevere with mathematical studies.

Arguments for the teaching of history of mathematics to adults need to be supported by a range of theoretical foundations. There is support from the history of mathematics itself for a philosophy of mathematics that sees it as fallible and socially constructed (Ernest 1991). Studies in the sociology of mathematics and mathematics education suggest the need for a broader view of mathematics than the traditional white, male, eurocentric version that commonly prevails. Walkerdine (1994) makes just this point. Ernest (1996) has pointed out that there is a strong relationship between the classroom experience of students and the general public image of mathematics. If someone's school experience has left them perceiving mathematics as fixed and absolute, exact and certain, and specified by rules, they are likely to be think of mathematics ever after as cold, inhuman, and rejecting. Recent philosophical analyses, however, have developed and enriched how mathematics is thought of. In a postmodernist analysis, mathematics is seen to be an outcome of social practices wherein people and history, among other things, play a vital constitutive role. Bishop (1988) addressed the values attributed to mathematics (viz. rationalism, objectivism, mystery, openness, control, and progress) each of which can be, and have been, valued highly by the mathematics community, but which have the potential to alienate members of the general public. In contrast to these potentially-alienating mathematical values, Bishop noted six universal activities (viz. counting, locating, measuring, designing, explaining, and playing) which underline the commonality of mathematics in most people's cultural activities, thereby providing justification for an ethnomathematical approach to teaching. Other work has built upon these ideas.   Skovsmose's critical approach to mathematics education for democratic competence (Skovsmose 1994) takes account of these values, and here the use of history of mathematics can illuminate how mathematical values and activities have been utilised in the past. The approaches of these recent educational thinkers is pertinent to both adult and vocational mathematics education.

Teachers in adult and vocational education make certain epistemological choices. They may choose, or be compelled, to operate variously within paradigms such as the traditional method of transmission, constructivism (radical, social), socio-cultural situated learning in a community of practice.   It is possible to incorporate the use of history of mathematics under each, but the methodologies utilised are dependent on various factors in the teaching situation such as, for example: the size, location, and heterogeneity of the class; access to various forms of multimedia, including print-based; the type of interaction between teacher and learner, whether personal or distance education modes of teaching, including self-paced learning; and the time allocated for lessons.   Within these parameters, some possibilities for using the history of mathematics are:

a) teaching *through* history and ethnomathematics;
b) teaching *about* history and ethnomathematics;
c) encouraging students' reflection on their own experiences of mathematics education—their personal history—to encourage metacognition; and
d) an integrated curriculum with a problem solving or project-based approach where the history of mathematics and ethnomathematical studies develop within the contextual setting for teacher and learner (e.g., FitzSimons, 1995; this volume).

Naturally the choice(s) will depend on the teacher's judgement of their appropriateness for the objectives of the session, and the teacher's ability to adjust the teaching style and content.

Much innovative work has been carried out by practitioners in the absence of funded research. Although the field of adult education in mathematics is burgeoning (FitzSimons 1997), there have been few dissertations noted to date, and the likelihood of formal research on the impact of using history of mathematics in adult education is even more remote. It is recognised that few busy teachers have time to document and analyse their teaching experiences, given the intensity of work pressures. However, it is in the interests of both government and industry to ensure that the best possible outcomes, according to their

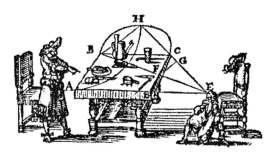

Figure 6.3: Problems with tables are of all ages (from the algebra text by A. de Graaf, Amsterdam 1672)

own criteria, are achieved. This section has asserted the benefits of using history of mathematics, based on the writer's personal experience and her reading of mathematics education and related literatures. Further, more formal, research is needed to justify this assertion and to provide documentary analysis of effective strategies according to the diverse needs of particular adult student groups.

References for §6.2.3

Bishop, A. J. 1988. *Mathematical enculturation*, Dordrecht: Kluwer.

Bourdieu, P. 1991. *Language and symbolic power*, Cambridge, UK: Polity.

Brown, C. A., & Borko, H. 1992. 'Becoming a mathematics teacher' in D. A. Grouws (ed.), *Handbook of research on mathematics teaching and learning*, New York: Macmillan, 9-23

Ernest, Paul 1991. *The philosophy of mathematics education,* Hampshire, UK: Falmer.

Ernest, Paul 1996. 'Popularization: myths, mass media and modernism', in: A. J. Bishop, K. Clements, C. Keitel, J. Kilpatrick, & C. Laborde (Eds.), *International handbook of mathematics education,* Dordrecht: Kluwer, 785-817

FitzSimons, G. E. 1995. 'The inter-relationship of the history and pedagogy of mathematics for adults returning to study'. Paper presented to the International Study Group for the Relations of History and Pedagogy of Mathematics, Cairns, Australia.

FitzSimons, G. E. 1996. 'Is there a place for the history and pedagogy of mathematics in adult education under economic rationalism?' in *Proc. HEM Braga* ii, 128-135

FitzSimons, G. E. (ed.) 1997. 'Adults returning to study mathematics', in *Papers from Working Group 18, International Congress on Mathematical Education*, Adelaide.

Knijnik, G. 1993. 'An ethnomathematical approach in mathematical education: a matter of political power', *For the learning of mathematics*, **13** (2), 23-25.

Maass, J., & Schloeglmann, W. 1997. 'The structure of the discipline of mathematics and its practical applications: two opposite orientations in mathematics education for adults'. In *Proceedings of the Third Annual Conference of Adults Learning Mathematics: A Research Forum*, London: Goldsmiths College, University of London, 158-165

Mayer, E. (chair) 1992. Report of the committee to advise the Australian Education Council and ministers of vocational education, employment and training on employment-related key competencies for post-compulsory education and training, Melbourne: Australian Education Council and Ministers of Vocational Education, Employment and Training.

Noss, Richard 1997. 'New cultures, new numeracies', Inaugural professorial lecture, London: Institute of Education.

O'Connor, P. 1994. 'Workplaces as sites of learning', in P. O'Connor (ed.), *Thinking work, vol. 1: theoretical perspectives on workers' literacies,* Sydney, NSW: Adult Literacy and Basic Skills Action Coalition, 257-295

Skovsmose, Ole 1994. *Towards a philosophy of critical mathematics education*, Dordrecht: Kluwer.

Walkerdine, Valerie 1994. 'Reasoning in a post-modern age', in P. Ernest (ed.), *Mathematics, education and philosophy: an international perspective*, London: Falmer Press, 61-75

6.2.4   Minority school populations

Webster's *New World College Dictionary* (1996) gives as its third definition of *minority* "a racial, religious, ethnic or political group smaller than and differing from the larger, controlling group in a community, nation, etc". Using a global perspective, 'minority' does not refer to any particular racial, ethnic, religious,

Figure 6.4: *This image from 16th century Peru, showing the Secretario del Inca with his quipu, reminds South American students that the ancient people of their pre-Spanish heritage had sophisticated means for recording and transmitting numbers.*

ethnic or political group. Within a school setting, a school district, or the schools in a particular country, minority students may number more than one half of the school population. Odd as this seems, what the word minority refers to, in terms of educational anthropology, is the population of students who have a culture (ideas, customs, skills, arts, etc.) which is different from the dominant school culture. For example, a school may use English as its language of instruction although many or even most students may not speak English as their first language. Another example would be in the cases of independent nations which, in the past, existed as colonies of other nations. In former times, and perhaps even now, the dominant school culture could be very different from the native culture. One need only look to the continent of Africa for such examples.

During the last half of the twentieth century, war, famine, and other turmoil has caused the emigration of many people to countries in Europe, in North America, and to the southern Pacific countries of Australia and New Zealand. Most of these immigrants have racial, religious or ethnic backgrounds different from the dominant culture of the countries to which they have emigrated. The immigrant children bring their culture to school with them. In some countries, such as Brazil, Australia and New Zealand, and parts of the United States and Canada, some of the minorities are not immigrants, but members of the aboriginal or indigenous peoples with their own culture and language.

While there is no universally dominant school culture, each country looks to educate its students in a way that it perceives as appropriate. Within the United States, because there is no national curriculum, policies of different states greatly differ about the best way to educate students from diverse cultural backgrounds. In one US state, highly populated by Spanish speaking immigrants from Central and South America, an ESL ('English as a second language') program in the schools does not exist. Yet in another state, populated by many Mexican immigrants, there are such ESL programs. In a number of countries in Europe, special language programs for non-native speakers in the schools is not an issue. These countries expect non-native speakers to enter the classroom, learn the national language and adapt to the school culture.

The purpose of this section is not to question how a particular country educates its minority students or criticise national educational policies. Rather it is to examine ways in which, globally, teachers of all students can provide learning environments which enhance the learning of mathematics. The reader should bear in mind that there is nothing deficient about a minority population; being different does not mean being of less value. Rather, educators need to realise that all students, in particular minority students, have a culture which may or may not conform to the dominant classroom expectations. The question remains how teachers can create within the classroom an equitable learning environment in which all students may learn and realise the value of their education.

At first glance mathematics would appear to be the curriculum area in which students from all diverse backgrounds have common ground. After all, except for a few minor differences in the algorithms for basic skills (such as the way in which calculations are written down on paper), computational skills throughout the world are much alike. None the less, students globally learn mathematics based on their familiar linguistic and cultural patterns (Trentacosta 1997). Many study school mathematics without understanding the use for it; many dislike school mathematics; many more exhibit great mathematics anxiety. Minority students appear to suffer the worst.

One way in which curriculum specialists encourage teachers to make their content area meaningful is by humanising the subject. What better way can mathematics be humanised then by the use of mathematics history in the classroom? Besides being entertaining, the history of mathematics provides the student with information about the global roots of mathematics. Mathematics history helps students realise that mathematics is not just the invention of the dominant school culture. Rather, it helps students realise that mathematics evolved from many sources and in many places. For example, minority students from Central American can learn that recent Olmec research suggests that the ancient people of their heritage developed the concept of zero perhaps earlier than any other ancient peoples. Students with an Ashanti African heritage can appreciate that the Ashanti mathematics bone is one of the oldest mathematics artifacts. Students with an Indian heritage can celebrate the contributions that their ancestors made to our present day numeration system and to concepts of negative numbers. Our female students can be inspired by the stories of courageous women mathematicians. The noted ethnomathematics scholar, Marcia Ascher, points out that when teachers emphasise the roles different cultures have played in the evolution of mathematics, students' pride in the accomplishment of their people is enhanced and they begin to value mathematics as a human activity (Ascher 1991). The history of mathematics, using both its European and non-European roots, makes mathematics relevant to the cultural heritage of all students.

To help the reader find information about and plan activities for students of diverse cultural backgrounds a resource bibliography is provided.

Resource bibliography

(* Indicates a work appropriate for primary and upper elementary or middle school teachers; ** mainly secondary; *** reference for all teachers)

***Ascher, Marcia 1991. *Ethnomathematics, A Multicultural View of Mathematical Ideas*, Pacific Grove, CA: Brooks/Cole Publishing

***Ascher, Marcia & Robert Ascher 1997. *Mathematics of the incas: code of the quipu*, New York: Dover

*Alcoze, T., et al. 1993. Multiculturalism in mathematics, science, and technology: readings and activities, Menlo Park, CA: Addison-Wesley, 1993.

*Kaleidoscope Series, 1994. *Count on it*, North Billerica, MA: Curriculum Associates

*Lumpkin, B. & Strong, D., 1995. *Multicultural science and math connections*, Portland, MA: Walch,1995.

**Johnson, A, 1994. *Classic math history topics of the classroom*, Palo Alto, CA: Dale Seymour Pub.

***Powell, A. B. and M. Frankenstein, Eds, 1997. *Ethnomathematics: challenging eurocentrism in mathematics education*, NY: State University of New York Press

**Smith, S. 1995. Agnesi to Zeno: over 100 vignettes from the history of mathematics. Berkeley, CA: Key Curriculum Press

**Swetz, F., 1994. Learning activities from the history of mathematics, Portland, MA: Walch

***Trentacosta, J. 1997. *Multicultural and gender equity in the mathematics classroom: the gift of diversity*, Reston, VA: National Council of Teachers of Mathematics

***Zaslavsky, C., 1996. *Fear of math*, New Brunswick, New Jersey: Rutgers University Press

*Zaslavsky, C., 1994. Multicultural math: hands-on math activities from around the world, New York: Scholastic Professional Books

*Zaslavsky, C., 1987, 1993. Multicultural mathematics: interdisciplinary cooperative-learning activities. Portland, MA: Walch

***Zaslavsky, C. 1996. The multicultural math classroom: bringing in the world, Portsmouth, NH: Heinemann

6.2.5   Students having educational challenges

Worldwide, students with educational challenges are either not schooled, or put into special classes, or are integrated into the regular classroom with their agemates. These students may be found in primary, secondary and alternative education programs. In some cases, those with learning disabilities reach the tertiary level. While the use of mathematics history can be an excellent pedagogical tool, resources for teachers are mostly unavailable. Therefore, teachers are left to develop their own resources. This is not an easy task because it requires a background in mathematics history, as well as the understanding of the cognitive level of the student and the student's special needs. If countries actually have special education training for teachers, one cannot expect that the training will include a strong background in mathematics, much less the history of mathematics.

None the less, students with educational challenges can enjoy and be inspired by mathematics history. For example, the abacus is an excellent manipulative resource for helping students develop number sense. Limiting addition and subtraction to examples which do not require regrouping can provide an opportunity for exceptional learners to experience an ancient, multicultural tool. It has also been found that some of the ancient computational algorithms are better to use with students with learning disabilities. Students enjoy learning about the history of these algorithms along with their success in the use of the algorithm. It can be helpful for students to realise that other scientists and mathematicians also had learning disabilities. Albert Einstein often mentioned his poor computational skills.

Certainly, using the history of mathematics with exceptional children is an area which is ripe for exploration. At this time there do not appear to be any historians or educational researchers who are working in this area. There are individuals who have shared their few experiences of using mathematics history. But these practices are little known to fellow practitioners.

6.2.6   Mathematically gifted and talented students

Coralie Daniel

The idea that almost everything can be done in a variety of ways is as true for mathematics as it is for anything else. Many mathematics teachers would respond to a remark conveying this idea, with examples and anecdotes from their classroom experiences that illustrate the point. Yet, while this is likely to be the case in a conversation, few would say that they actually begin their lessons with the idea of finding the widest variety of solutions, or that they prepare their lesson plans with the specific intention of developing a particular concept through the presentation of a variety of widely different strategies or processes. Many classroom lessons are based on an assumption that a teacher's task lies in introducing an idea and then giving an example, showing a formula, teaching a method that proves the idea. Most students seem content that the single proof approach is a reasonable way for a teacher to stimulate the learning process and transmit knowledge, and because they have limited mathematical experience, they are unlikely to ask whether or not there are alternative methods that could be learned or thought through. But this is not true of gifted and highly talented mathematics students.

From a very early age gifted and talented children make an impression through the complexity of the *why*? and *how*? questions they ask. Most people are familiar with children's *why*? questions which follow on in a sequence derived from the adult responses, without thought on the child's part. Compare, though, the following pre-schoolers' *why*? and *how*? questions (Daniel 1995):

"Why are the clouds gray if what you said yesterday about why the sky is blue, is true?"

"I can see how these blocks go together on the floor, so how can I write that on paper?"

"How can I make these spaces [between telegraph poles] seem the same when they are not?"

The children and students who ask such questions retain information and think about obvious links between things, but they also experiment, think laterally, notice detail,

John Fauvel, Jan van Maanen (eds.), *History in mathematics education: the ICMI study*, Dordrecht: Kluwer 2000, pp. 188-195

- He tuhi pūrongo e pā ana ki te whakamahia o tētahi inenga whaitake e whai wāhi mai ana ētahi āhua ōrite. Hei whakatauira, ko te rapanga e whakaahuatia ana i raro nei:

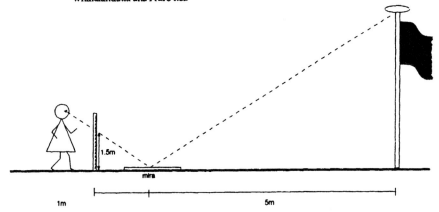

L V I I. Zeker Perzoon, ſtaande in V, ziet in de Spiegel P (. Horizontaal met A zijnde) de top des To-rens B, en 4 Voet te rug gaande, tot in D, en de Spiegel in V leggende, bevint het zelvige: Vra-ge na de hoogte des To-rens? Zoo P V is 3 ½, en V O, of D O, 5 Voe-ten.

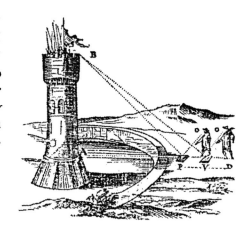

Dewijl de hoek O P V gelijk is aan de hoek B P A, en O V D gelijk aan B V A, door de natuur van de Weêr-kaatzing, zoo volgt dat

PV tot VO, als PA tot AB

$a \text{—} b \text{—} x /$ $\dfrac{bx}{a}$

VD tot DO, als AV tot AB

en $c \text{—} b \text{—} a + x /$ $\dfrac{ab + bx}{c}$

AP ∞ x
PV ∞ a
VO, of DO ∞ b
en VD ∞ c.

Figure 6.5: Problems found in many cultures offer gifted students opportunities to discover mathematics for themselves. Both Maori and Dutch students have the opportunity to work out, from the hint that a mirror on the ground provides a clue, how to measure the height of an object. The first case is from the New Zealand curriculum in Maori (1994), the second, a bit more complicated since the foot of the tower is out of reach, from a textbook in algebra by Abraham de Graaf (1672).

conceptualise, and allow prior learning to influence other ideas. They develop one idea from another, but in doing so they make value judgements, adapt ideas, enjoy discussing how things interact, are curious to see how knowledge and ideas can be applied to the world as they know it, and are stimulated by considering (but not necessarily adopting) other people's ideas. They are keen to proceed from one point of view to another by discussion and conversation, rather than by an example taught, copied down, practised and thus (presumably) learned. Teaching and learning, as the most important aspects of the educational process, have already been superseded by thinking, in the minds of gifted and talented students, usually before they have even begun formal schooling. This is often threatening to teachers; but it need not be.

It is not difficult to see a link between these aptitudes and the history of the development of mathematical ideas. As with all knowledge and philosophies, it was the application of just these attitudes—by different people over centuries of time—that enabled mathematical ideas, concepts, and proofs to be developed at all. This understanding in itself gives a reason for the inclusion of ideas from the history of mathematics in teaching gifted and talented mathematicians at any level.

The inclusion of historical material is not simply the adding of whimsical stories or biographical extras to basic mathematics lessons, to make students feel relaxed and add a level of fun to the mathematics classroom. Inclusion of the history of mathematics should also be undertaken for the way in which such an inclusion opens up different areas of mathematics and broadens the possibilities of helping to satisfy the curiosity and thinking needs of gifted students. For example, it is almost certain that many teachers have told their students something about Pythagoras (even if it is merely his name) when they have been teaching anything about the characteristics of the various squares on the sides of right-angled triangles. But how many teachers will have said (or even been aware) that there are at least 370 (Loomis 1968) known ways to show that the square on the hypotenuse of a right-angled triangle is equal to the sum of the squares on the other two sides? How many teachers have shown a number of those solutions in the same lesson series? This is a good way of illustrating to students that Pythagoras was one among many who was on to something very useful that could be thought about in many different ways; the alternative, teaching that Pythagoras discovered a theorem for which the proof is such-and-such, is very confining, not to say boring, by comparison. A teacher is likely to choose a presentation that seems to be the most logical, or choose the version remembered from some level in their own education and presume that is the most logical, the most universally known, and the easiest to follow.

There are ample indications, both in research and in anecdotal evidence, that there are real differences in the ways people approach thinking, in the ways that they relate to the various branches of a subject area, both aesthetically and functionally, and in the sort of information they are most likely to be able to follow and use again (Krutetskii 1976; Gelman 1993; Gross 1993; Holton & Daniel 1996). These genetic differences and intuitive preferences can be broadly described as placing any one individual in a position where they will use best an approach which principally uses their own style, whether analytical and reasoning skills, or reasoning and geometric skills, or geometric and pictorial skills (Krutetskii 1976; Holton & Daniel 1996).

Gifted students especially need to know this, so that their view of their own ability is not impaired by believing that one type of thinking is better than another. When one looks at various mathematical proofs that have been offered for particular ideas down through the ages, one can begin to identify the sorts of differences in approach that have influenced each of the mathematicians working on the idea. Looking at proofs from this point of view, and placing them in their historical context, also makes one aware of another interesting aspect of the contribution that a regard for the history of mathematics can make to classroom presentation and to the confidence of both teachers and gifted students in exploring various solutions to problems.   It becomes clear that more recent proofs of a proposition are not necessarily the better proofs simply because they are more recent. In Durell (1952), a text that was used extensively in New Zealand schools, the proof given as that of the Pythagorean theorem is an analytical proof attributed to Euclid.   Although Durell's text (p. 268) states that Pythagoras' method of proof is not known, the more geometrical proof commonly attributed to Pythagoras (Eves 1992) is given as an introductory comment earlier in the chapter.   The geometrical proof would be considered to be the more self-evident and the more easily accessible of the two, and many proofs which are over fifteen hundred years more recent cannot be considered to throw greater light on the truth of the idea. The Euclidean proof would certainly interest most gifted mathematics students but it would be less immediately accessible to students with a geometric and a pictorial intuitive approach.

Realisation of the fact that proofs may well reflect differences in thinking styles, as well as differences in the fashions of thought patterns of any given period in time, should encourage teachers and students alike to have confidence in exploring and investigating one proof or another, and to openly allow different personal assessments of which proof appeals and which does not.

Recognising that different students in one's classroom will intuitively access mathematical concepts more easily through one or another of these reasoning, or geometric, or pictorial approaches—but not through all—creates a new challenge for teachers. The link between different intuitive ways of seeing things and perceptions of what to put together to answer the question is strong not only in students but also in teachers themselves.

Few educators hold any longer to the idea of people being empty vessels who can be filled up with learning or expertise. Usually the phrase is used in describing students' roles in the educational process, but it is as true in relation to defining teachers' capabilities as it is in relation to describing students' potential as learners. Osborn (1983) has categorised aspects of mathematical ability slightly differently from some other writers but he recognised that teachers (albeit unwittingly and unintentionally) will teach, and evaluate and assess, principally from the perspective of their own thinking type.   While it is reasonably easy to develop skills which enable one to recognise solutions that come from a thinking style which is different from one's own, it is difficult (and in the long-term perhaps impossible) to be able to memorise, or easily recall, solutions which do not come from one's own approach. This is not because people are unwilling to do it, but because the neural system in their brain is programmed not to do it (Edelman 1994; Csikszentmihalyi, Rathunde & Whalen 1997; Dehaene 1997).

Not only do different things trigger off different responses, but different brains encode different things. Repetition does not go far towards changing this. Many mathematics programmes used in the past have assumed that repetition will change this, but most mathematicians would see that we do not have the same expectations in regard to other fields of capability. For example, we accept that just because one is an outstanding singer, does not mean that one will necessarily be able to become a concert pianist. Practice will improve what one can do; but it will not automatically turn a high-achieving singer into a superb pianist. In the field of music we would not expect effort alone to change the teacher or the student, and so we should not expect it in mathematics, either, even among gifted students and highly talented teachers.

In mathematics we need to help teachers find non-threatening ways to recognise and declare their own intuitive preferences in approach and to access and share examples of other methods of achieving solutions. Using the history of mathematics as an intrinsic part of one's teaching can help immensely in providing such an approach. For teachers, their own memory and initial response can become less influential in determining what is presented in the classroom, without fear of losing their credibility. Students see that the broadness of solutions presented in history proves in itself that one person is unlikely to be able to do everything from every point of view. It allows students and teachers to see that differences are acceptable; it offers them opportunities to make judgements about the types of solutions which increase their understanding; and it potentially offers the social and educational advantage of allowing other students to identify some of their skills as being similar to some of the skills of gifted students in their midst.

Whether or not a teacher merely gives a precis of the historical context of the mathematics being studied in class, does not matter. Finding out what will be acceptable to the teacher is a very important aspect of survival for gifted and talented students in a mathematics classroom. References to the history of particular mathematical ideas will have opened up avenues for mental exploration and actual research which the student can follow up with the knowledge that what is found out will be acceptable to the teacher. Adding to teachers' repertoires the resource of different examples, approaches, and proofs from history can give them more confidence in using discussion and opinion as a part of their methods in teaching mathematics, and more confidence in accepting that gifted and talented students may well know more mathematics and be able to solve more complex problems, than many teachers.

Problems for gifted and talented students in mathematics classrooms include boredom, a lack of a sense of stimulation, and isolation from others because they wish to explore mathematical concepts rather than merely learn specific methods of proof. Comments from teachers indicate that it is not only difficult to find enough extension work for gifted students, but also difficult to find time to concentrate on locating material when the whole class does not need to be catered for in this way. Inclusion of the history of mathematics not only increases access to possible mathematical extension for gifted students, but also enables teachers to show that they are not afraid of exploring different solutions. This benefits the whole class as well as providing a discussion environment which can increase the contribution gifted mathematics students can make and decrease their sense of working alone.

There are college texts available, such as Eves' *An introduction to the history of mathematics* (1992), which will provide teachers with a quick reference book for both practical examples and references to other texts available. Such books often also give the story of the development of mathematical ideas a general historical context by linking the patterns of mathematical developments with those of other academic disciplines and with particular historical periods and events.

Research shows that gifted and talented mathematicians are most likely to be advanced and voracious readers from an early age, and also frequently show a real interest in taking courses and papers in history and in philosophy (Daniel 1995). The books read by mathematically gifted students often include books by writers such as Tolkein, C.S.Lewis, Penrose, Sagan, Adams, and Asimov, and this in itself gives an indication of the way in which gifted and talented mathematics students seek themselves to link the worlds of mathematics, science, and philosophy (Daniel and Holton 1995). The inclusion of material in the mathematics classroom which increases the interdisciplinary connections which students can make increases the reward received from their interest in mathematics. The embedding of mathematics in its historical context helps to encourage able mathematics students to use, and see value in using, other skills and interests as a part of their progress towards being able mathematicians. And again, it increases the sense of inclusiveness with the class, as less able students will also be motivated and stimulated by being able to make links with other subject areas.

History has many examples of mathematicians who were variously misunderstood, under-appreciated, acclaimed and then never heard of again, or penalised for their skills and ideas. At the very least, a study of the history of mathematicians and their ideas will be supportive for gifted and talented mathematics students (who are self-conscious about that), for it offers models of differences and of intellectual fortitude in the face of criticism, examples of thoughtfulness and experimentation, and evidence that those who did not follow the contemporaneously accepted approach to

Figure 6.6: In museum visits and elsewhere the ethnomathematics of cultures from across the world is very evident to the alerted eye. Here patterns from the Bakuba people of the Congo demonstrate what mathematicians think of as the seven one-colour one-dimensional patterns.

solving a problem were not mad. Without the encouragement to think independently and the freedom to contribute unashamedly, many famous mathematicians would not have made the contribution they have to our understanding of mathematics. Similarly, for our own society to gain the greatest benefit from those who are mathematically able, it is important that as large a number of mathematically gifted and talented people as possible are helped to develop in ways which foster their abilities and encourage them to use their talent openly.

Last winter I carried out an experiment on perception and language. I asked people who happened to come into my office (about a dozen in all) to tell me what they could see when they looked out of the window. Basically, three different kinds of answers were given. Some named first the objects that lay on the horizon then worked backwards, with details of things between the horizon and themselves, and giving opinions in their descriptions through the use of words such as 'pretty', 'small', 'purple'. A second group named something in the middle distance, described in detail the things to its left and right, and used functional and emotional words such as 'useless', 'benign', 'dangerous' to describe what they saw. Three people named first the building immediately across the street, and stopped there. When asked "What else?", they described its colour and texture, and then the small tree and the cars parked on the road between the wall and the window.

At least three things became evident from listening to the descriptions and discussing the viewers' interpretations of what they saw. First, the question of what was 'good visualising', 'logical', or 'a fair description' suddenly took on new meaning—or no meaning at all! Such words were themselves subjective and were defined differently by different viewers. Second, it was significant that it was winter, because large deciduous trees obstruct the range of views in summer, and hence would have had an influence on what was described. Third, the total picture of what could be viewed from my window was actually most richly described by combining all three types of description. If one thinks about these three things from the point of view of teaching mathematics, then one can see an immediate advantage to teaching within a framework which

1.   identifies different views of what it is important to know,
2.   provides various contexts in relation to areas of knowledge and world views, and
3.   shows that there are many ways of fitting differences together and still making sense of them.

The use of history offers such a framework in the teaching and understanding of mathematics, just as much as it does in other subjects.

References for §6.2.6

Csikszentmihalyi, M., Rathunde, K., Whalen, S. 1997. *Talented teenagers: roots of success and failure*, Cambridge: University Press
Daniel, C. 1995. 'The identification of mathematical ability and of factors significant in its nurture'
Daniel, C., Holton, D. 1995. 'Aspects of studies of talented mathematics students and implications for the secondary school education of the mathematically able', paper presented at First National Conference for Teaching Gifted Students at Secondary Level, Palmerston North, N.Z.

Dehaene, S. 1997. *The number sense: how mathematical knowledge is embedded in our brains*, Oxford: University Press

Durell, C.V. 1952. *A new geometry for schools*, London: G. Bell and Sons

Eves, H., 1992. *An introduction to the history of mathematics*, 6th ed. Orlando, U.S.: Saunders College Publishing

Edelman, G., 1994. *Bright air, brilliant fire: on the matter of the mind*, London: Penguin

Gelman, R. 1993. 'A rational-constructivist account of early learning about numbers and objects', *The psychology of learning and motivation* **30**, 61-96

Glotov, N.V. 1989. 'Analysis of the genotype-environment interaction in natural populations', pres. at Symposium on Population Phenogenetics, Moscow, Russia

Gross, M.U.M. 1993. *Exceptionally gifted children*, London: Routledge

Holton, D.A., Daniel, C. 1996. 'Mathematics', in Don McAlpine and Roger Moltzen (eds.), *Gifted and talented: New Zealand perspectives*, Palmerston North: ERDC Press, 201-218

Krutetskii, V.A. 1976. *The psychology of mathematical abilities in schoolchildren*, Chicago: University Press

Loomis, E.S. 1968. *The Pythagorean proposition*, Washington, DC: NCTM

Osborn, H.H., 1983. 'The assessment of mathematical abilities', *Educational Research* **25**, 28-40

Sternberg, R.,1995. Interview, *Skeptic* **3** (3), 72-80

6.3   Opportunities for change

6.3.1   Teacher education

The educational, cultural, economic and social diversity which is found in the history of mathematics can be used to help students and teachers learn more effectively.   It follows that teachers, at all levels, would benefit from some experience and background in the history of mathematics.   A major question remains, of how teachers are to gain such experience.   This was discussed more fully, in a general context, in chapter 4.   Here we suggest a few ideas in relation to teacher training, in the context of the concerns of this chapter. Training may be available either while teachers are in their initial pre-service training or as in-service provision for practising teachers, and we make some brief remarks on each.

There are several possibilities for making the history of mathematics part of pre-service teacher education.

a) Integration for non-specialist teachers of mathematics.

Easy-to-find references and resources for teaching mathematics under topic headings would enable teacher-trainers to give a context for whatever style they adopt with non-specialist teachers, mainly those preparing to teach in primary or elementary schools. So, under the heading of number could be listed such topics as place value, examples of counting in Babylonian numerals or counting with sticks and stones. Some schools with non-specialist teachers have theme or topic-based lessons. If the topic were Ancient Egypt, for example, mathematics can be cross referenced through finding angles, triangles and measures under pyramid construction, or indeed pursuing the way loaves of bread were divided for distribution in the Rhind Papyrus.

Benjamin Banneker's
PENNSYLVANIA, DELAWARE, MARY-
LAND, AND VIRGINIA
A L M A N A C,
FOR THE
YEAR of our LORD 1795;
Being the Third after Leap-Year.

—PRINTED FOR—
And Sold by JOHN FISHER, Stationer,
BALTIMORE.

Figure 6.7: Benjamin Banneker (1731-1806), the first African American to be recognised for his mathematical abilities, can be an inspiration for teachers of minority students in north America today. Teachers of gifted students everywhere can encourage projects into his life, to share his work in surveying and almanac-making with the rest of the class.

b)   Separate modules on the history of mathematics for specialist mathematics teachers.

Those training to teach at primary, elementary, secondary and further levels in various settings are often college or university students. Sometimes a 'standard' history of mathematics course is offered to give the trainee teachers an overview of the subject. Even, or especially, in such a course it is important that the balance be wide ranging and specifically address the requirements of the gifted, different cultures and social groupings, the educationally disadvantaged, and other individuals or groups along the lines discussed in this chapter. This will enrich teachers' ability to meet challenges that they may face in the course of their professional life.

c)   Integrate the history of mathematics into teacher training

It is particularly appropriate for those training as specialist secondary mathematics teachers (often post-graduates) that they be provided with references and resources under mathematical topic headings, and guidance in a range of uses of history in the classroom. The same principle holds as in the previous case, that potential teachers need to acquire confidence in making use of resources for a wide range of students. For teaching material under the heading of Pythagoras' theorem, for example, the possibilities might include: different approaches to its proof through history (appropriate for some gifted students); uses and applications of the theorem through the ages (particularly for adults and workplace learners); using ready prepared jigsaw pieces which fit to show the Pythagorean relationship (good for students with a mental or physical handicap); different representations of the theorem to reflect the cultures represented amongst the students and beyond (to help relate minority group students to the world-wide nature of the history of mathematics).

In the case of current teachers of mathematics who are receiving up-dating or other in-service training, the issues are slightly different because of their greater classroom experience. Such teachers can be helped to cater for the needs of a diversity of students, by suggesting ways for using material from the history of mathematics in their teaching. Summarising broadly, most such uses will be by integrating, adding or substituting historical material. There are several possibilities by which the HPM community can contribute to in-service provision (we draw attention here to some possibilities especially appropriate for the themes of this chapter).

— Case studies of teachers using the history of mathematics in various ways. Each such study needs to include the context—country, place, language, students, teacher, resources, activity and student response. Such a collection of case studies would provide a vision of alternatives, both inspirational and occasionally as a warning. Each would have to be brief and concise to be accessible to busy teachers across the world.

— Classroom-ready resources like photocopiable sheets, press-out models, outline lesson plans.

— Multi-lingual resources. Particularly in societies in which several languages co-exist in the community, it will be useful for translations to be made available of key identified materials.

— Day courses funded internationally, with a local and international flavour. These could be run by local advisors or advisory teachers who would show how the history of mathematics is part of each country's mathematics and history curriculum. Resources and Activities

The following are specific resources and activities which are known to have been used by at least one practising teacher in at least one country, who claims it to have 'worked' for them. These :fall into four main categories of activity or resource, which may be described as researching, presenting, visiting, and experiencing.

Researching   These activities for the students have a variety of names such as 'researching a topic', 'doing a piece of coursework', 'doing a project'. They can range from small, short, relatively closed exercises in finding out using a given resource, to an open-ended whole class investigation lasting several weeks, and involving many different aspects. Examples would include finding out about a

---

Biography Project - Poster

In this project you will learn about a mathematician and create a poster display. This project will be due on the first day your class meets, during the last week of classes, this semester. The posters will be displayed for other students to critique and to ask you questions about your mathematician.

You must include the following elements in your display:

- picture of your mathematician - credit the source (10 points)
- one page curriculum vita or resumé of your mathematician. You may be creative, but you must be factual concerning: schooling, papers/books written, mentorship - note this is not a biography (10 points)
- timeline portraying significant events of his/her life interwoven with significant world events and contemporary lives (20 points)
- one page summary of an important result attributed to this mathematician, including relevant graphs and mathematics as necessary (20 points)
- explanation of how other mathematicians/mathematics and scientists/science directly influenced your mathematician and his/her work (10 points)
- brief anecdote about your mathematician (10 points)
- bibliography (at least 3 references, and at least one of these references must be a book) (5 points)
- your poster should be easily read, neatly organized, and creatively displayed. (10 points)
- critique of your display by the other students in the class. Five points would be an "A" grade, four points a "B", etc. (5 points)

Joanne Peeples <joannep@epcc.edu>
(adapted from an assignment by Lynn Fosbee Reed, The Governor's School for Government and International Studies, Richmond, VA)

Figure 6.8: the US teacher who devised this project for her math class combined 'researching' and 'presenting' activities for the pupils, who responded with enthusiasm to the challenges of a carefully guided set of instructions.

mathematician; the evolution of the subtraction algorithm; early symbolism around the world; some 'other' number system; who 'invented' calculus?; writing your personal history of learning and doing mathematics.

Presenting There are many different ways of presenting, communicating and disseminating the findings and discoveries made by students of all ages. Examples would include: play-writing and acting; role play in costume; simulation of a mathematical discovery by re-enacting the process; spoken presentation; video; slides; projected transparencies; photos; posters (cf. figure 6.8); structured discussion or debate; drawing; a written essay or paper; building a model; playing a game; creating a stereogram. We may note two things in particular about the process of presenting. One is that such experiences develop and deepen student competencies across the range of the skills they are learning at school. The other point to notice is that some students will flourish particularly strongly in such activities and show talents that were hidden in the usual class contexts.

Visiting   Visits outside the classroom can bring history to life. The events, people and objects which illustrate the history of mathematics can be found in a wide variety of places. Many teachers have learned to make the most of what is available in the school locality. It is often surprising how much there is. Many schools are not too far from one or all of: museum; exhibition; church; site of historic interest; sundial; palace; cemetery; building; art gallery; play; concert; park; countryside; city; boat trip; river bank; coast line; talks; lectures; demonstrations; historical mathematics tours; historical mathematics trails. One pre-college teacher from the United States has for the last 10 years taken students on a 'Math Tour of England'. Among the places and artifacts which can be visited are the Rhind mathematical papyrus at the British Museum, Babylonian mathematical tablets also at the BM, ancient scientific instruments in the Science Museum (London), letters of Newton at the British Library, astrolabes at the Science Museum in Oxford, Newton's birthplace in Lincolnshire, and so on.

Experiencing   The objects and artifacts which make history and are its primary evidence need to be seen, heard, touched, played with and experienced at first hand. Some things will have to be copies, models, films or photocopies of the real thing. Objects fall into several categories. *Measuring instruments* like a sextant, water clock, sundial, dividers, compasses, weights and balances. *Calculating devices* like abaci, quipus, counting boards, early computers, Napier's bones. *Written material* in the form of manuscripts, early books and printing, stone inscriptions, papyri, clay tablets, diaries, text books. *Natural objects* which have inspired the creation of mathematics in the past like spirals on shells and fir cones, the movement of the stars and planets, the rhythm and beat in music. *Artifacts,* objects made by people, often embody mathematics and its history: woven baskets, furniture, building design, tiling, friezes, wrought-iron work, machines of all sorts, games and puzzles.

References for §6.3

Sheath, Geoff, Muriel Seltman and Wendy Troy 1996. 'The history of mathematics in initial
teacher training' in *Proc. HEM Braga* **ii**, 136-143
Singmaster, David 1996. *A mathematical gazetteer*, London: South Bank University
Singmaster, David 1999. Mathematical gazetter of Britain,
http://www.dcs.warwick.ac.uk/bshm/
Tanford, Charles & Jacqueline Reynolds 1992. *The scientific traveller: a guide to the
people, places and institutions of Europe*, New York: Wiley
Tanford, Charles & Jacqueline Reynolds 1995. *A travel guide to scientific sites of the British
Isles*, Chichester: Wiley

6.4   Conclusion

The aim of this chapter has been to help the reader become aware of the pedagogical opportunities of using the history of mathematics in institutional settings other than the typical university environment. Because of their own experiences, the authors know and are keen to share the knowledge that students can enjoy, learn from, and be enriched with experiences in mathematics history long before they are in the university. Research indicates that early adolescence is the time that students begin defining their attitudes toward mathematics. It therefore seems imperative that opportunities for incorporating mathematics history into the classroom begin early and continue throughout the educational careers of students. Universities and teacher education programmes should find a place for history in the curriculum from two perspectives: the curriculum of the teacher in training and the curriculum of the classroom student.

When students receive a limited education, as many do, it is for a variety of reasons. Some for lack of economic resources, some because of cognitive

Figure 6.9: The 'diverse educational requirements' team working on this chapter at the ICMI Study Meeting: Coralie Daniel, Wendy Troy, Gunnar Gjone, Gail FitzSimons, Karen Dee Michalowicz, Vicky Ponza

limitations, some because higher education is not readily available or appreciated; and some, unfortunately, because of the quality of their teachers. Even students from affluent backgrounds can find themselves studying in mathematics classrooms where the quality of the instruction causes anxiety, frustration, and negative attitudes. But, more generally, the needs of students of diverse educational backgrounds are increasingly being appreciated, and the availability of resources to help in their mathematical learning is more apparent. It has been argued here that a historical component, or the possibility of historical resources, can help teachers support students in such situations.

Chapter 7

Integrating history of mathematics in the classroom: an analytic survey

Constantinos Tzanakis and Abraham Arcavi

with Carlos Correia de Sá, Masami Isoda, Chi-Kai Lit, Mogens Niss, João Pitombeira de Carvalho, Michel Rodriguez, Man-Keung Siu

Abstract: *An analytical survey of how history of mathematics has been and can be integrated into the mathematics classroom provides a range of models for teachers and mathematics educators to use or adapt.*

7.1 Introduction

Mathematics is often regarded as a collection of axioms, theorems, and proofs. Organised and presented as a formal deductive structure, this assumes, at least implicitly, that the logical clarity of such a presentation may be sufficient for understanding mathematics. Under this view, strongly influenced by formalism as a philosophical trend, mathematics seems to progress by a more or less linear accumulation of new results (Davis and Hersh 1980, Ch.7; Brown 1977, Ch.4). Publicly, it consists of polished products of mathematical activity, which can be communicated, criticised (in order to be finally accepted or rejected) and which may serve as the basis for new work. Increasingly, though, it is recognised that this view of mathematics is just one aspect of what constitutes mathematical knowledge. The process of *doing mathematics* is equally important, especially from a didactical point of view. This process includes using heuristics, making mistakes, having doubts and misconceptions, and even retrogressing in the development and understanding of a subject (Lakatos 1976, Introduction; Courant and Robbins 1941, Introductory comments; Stewart 1989, 6-7; Schoenfeld 1992; Barbin 1997). In this understanding, the meaning of mathematical knowledge is determined not only by the circumstances in which it becomes a deductively-structured mathematical theory, but also by the procedure that originally led to it and which is indispensable for its understanding (cf. Brousseau 1983, 170; Hadamard 1954, 104).

John Fauvel, Jan van Maanen (eds.), *History in   mathematics education: the ICMI study*, Dordrecht: Kluwer 2000, pp. 201-240

To learn mathematics, then, is not only to become acquainted with and competent in handling the symbols and the logical syntax of theories, and to accumulate knowledge of new results presented as finished products. It also includes the understanding of the motivations for certain problems and questions, the sense-making actions and the reflective processes which are aimed at the construction of meaning by linking old and new knowledge, and by extending and enhancing existing conceptual frameworks (Hiebert and Carpenter 1992, 67; Schoenfeld *et al.*, 1993). Teaching mathematics then becomes a much more complex enterprise than just the mere exposition of well organised mathematical developments. It should include giving opportunities to *do* mathematics, in the sense described above. In this respect history of mathematics seems a natural means for exposing mathematics in the making, and thus it may play a very important role in mathematics education.

This chapter is intended to review how the history of mathematics can be and has been harnessed and integrated in mathematics education. More specifically, in section 7.2 the above general argument for the relevance of history is analysed in more detail. The analysis provides several reasons why the history of mathematics may be relevant to the teaching and learning process, both for the teacher and the learner. In this process, some arguments questioning the use of history in mathematics education are raised and dealt with. In section 7.3, we elaborate on the important question of how integration of history can be effected, and in section 7.4, the longest section of the chapter, we survey and exemplify a wide spectrum of different possible implementations of history in the mathematics classroom.

7.2   Why should history of mathematics be integrated in mathematics education?

Integrating the history of mathematics in mathematics education has been advocated for a long time (De Morgan 1865; Glaisher 1890; Poincaré 1908; Barwell 1913; Miller 1916; MAA 1935; Klein 1914/1945, 268; British Ministry of Education 1958; Lakatos 1976, Introduction and Appendix 2; Leake 1983; see also Kline 1973, Ch.4). In 1969, the US NCTM (National Council for the Teaching of Mathematics) devoted its 31st Yearbook to the history of mathematics as a teaching tool (NCTM 1969). On the other hand, several difficulties have been raised, challenging the desirability or feasibility of seeking to integrate history of mathematics in mathematics education. In this section, we first summarise these objections (in a list extended from that given in Siu 1998) and then we classify and discuss the different arguments that have been or may be proposed in favour of integrating history in mathematics education, dealing implicitly, in the process, with the objections.

Some objections

Arguments against the incorporation of history are based on at least two sources of difficulty: philosophical and practical. Among the former we hear that:

(*O1*) History is not mathematics. If you must teach history, then you need to teach mathematics itself first: teach the subject first, then its history.

(*O2*) History may be tortuous and confusing rather than enlightening (e.g. Fowler in Ransom 1991, 15; Fauvel 1991, 4).

(*O3*) Students may have an erratic sense of the past which makes historical contextualisation of mathematics impossible without their having had a broader education in general history (e.g. Fauvel 1991, 4).

(*O4*) Many students dislike history and by implication will dislike history of mathematics, or find it no less boring than mathematics.

(*O5*) Progress in mathematics is to make the tackling of difficult problems a routine, so why bother by looking back? (e.g. Le Goff 1996, 13)

(*O6*) History may be liable to breed cultural chauvinism and parochial nationalism.

Some of the practical objections to incorporating history within mathematics teaching and learning are:

(*O7*) *Lack of time*: there is not enough classroom time for mathematics learning as it is, still less when it is proposed to teach history of mathematics as well (e.g. Buhler 1990, 43).

(*O8*) *Lack of resources*: there are not enough appropriate resource materials to help even those teachers who may want to integrate historical information (e.g. Fauvel 1991, 4; Le Goff 1996, 13).

(*O9*) *Lack of expertise*: the teacher's lack of historical expertise (e.g. Fowler in Ransom 1991 p.16) is a consequence of the lack of appropriate teacher education programmes; indeed, not only historical but also interdisciplinary knowledge is required, which is far beyond what mathematics teachers are equipped for. The lack of expertise leads to an even more debilitating lack of confidence.

(*O10*) *Lack of assessment*: there is no clear or consistent way of integrating any historical component in students' assessment, and if it is not assessed then students will not value it or pay attention to it.

Some arguments in support of integrating history

There are five main areas in which mathematics teaching may be supported, enriched and improved through integrating the history of mathematics into the educational process:

a)   the learning of mathematics;

a)   the development of views on the nature of mathematics and mathematical activity;

b)   the didactical background of teachers and their pedagogical repertoire;

c)   the affective predisposition towards mathematics; and

d)   the appreciation of mathematics as a cultural-human endeavour.

(cf. references at the beginning of this section and also, Arcavi 1985 Ch.1; Fauvel 1991; Ransom et al. 1991; Lefort 1990, 87-88; Grugnetti 1998, 1-2).

In the following we elaborate on these arguments, and by implication we deal with some of the aforementioned objections (citing them by number at the end of each argument to which they are related).

(a) The learning of mathematics

1. *Historical development vs. polished mathematics*: mathematics is usually taught in a deductively oriented organisation. However, the historical development of mathematics shows that the deductive (or even strictly axiomatic) organisation of a mathematical discipline comes only after this discipline has reached maturity, so that it becomes necessary to give an *a posteriori* presentation of its logical structure and completeness. Freudenthal (1983, ix) describes this as follows:

"No mathematical idea has ever been published in the way it was discovered. Techniques have been developed and are used, if a problem has been solved, to turn the solution procedure upside down . . . [and turn] the hot invention into icy beauty."

Thus, mathematics is usually globally and retrospectively re-organised. On the one hand, it would seem that this re-organisation is needed to avoid possible tortuous and long-winded accounts. On the other hand, questions and problems which constituted basic motivations for the development of an idea, as well as any doubts along the way, remain hidden under a linearly organised, deductive body of knowledge, in which new results seem to be simply added in a cumulative way.

In this connection, the proper integration of history into mathematics education can play an important role by helping to uncover how "our mathematical concepts, structures, ideas have been invented as tools to organise the phenomena of the physical, social and mental world" (Freudenthal 1983, ix). In this way the learning of a mathematical concept, structure or idea may gain from acquaintance with the motivation and the phenomena for which it was created (Barbin 1996, 196; Nouet 1996, 125; Tzanakis 1996, 97). This fact has been recognised and advocated by many (Klein 1926-7/1979, 316; Polya 1954, 1968; Lakatos 1976 Introduction and Appendix 2). However, it implies neither that there is a uniquely specified presentation of a subject that follows exactly the usually complicated historical development, nor that the learning of mathematics should be guided by "ontogenesis recapitulates phylogenesis" (Fauvel 1991, 3-4; Sierpinska 1994, 122; Rogers 1998 §2, 3). History could at best suggest possible ways to present the subject in a natural way, by keeping to a minimum logical gaps and *ad hoc* introduction of concepts, methods or proofs. In this way the historical record could inspire teachers and help them in their teaching (cf. §3.2 and Menghini 1998 §2, Tzanakis and Thomaidis 1998 §3.3). (*O1, O2, O5*)

2. *History as a resource*: the history of mathematics provides a vast reservoir of relevant questions, problems and expositions which may be very valuable both in terms of their content and their potential to motivate, interest and engage the learner (Van Maanen 1991, 47; Arcavi in Ransom et al. 1991, 11; Friedelmeyer 1990, 1; 1996, 121; Ransom *et al* 1991, 8; Ernest 1994, 237-238). In this connection, historically inspired exercises may stimulate the student's interest and contribute to

curricular enhancement alongside those exercises and problem which may seem more artificially designed.   Through such exercises, aspects of the historical development of a subject become a working knowledge for the student; in this way history no longer appears as something alien to mathematics proper (cf. §3.2 and Tzanakis 1996, 97). (*O1, O7, O4, O10*)

3. *History as a bridge between mathematics and other subjects*: history exposes interrelations among different mathematical domains, or, of mathematics with other disciplines, (for example, physics: Tzanakis 1999, 2000).   It also suggests that mathematical activities and results may be interdependent (Jozeau 1990 p.25). Thus, integration of history in teaching may help to bring out connections between domains which at first glance appear unrelated.   It also provides the opportunity to appreciate that fruitful research in a scientific domain does not stand in isolation from similar activities in other domains.   On the contrary, it is often motivated by questions and problems coming from apparently unrelated disciplines and having an empirical basis. (*O9*)

4. *The more general educational value of history*: students involved in historically oriented study projects may develop personal growth and skills, not necessarily associated only with their mathematical development, such as reading, writing, looking for resources, documenting, discussing, analysing, and 'talking about' (as distinct from 'doing') mathematics (Ransom *et al* 1991, 9). (*O10*)

(b) The nature of mathematics and mathematical activity

1. *Content*: a more accurate view of mathematics and mathematical activity may be provided by historically important questions, problems, and answers (whether provided directly by primary sources or reconstructed in a modern language). Students may learn that mistakes, heuristic arguments, uncertainties, doubts, intuitive arguments, blind alleys, controversies and alternative approaches to problems are not only legitimate but also an integral part of mathematics in the making (see for example, Arcavi et. al, 1982, 1987; and in this chapter, §7.4.6). They may become more able to understand why conjectures and proofs, which have been put forward in the past, do or do not supply satisfactory answers to already existing problems.   Indirectly, students may be encouraged to formulate their own questions, make conjectures and pursue them (Friedelmeyer 1996, 121; Rodriguez 1998, 4; Tzanakis 1996, 97).   History also makes more visible (to both teachers and students), the evolutionary nature of mathematical knowledge and the time-dependent character of fundamental meta-concepts, such as proof, rigour, evidence, error etc. (Ransom *et al* 1991, 12; Barbin 1996, 198-202; 1997; Nouet 1996, 126). (*O5, O4*)

2. *Form*: mathematics is evolving not only in its content, but also in its form, notation, terminology, computational methods, modes of expression and representations.   History helps students to understand this as well as the mathematical (verbal, or symbolic) language of a given period, and to re-evaluate the role of visual, intuitive and non-formal approaches that have been put forward in the past (van Maanen 1991, 47).   Then, with the aid of original material, or even simple extracts from it (see §7.3, below) both the teacher and the learner may become aware of the advantages and/or disadvantages of modern forms of

mathematics. (*O2*). (Think for instance of vector analysis without vector notation as it appeared in the second half of the 19th century in Maxwell's electrodynamics; classical mechanics in Newton's euclidean geometric form; Diophantus's algebraic notation; the advantages and disadvantages of the older formulation of differential geometry by using indices to describe tensor quantities, compared to its modern, coordinate-independent formulation.)

(c) The didactical background of teachers

By studying history and trying to reconstruct aspects of the historical development of specific mathematical topics in a didactically appropriate manner, teachers may:

1. Identify the motivations behind the introduction of (new) mathematical knowledge, through the study of examples that served as prototypes in its historical development and which may help students to understand it (cf. (a1) above). (*O5*)

2. Become aware of:

(i)   the difficulties, or, even obstacles, that appeared in history and may reappear in the classroom;

(ii)   how 'advanced' a subject may be—namely, even when a subject may appear simple, it may have been the result of a gradual evolution. In general, this evolution was based on concrete questions and problems which are not evident if the subject is presented in its modern form right from the beginning. But these questions and problems may presuppose a mathematical maturity on the part of the student that may not exist yet. In this sense, the history of mathematics may help the teacher to become aware of the pros and cons of presenting a subject at a particular level of education (Arcavi in Ransom 1991, 11; Tzanakis 1996, 97; Horng 1998, 1; Rodriguez 1998, 4-5). (*O5*)

3. Get involved into, hence become more aware of, the creative process of 'doing mathematics' (Barbin 1997). Thus, teachers (and in this connection, students as well), can not only enrich their mathematical literacy, but also appreciate better the nature of mathematical activity.

4. Enrich their didactical repertoire of explanations, examples, and alternative approaches to present a subject or to solve problems (cf. (a2) above). (*O1*).

5. Participate in a situation in which they have to decipher and understand a known piece of correct mathematics but whose treatment is not modern (see also Ch. 9, about working with primary sources), and thus they can exercise sensitivity, tolerance and respect towards non-conventional or idiosyncratic ways to express ideas or solve problems. This argument is valid for students as well. (*O2*)

(d) The affective predisposition towards mathematics

History can provide role models of human activity, from which several things can be learned, among them the following:

1. That mathematics is an evolving and human subject rather than a system of rigid truths. It is a human endeavour which requires intellectual effort and it is determined by several factors, both inherent to mathematics itself and external to it (cf. (e) below). In particular, it is not a God-given finished product designed for rote learning.

2. The value of persisting with ideas, of attempting to undertake lines of inquiry, of posing questions, and of attempting to develop creative or idiosyncratic ways of thought (cf. (b1) above).

3. Not to get discouraged by failure, mistakes, uncertainties or misunderstandings, appreciating that these have been the building blocks of the work of the most prominent mathematicians. (*O2, O5*)

(e) The appreciation of mathematics as a cultural endeavour

As stated above, mathematics is not a rigidly structured system of results, but a continuously evolving human intellectual process, tightly linked to other sciences, culture and society (cf. Rickey 1996, 252; Ernest 1994, 238; Van Maanen 1991, 47). For example:

1. Through the detailed study of historical examples, students can be given the opportunity to appreciate that mathematics is driven not only by utilitarian reasons (a currently prevailing view), but also developed for its own sake (Hallez 1990, p.97), motivated by aesthetic criteria, intellectual curiosity, challenge and pleasure, recreational purposes etc. (Chandrasekhar 1987, Ch.4; Kragh 1990, Ch.14; Tzanakis 1997). (*O5*)

2. History can provide examples of how the internal development of mathematics, whether driven by utilitarian or 'pure' reasons, has been influenced, or even determined to a large extent, by social and cultural factors. (*O9*)

3. Mathematics in its modern form is mostly viewed as a product of a particular (western) culture.   Through the study of history of mathematics, teachers and students have the opportunity to become aware of other, less known, approaches to mathematics that appeared within other cultures, and the role it played in them.   In some cases, these cultural aspects may help teachers in their daily work with multi-ethnic classroom populations, in order to re-value local cultural heritage as a means of developing tolerance and respect among fellow students (Nouet 1996, 126; Ch. 6, above). (*O6*)

The discussion in this section illustrates the many roles which history may play in mathematics education, varying according to both the intended purposes and the beneficiaries. Both students and teachers benefit; the latter may profit, not only as practitioners, but also as students themselves, both in their pre-service education and in in-service development programs. Above all, the present discussion brings out the need for:

(i)   easily accessible, comprehensible resources, available to teachers and students (e.g. Fauvel and Gray 1987). (*O8*)

(ii) a systematic preparation of future teachers both during their initial training and through in-service studies (cf Ch. 4). (*O9*)

The present volume in general, and the rest of this chapter in particular, is intended as a contribution to the fulfilment of these needs. In the next sections, we survey ways in which history may in practice be integrated with educational experiences.

7.3   How may history of mathematics be integrated in mathematics education?

Making explicit some reasons for integrating history in mathematics education, as we did in the previous section, still leaves open the question of how this integration may be accomplished. In this section, we distinguish and analyse three different yet complementary ways in which this may be done. In broad terms they may be characterised as:
1. Learning *history*, by the provision of direct historical information.
2. Learning *mathematical topics*, by following a teaching and learning approach inspired by history.
3. Developing *deeper awareness*, both of mathematics itself and of the social and cultural contexts in which mathematics has been done.

7.3.1   Direct historical information

By direct historical information, we mean both
a)   isolated factual information, such as names, dates, famous works and events, time charts, biographies, famous problems and questions, attribution of priority, facsimiles etc., and
b)   full courses or books on the history of mathematics. These may be a simple account of historical data, or a history of conceptual developments, or something in between.
In both cases the emphasis is more on resourcing history than on learning mathematics (in contrast to what is described in the following subsections). Given that emphasis, this is an auxiliary way of integrating history; by itself it does not directly change the intrinsic teaching of particular mathematical content (although it will surely affect the learning experience).

In section 7.4 we describe in some detail different implementations of this emphasis, such as historical 'snippets' (§7.4.1), parts of packages 'ready to use' in the classroom (§7.4.5), becoming acquainted with famous problems (§7.4.7), certain kind of plays (§7.4.10), certain visual displays (§7.4.11), visits to museums (§7.4.12) and databases in the WWW (§7.4.13). Although direct historical information may not be the main emphasis of the remaining implementations described in section 7.4, it can be an integral part of them.

Most of the arguments analysed in section 7.2 can be partially supported by such integration, depending on the form, the scope and the chosen depth.

7.3.2   A teaching approach inspired by history

This is essentially what may be called a genetic approach to teaching and learning. It is neither strictly deductive nor strictly historical, but its fundamental thesis is that a subject is studied only after one has been motivated enough to do so, and learned only at the right time in one's mental development. This means that those questions and problems which the subject at that stage may be addressing have been sufficiently elucidated and appreciated (cf. Toeplitz 1963, Edwards 1977). Thus, the subject (e.g. a new concept or theory) must be seen to be needed for the solution of problems, so that the properties or methods connected with it appear necessary to the learner who then becomes able to solve them. This character of *necessity of the subject* constitutes the central core of the meaning to be attributed to it by the learner. In this sense, in a genetic approach the emphasis is less on how to use theories, methods and concepts, and more on why they provide an answer to specific mathematical problems and questions, without however disregarding the 'technical' role of mathematical knowledge (Sierpinska 1991 §II). From such a point of view, the historical perspective offers interesting possibilities for a deep, global understanding of the subject, according to the following general scheme (Tzanakis 1996; Tzanakis 2000, §1; cf. Kronfellner 1996, 319; Lalande *et al* 1993):

(1) Even the teacher who is not a historian should have acquired a basic knowledge of the historical evolution of the subject.

(2) On this basis, the crucial steps of this historical evolution are identified, as those key ideas, questions and problems which opened new research perspectives.

(3) These crucial steps are reconstructed, so that they become didactically appropriate for classroom use.

(4) These reconstructed crucial steps are given as sequences of historically motivated problems of an increasing level of difficulty, such that each one builds on some of its predecessors. The form of these problems may vary from simple exercises, of a more or less 'technical' character, to open questions which probably should be tackled as parts of a particular study project to be performed by groups of students.

Concerning this scheme, we make some further remarks.

(i)   Both the teacher and the students may well make use of original and secondary sources (§ 7.4.2, 7.4.3, Ch. 9).

(ii)   Mainly in stage (2) above (and partly in (3)), the teacher makes an effort to grasp the difficulties inherent in the subject and to gauge possible obstacles in its understanding. Then the selection of questions and problems can be made, motivated by history, so as to activate the curiosity of the learner and smooth the learner's path, by creating and/or explaining the necessary motivations for studying new theories, methods and concepts. In this way, one could have an answer to the important question put forward by Brousseau (1983, 167; our translation):

A pupil doesn't do mathematics if he is not given problems and does not solve problems. Everybody accepts this fact. The difficulties arise once it is required to know, which problems must be given to him, who puts them and in what way.

At this level, inductive reasoning and analogies dominate as creative and discovering patterns, emphasising the mathematical activity itself rather than the well-organised arrangement of its results (Polya 1954, Polya 1968, Tzanakis 1997 §7, Tzanakis 1998).

(iii) In the reconstructions of stage (3), history may enter either explicitly or implicitly. There is a duality here, as has been stressed by several authors, from Toeplitz's work to recent researches (see e.g. the distinction between direct and indirect genetic approach (Toeplitz 1927, 1963; Schubring 1978, 1988), 'forward and backward heuristics' (Vasco 1995, 61-62), explicit and implicit use of history (§ 3.1; Menghini 1998, §2)). In a reconstruction in which history is explicitly integrated, mathematical discoveries are presented in all their aspects. Different teaching sequences can be arranged according to the main historical events, in an effort to show the evolution and the stages in the progress of mathematics by describing a certain historical period (Menghini 1998, 3; Schubring 1978; Schubring 1988; cf. Hairer & Wanner 1996; Friedelmeyer 1990; Martin 1996). In a reconstruction in which history enters implicitly, a teaching sequence is suggested in which use may be made of concepts, methods and notations that appeared later than the subject under consideration, keeping always in mind that the overall didactic aim is to understand mathematics in its modern form. In such an approach, the teaching sequence does not necessarily respect the order by which the historical events appeared; rather, one looks at the historical development from the current stage of concept formation and logical structuring of the subject (Kronfellner 1996; Siu 1997; Stillwell 1989; for examples see e.g. Radford and Guérette 1996 §§ 2, 3; Tzanakis 1995; Tzanakis 1999). At this point, it is important to stress that the above two possible types of reconstructions of the historical development are not mutually exclusive. They have a dual character with respect to each other and both may be used in teaching a subject in complementary ways (cf. Ofir 1991, 23; Flashman 1996): in an explicit integration of history, emphasis is on a rough but more or less accurate mapping of the path network that appeared historically and led to the modern form of the subject; in an implicit integration, the emphasis is on the redesigning, shortcutting and signalling of this path network (Vasco 1995, 62). In both cases, historical aspects of famous problems, intuitive arguments, errors, and alternative conceptions may be incorporated in teaching (§7.4.6, §7.4.7).

(iv) A reasonable concern about such an approach might be the fear that it takes too much time, or leads to over-voluminous textbooks. Such a fear is not well-founded. The sequence of problems (devised in the sense of (4) above) can give compact opportunities to the learner to arrive at constructive results, starting from easy corollaries of the main subject and often following the main steps of the historical path. In this way, the solution of exercises becomes an essential ingredient of learning, leading to the construction of the necessary technical knowledge on the basis of interesting problems and not on the basis of exercises artificially constructed and often devoid of interest (§§7.4.4, 7.4.2). One must be careful, of course, not to seem to abuse this strategy by presenting fundamental aspects of the subject (e.g. basic concepts, or difficult theorems) in the form of exercises, or problems.

The approach outlined as (1) to (4) above has distinct advantages, some of which are the following:

– Reconstructions of examples (point (3) and remark (iii) above) make it possible for students to understand the motivation for the introduction of a new concept, theory, method, or proof, and to grasp their content more profoundly (§7.2, a2, c1).

– The learner and the teacher are thus encouraged to think of themselves as pursuing their own researches (§7.2, b1, c3, d2).

– Point (2) above (identification of the crucial steps) often reveals interrelations between different mathematical and non-mathematical domains, which have a great didactic interest (§7.2, a3).

– It is possible to make the solution of problems and exercises an essential ingredient of the presentation, very helpful for a complete understanding of the subject (see point (4) and remark (iv) above). Often the interest is naturally induced by historically important and mathematically fruitful questions, without however neglecting their role as a means to improve one's knowledge of 'mathematical techniques' (§7.2, a2).

– The approach suggests several possibilities for teaching a subject, according to the specific needs of the classroom and the curriculum; e.g. emphasise the historical aspects, or specific mathematical ideas, or interrelations between different mathematical or nonmathematical domains etc (§7.2.a1).

– By points (1) and (2) above, the teacher has the opportunity to compare modern mathematics with its form in the past (notation, terminology, methods of proof and of computation, etc). Presentation of aspects of this comparison may be beneficial for the students (§7.2, b2).

– Also by points (1) and (2) above, it is possible for the teacher to look for and recognise difficulties and obstacles to the learner's understanding (§7.2, c2).

7.3.3   Mathematical awareness

We propose that mathematical awareness should include aspects related to (a) the intrinsic and (b) the extrinsic nature of mathematical activity. In this connection, history offers interesting possibilities, which are outlined in the rest of this subsection.

(a) Awareness of the intrinsic nature of mathematical activity

The history of mathematics provides opportunities to unfold, analyse and emphasise important aspects of doing mathematics, such as:

(i)   The role of general conceptual frameworks and of associated motivations, questions and problems, which have led to developments of particular mathematical domains (e.g. Tzanakis 1995; cf. §7.2, d1, e1, e2).

(ii)   The evolving nature of mathematics, both in content and in form; notation, terminology, favourite computational methods, modes of expression and representations, as well as metamathematical notions such as proof, rigour and evidence, in comparison with mathematics of today (see e.g. Barbin 1996; Kleiner 1996; cf. §7.2, b).

(iii)   The role of doubts, paradoxes, contradictions, intuitions, heuristics and difficulties while learning and producing new mathematics in the context of specific questions and problems, and the motivations for generalising, abstracting and formalising in such a context (e.g. Lakatos 1976, appendices 1 and 2; Friedelmeyer 1996; cf. §7.2, d3, c2, a1).

(b) Awareness of the extrinsic nature of mathematical activity

Mathematics is often regarded as a discipline which is largely disconnected from social and cultural concerns and influences. Its history may illustrate the superficiality of such a view. For example:

(i)   Aspects of mathematics may be seen as closely related to philosophical questions and problems, the arts (music, architecture etc), other sciences and also humanities (e.g. Montesinos Sirera 1996; Pérez 1996; cf. §7.2, a3, e3).

(ii)   The social and cultural milieu may be seen to influence the development, or delay the development, of certain mathematical domains (e.g., references in (i) above; Brin *et al.* 1993; cf. §7.2, e2).

(iii)   Mathematics is recognisably an integral part of the cultural heritage and practices of different civilisations, nations, or, ethnic groups (e.g. Cousquer 1998, Horng 1996; cf. §7.2, e3).

(iv)   Currents in mathematics education throughout its history reflect trends and concerns in culture and society (e.g. Gispert 1997; FitzSimons 1996; cf. 7.2 e).

The emphases listed above can serve as a general outline of how to start to translate some of the arguments supporting the integration of history into mathematics education, detailed in §7.2. More specific guidelines for practical implementation are detailed in the next section. For example: research projects on history texts (§7.4.2), primary sources (§7.4.3), taking advantage of errors, alternative conceptions, change of perspective, revision of assumptions, intuitive arguments (§7.4.6), famous historical problems (§7.4.7), mechanical instruments (§7.4.8), experiential mathematical activities (§7.4.9) and outdoors experience (§7.4.12).

Ways of integrating history into mathematics education clearly involve the use of sources of reference material. These materials can be roughly categorised into three types:

a) Primary source material (excerpts from original mathematical documents).

b) Secondary source material (textbooks with history narratives, interpretations, reconstructions etc).

c) Didactical source material (see below).

Historians of mathematics are, by their profession, mostly interested in the evidence supplied by primary sources, and contribute to the progress of knowledge by writing secondary materials. Teachers of mathematics (at all levels) may benefit from both primary and (perhaps more from) secondary materials and they particularly welcome the third category of didactic materials. By didactic source materials, we mean the body of literature which is distilled from primary and secondary writings with the eye to an approach (including exposition, tutorial, exercise etc) inspired by history. Of the three categories, the didactic resource material seems to be the most lacking in the public domain. Teachers of mathematics and mathematics educators are encouraged to develop, individually, or, in collaboration, their own material in this category and to make it available to a wider community.

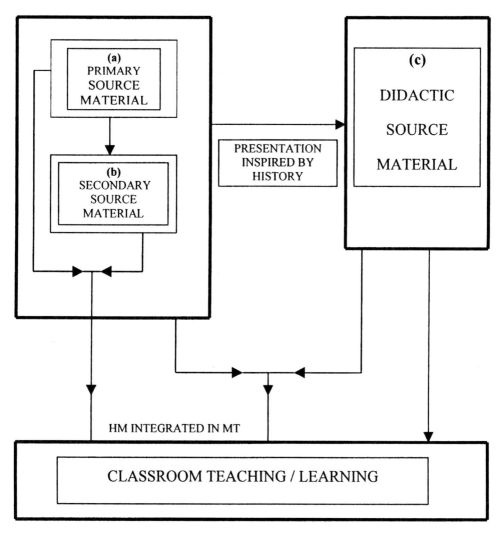

Figure 7.1: Reference materials that play a role when history of mathematics (HM) enters the classroom for mathematics teaching (MT).

The diagram above (figure 7.1) illustrates the kinds of reference material. The arrows indicate possible interconnections between the materials.

7.4   Ideas and examples for classroom implementation

In this section we survey a wide range of possible ways of implementing history in the mathematics classroom, through giving examples under each of the following headings:

1. Historical snippets
2. Research projects based on history texts
3. Primary sources
4. Worksheets
5. Historical packages
6. Taking advantage of errors, alternative conceptions, change of perspective, revision of implicit assumptions, intuitive arguments
7. Historical problems
8. Mechanical instruments
9. Experiential mathematical activities
10. Plays
11. Films and other visual means
12. Outdoors experience
13. The World Wide Web

7.4.1   Historical Snippets

Many mathematical textbooks, at all levels, have incorporated in their exposition historical information, which we call historical snippets. It is beyond the scope of this chapter to survey a representative sample of snippets from textbooks around the world. Instead, we propose a way of characterising and categorising them according to their format and their content, based on surveying a range of textbooks from various countries.

Under *format*, we consider

(a) whereabouts it is inserted in the text in relation to the mathematical exposition to which it refers: is the snippet before, during (interspersed within the text or as footnotes), alongside (in parallel to the main text but separated from it), or after the mathematical exposition?

(b) the didactical approach: is the snippet merely expository, or does it invite active involvement (a problem to solve, a notation to decipher, or proposed activities and projects)?

(c) how substantial it is: how much attention is devoted to the historical side, in comparison to the mathematical exposition? is a mathematician given just his dates, or are further and more helpful details of his life provided?

(d) style and design of the snippet: is the narrative informal, friendly, easy to read? Is it salient and distinguishable from the main text (using different colours, backgrounds, fonts)? Is it visually appealing?

Under *content*, we consider what the snippet consists of and what aspects of history it emphasises:

(e) Factual data: the snippet may consist, for example, of photographs, facsimiles of title pages or other pages of books, biographies, attribution of authorship and priorities, anecdotes, dates and chronologies, mechanical instruments, and architectural, artistic, or cultural designs.

(f) Conceptual issues: the narrative may touch upon motivation, origins and evolution of an idea, ways of noting and representing ideas as opposed to modern ones, arguments (errors, alternative conceptions etc), problems of historical origin, ancient methods of calculation, etc.

7.4.2   Student research projects based on history texts

We outline here an experience from Denmark. While this example is at university level, the principle is applicable at any level when appropriate changes are made. In the master's degree studies in mathematics at Roskilde University, project work occupies a central position. Students spend half of their time working in small groups on projects dealing with various aspects of mathematics. Each project aims at posing and answering a few research-type questions. A project typically takes 1-2 semesters to be carried out to completion, in parallel with more traditional course work. The main product of a project is a 70-150 page report written by the students in the group and defended at an oral examination with internal and external examiners. Students have to make three projects in mathematics, in one of which they consider aspects of the nature and structure of mathematics as a science with particular regard to its methods, theories and organisation so as to elucidate philosophical issues, historical developments, or the social role of mathematics. The underlying philosophy behind such projects is that every mathematics graduate, irrespective of his or her future career as a researcher, teacher, or user of mathematics, should have at least an impression of mathematics as a discipline situated in human culture and society (§7.2, e2), having a history and being related to other disciplines (§7.2, a3). In general, the research questions, investigations, and processes bear a strong resemblance to what is encountered in original and publishable research projects (§7.2, c3, b1, a4).

Examples of projects with a substantial historical component are 'Angle trisection: a classical problem', 'Euler and Bolzano: mathematical analysis from a philosophical perspective', 'The history of the theory of complex numbers', 'The genesis of non-euclidean geometry and its impact on the development of mathematics', 'The influence of Galois on the development of abstract algebra', 'The standard methods of mathematical statistics: internal and external factors in their genesis and development' and 'The early development of game theory'.

As an illustration, we outline the genesis and development of one project, on 'Cayley's problem and the early development of what later became fractals', carried out by six students, whose interest in chaos and fractals led to the project. Advised by active researchers in the field of dynamical systems with a serious interest in its history, the students decided to look at the history of 'Cayley's problem' (the problem of determining the domains of convergence for Newton's iteration method applied to a complex polynomial function). Initial investigations showed that Cayley and Schröder had studied variants of this problem independently of one another in the late 19th century.

Against this background, the following research questions were formulated for the project: How did Cayley and Schröder actually solve Cayley's Problem for a quadratic polynomial? and: Why is it so much more difficult to treat the cubic case than the quadratic case, and what was the historical evolution that led to results regarding the former case? To answer these questions, the group set out to identify and read relevant parts of secondary sources (with a historical and/or strictly mathematical content) and of original articles in the period 1870-1920 by Schröder, Cayley, Koenigs, Fatou and Julia, so as to acquire the mathematical knowledge of complex iteration necessary to understand what was going on. Based on this material, the group gave an interpretation of the historical evolution in the area under consideration and found that the reasons why it took three decades to go from degree 2 to higher degrees were essentially two-fold. Firstly, Schröder's and especially Cayley's methods were so tightly tailored to deal with quadratic polynomials that they were not generalisable to higher degrees, and, secondly, the general mathematical developments in set theory and topology of the early 20th century were a key prerequisite for the new results obtained by Fatou and Julia.

The project was reported in a 78 page report and was defended by the group at an oral examination by the supervisor and an external examiner from another university. While this represents one example of project work, appropriate for its institution and the students concerned, something similar may be devised for pupils of other ages and levels of experience.

7.4.3   Primary Sources

The centrality and importance of the use of primary sources was addressed by the Discussion Document (Question 8) which convened this ICMI study (Fauvel and Van Maanen, 1997). We do not give an example here, but refer the reader to chapter 9 for detailed discussion.

7.4.4   Worksheets

The use of worksheets is widespread in many mathematical classrooms around the world. They are meant for students to work either individually or in groups, and are of two main kinds:

(a) Worksheets which contain a collection of exercises in order to master a procedure, or consolidate a topic which was learned in the classroom and which can be worked, either in class, or at home.

(b) Worksheets which are designed as a structured and guided set of questions to introduce a new topic, a set of problems, or issues for discussion. The design usually takes into account the student's previous knowledge and by gradual questioning leads to the development of the basics of a previously unknown topic (cf. 7.3.2, point (4) and remark (iv)). These worksheets are usually meant to be used in the classroom, often in pairs or groups of students, the teacher acting as a consultant and guide. Worksheets of this kind are also used in teacher education courses.

It is this second type of worksheet that we are especially concerned with here, as they can be especially suitable for the development of mathematical understanding through the integration of history. They are appropriate at any level of schooling, including tertiary level. Arcavi 1987, Arcavi and Bruckheimer 1991, Bruckheimer and Arcavi 2000 describe single worksheets, as well as collections of worksheets, especially designed to integrate historical topics, both in classrooms and in teacher education courses.

These worksheets are usually structured around short historical extracts, accompanied by historical information to describe their context, followed by questions aimed at supporting the understanding of the contents (§7.2, a1), the discussion of the mathematical issues involved, comparison between mathematical treatments of then and now (§7.2, b), and solving problems in the extracts, or similar ones inspired by it (§7.2, a2). If the extract contains ancient notation, or notation which is foreign to the students, the questions support the translation into modern notation, by supplying partial 'dictionaries' to be completed by relying on the context.   When the extract contains more than one piece of information, the questions support the parsing of the text to cope with its apparent complexity (§7.2, c5). Critical reading of extracts is encouraged by asking students to check a calculation, discuss an argument, or simply by completing a mathematical sentence from which small pieces were deliberately omitted for this purpose (§7.2, a4). Sometimes, the worksheets bring out arguments about mathematics and its nature, or about the nature of a certain mathematical topic; the questions ask the students to elaborate arguments to support or oppose what is presented (§7.2, b1). In the case of worksheets designed for teacher education courses, the questions also ask students to address didactical issues raised by the text, their potential value for the teacher's own practice, and their similarities and differences with modern didactical approaches (§7.2, c).

The authors report on the ways they have used the worksheets, as well as on the design of the accompanying answer sheets, in which not only answers, but also further historical and mathematical consolidation are provided. Worksheets of this kind can be designed to study the history of a mathematical topic, or simply to be used as needed in the mathematical classroom for single class periods, where the topic is relevant to the curriculum.

7.4.5   Historical packages

Bruckheimer and Arcavi (2000) define 'historical packages' (or historical 'happenings') as a collection of materials narrowly focused on a small topic, with strong ties to the curriculum, suitable for two or three class periods, ready for use by teachers in their classrooms. These packages are more than historical asides, but less than a comprehensive historical approach to a large topic. Where possible, they are built around short fragments of primary sources (usually a 3-4 line quotation) and even though they are meant to be driven by the teacher, they are mostly based on the active participation of the students (§7.2, d2, b1). The role of the teacher consists of presenting the historical background needed, proposing the questions and the problems and guiding the discussion.

The package is meant to be self-contained: it provides the teacher with a folder including the detailed text of the activity, historical and didactical background, guidelines for classroom implementation, expected student reactions (based on previous classroom trials) and all the illustrative material needed in the form of pre-prepared overhead projector transparencies.   The transparencies contain (i) reproductions from original texts, pictures of mathematicians, etc. and (ii) the quotations, the problems and issues for discussion.

The aforementioned authors report on the development of five such packages: 'Ancient numerals and number systems', 'Arithmetic in ancient Egypt', 'π and the circumference of the circle', 'Word problems and equations' (described in Ofir and Arcavi, 1992), and 'Casting out nines' (described briefly below and in more detail in Bruckheimer *et al* 1995).

'Casting out nines' tells the story of a topic which in some countries has dropped out of the curriculum, but which nevertheless contains mathematically rich and relevant issues for discussion with junior high and high school students. At the same time, it gives an opportunity for teachers to enhance their mathematical literacy. The activities are organised around quotations from old textbooks, on ways to check multiplication calculations based on properties of numbers modulo 9. The quotations include statements which the students are asked to test, discuss, justify, or reject (§7.2, a4). Once the method is understood and justified, students are asked to discuss the issue of simplicity versus reliability (casting out nines provides a necessary but not sufficient condition for the correctness of the calculation; §7.2, b1). Then the discussion is moved towards checks by means of other numbers as they appear in an 18th century textbook.

As another example, we mention a package on the Pythagorean theorem, prepared mainly in English for Grade 8 students (age 14) in Hong Kong (Lit 1999). The textbook in use at school was analysed and improved by taking into account historical aspects of the subject. The package includes activities, manipulations, proofs of the Pythagorean theorem in various cultures, related problems taken from ancient scriptures, and historical narratives, as detailed below:

A. (i) Presentation of the Chinese, Egyptian, Greek and Babylonian origins of the Pythagorean theorem. (ii) Presentation of the Chinese 'water weed problem' (Bai 1990, 419-420), and the Indian 'lotus problem' (Du *et al* 1991, 572-573; for both problems see Swetz and Kao 1977, 30-33). (iii) The original writings for reference, to be used by the students.

B. Seven different proofs of the theorem by methods that appeared in various cultures in order to see the same problem in different perspectives (§7.2, e3)

C. A diagrammatic proof from ancient Chinese mathematics which illustrates that the problem can be treated not only algebraically.

D. A simplified account of an early crisis in the foundations of mathematics, concerning the discovery of irrationals and its relationship to the Pythagorean theorem (§7.2, a3).

E. Historically motivated activities such as (i) making a right-angled triangle from a string with 11 knots; or (ii) deciphering the relations between number columns of the Babylonian tablet 'Plimpton 322', given on a worksheet in Hindu-Arabic numerals (Boyer 1968; Eves 1990; cf. §7.2, a2, c4).

F. On the basis of *E*(i) above, students are asked to explain the proof of the Pythagorean theorem by geometrical dissection, which illustrates the use of manipulations.

The package contains all the necessary written documents and materials for items *A* to *F*, four articles (in Chinese) for the students to read, and brief guidelines to teachers. Before the actual implementation, the original design of the package was pre-tested in two Grade 8 classes in October 1996, leading to its revision and final design.

7.4.6   Taking advantage of errors, alternative conceptions, change of perspective, revision of implicit assumptions, intuitive arguments etc.

An advantage of implementing history in the presentation and learning of mathematics is the opportunity it presents to appreciate and make explicit use of the constructive role of (i) *errors*, (ii) *alternative conceptions*, (iii) *changes of perspective concerning a subject*, (iv) *paradoxes, controversies and revision of implicit assumptions and notions*, (v) *intuitive arguments*, that appeared historically and may be put to beneficial use in the teaching and learning of mathematics, either directly or didactically reconstructed (in the sense of §7.3.2).

Some of these have to do with a broad theme and usually span a long period, perhaps in different guises. These can help to give a perspective on the issue concerned (cf. §7.2, b1). An example of *(ii)*, for the sake of illustration, is early belief in the. commensurability of all magnitudes; of *(iii)*, rejection of negative numbers, even as late as the beginning of the 19th century; of *(iii)* and *(iv)*, 'proofs' of the 'Parallel Postulate'; of *(i)* and *(v)*, controversy over the infinitesimals and the many strange results about infinite series; of *(iv)*, paradoxes in probability theory ever since the 17th century. Here one may think of D'Alembert's calculation of 2/3 for the probability of a coin falling at least once head if tossed twice, in the French *Encyclopédie* (1754), or Bertrand's different answers, by using different methods for estimating the probability that a chord chosen at random will be larger than the side of an equilateral triangle inscribed in the circle, published in *Calcul des probabilités* (1889); see Székely 1986)). References for and fuller discussion of these examples abound in the literature.

Other examples afford an opportunity to look into the methodology of mathematical invention, e.g. exploration of the Euler-Descartes formula *V-E+F=2* in solid geometry (see Biggs *et al.* 1976, Ch.5; Lakatos 1976; Siu 1990, Ch.4).

Finally, some others may contribute to enhance understanding by (all examples here are of case (*i*), ie the historically constructive role of errors):

a)   leading to specific significant notions. An example here would be Cauchy's 'proof' of the convergence of a sequence of continuous functions; see Rickey 1995, 130-132; Siu 1997, 147-148;

b)   leading to a correct proof of a specific theorem. An example here would be Kempe's 'proof' of the Four Colour Map Conjecture and Heawood's proof of the Five Colour Map Theorem in the late 19th century; see Biggs et al. 1976, Chap 6; Siu 1990, Chap. 6.

c)   leading to both significant notions and correct theorems. An example here would be Lamé's 'proof' of Fermat's Last Theorem and Kummer's counterexample and subsequent work on cyclotomic integers, from which the notion of ideal in commutative ring theory arose; see Edwards 1977, Ch.4; Siu 1990, Ch. 6; Siu 1997, 148-149.

Now we give some details of specific examples.

(i) Errors

Here are five short examples, useful for undergraduate and/or senior high school students (§7.2, a, c4):

(1) A surviving deed from Edfu in Egypt, dating back to the 2nd century BC, gave the area of a quadrilateral as the product of the pairs of arithmetic means of opposite sides. From this the area of a triangle was deduced, as the product of the mean of two sides and one half of the third side. Students can be asked to investigate how good the formula is, when it will give a correct answer and what some special cases yield (see Eves 1990, 63, exercise 2.13(f)). They can also be asked to discuss the hypothetical historico-mathematical issue of whether the ancient Egyptians were aware of the mistake. An argument here could be that their awareness or otherwise of this fact is irrelevant, if the areas they calculated were approximately rectangular (in which case the mistake is small and the method is convenient).

(2) Archimedes (3rd century BC) obtained the formula $A = \pi ab$ for the area of an ellipse with semi-axes a, b in *On conoids and spheroids*. Hence, the ratio of the area of an ellipse to that of its circumscribed rectangle is $\pi/4$. Based on analogy, Fibonacci (13th century) argued that since the ratio of the area of a circle (which is the special case with $a=b$) to that of its circumscribed square (which is of course $\pi/4$) is equal to the ratio of the perimeter of a circle to that of its circumscribed square, the same held true for an ellipse and its circumscribed rectangle. This would yield the formula $P = \pi (a+b)$ for the perimeter of an ellipse. Students can be asked to comment on its validity, and more generally (and probably on the basis of more examples) to discuss the method of analogy in mathematics (see Siu 1990, Ch. 2; cf. Polya 1954, 77-79; and here §7.3.2, remark (ii)).

(3) In Chapter Four of *Jiu Zhang Suan Shu* ('Nine Chapters on the Mathematical Art', c.100 BC – AD 100), the volume of a sphere was said to be 9/16 that of its circumscribed cylinder (Bai 1990). In his commentary, Liu Hui (c. 250) pointed out that this is incorrect and gave further elaboration, which led to a correct formula, derived through an ingenious means by Zu Chong-Zhi and his son Zu Geng in the late 5th century. Students can be asked to compare the incorrect formula with the correct one and to guess how the 9/16 might have come about. This can lead to a discussion of the interesting principle known in the West as Cavalieri's principle, stated in 1635 (see Siu 1993, 353-354; Wagner 1978).

(4) Galileo conjectured that a heavy rope suspended from both ends hangs in the shape of a parabola (which is indeed the case if the rope suspends a plank, in the manner of a suspension bridge). This problem of the curve formed by a hanging rope was later posed by Jakob Bernoulli and solved by mathematicians of the 17th and 18th centuries, including Huygens, Leibniz and Johann Bernoulli. Huygens coined the word 'catenary' (*catena*) for this curve. Students can be asked to find the equation of a catenary and compare it with a parabola. This example can also be

used to motivate and predispose students to the study of differential equations (Rickey 1995, 127-129; cf. 7.2.a2, c1).

(5) In a 3-page paper of 1878 which appeared in the first issue of the *American journal of mathematics*, Arthur Cayley claimed that there were three groups of order 6, characterised by generators and relations:

(i)   a; $a^6 = 1$

(ii)   a, b; $a^2 = b^3 = 1$, $ab = ba$

(iii)   a, b; $a^2 = b^3 = 1$, $ab = b^2 a$, $ba = ab^2$

Students can be asked to find whether this claim is valid, and if not, to explain why some among the three are isomorphic (see Lam 1998, 363).

(ii) Alternative conceptions

The history of the notion of function, as a rule by which an element of a set is associated to exactly one element of another set, is relatively late. It came after more restricted, but intuitive conceptions were found to be insufficient (e.g. the function as a formula, cf. Euler's definition, Boyer 1959, 243; for a comprehensive account of the history of the function concept, see Youshkevitch 1976). At the high school level, this fact may help the teacher to appreciate the difficulties of his students to understand the abstract definition in depth (§7.2, c2). Even where they have been taught the rule-based definition over several years of education, students may continue to identify a function by its mode of representation, in effect a formula (Bakar and Tall 1991; Grugnetti 1994; Vinner and Dreyfous 1989).

Gottlib (1998) developed an activity for use in teacher workshops or courses, in which participants follow some of the stages of the historical evolution of the function concept, which were accompanied with details of mathematicians' struggles and rejection of new ideas. Cognitive studies that explore student conceptions and difficulties when learning functions, are also examined. Then, the teachers compare the past developments with the experiences of students who learn the concept, in order to develop appreciation of the complicated process of learning a complex concept, such as the function concept. Finally, the teachers consider didactical implications (for another such activity, see Lycée Group 1996). At a higher level, the history of the abstract definition of a function may be given in more detail, based on historically motivated questions, like Fourier's assertion that any periodic 'function' may be represented by a trigonometric series and Dirichlet's function as a famous counterexample to this assertion (Boyer 1959, 599-600; Struik 1948, 148; Kronfellner 1996; Siu 1995; §7.2, a2).

Incidentally, this example may also help teachers to see how alternative conceptions (of a function) and errors can give the problem background from which the concepts of uniform convergence of functions, Riemann integrability, and functions of bounded variation emerge as proof-generated concepts (see Lakatos 1976, 146-148; Siu 1997).

(iii) Change of perspective

The distinction between synthesis and analysis in geometry is an interesting illustration of how mathematical methods and points of view concerning the same subject are not unique. Euclid's *Elements* is the historical paradigm of synthetic exposition in geometry. According to Pappus' *Mathematical collection* ('Synagoge'), book vii, the ancient mathematicians used not only synthesis, but also analysis in their geometrical research (see Thomas 1941, 596-600; Ver Eecke 1982, 477-512; Heath 1981, 400-401; Fauvel and Gray 1987, 208-209). In the late 16th century, Viète impelled an 'analytic program', quite distinct from the Greek method of analysis, that eventually led to the creation of analytic geometry by Fermat and Descartes (Boyer 1956, Ch. 5; Mahoney 1994, Chs.2, 3; Mancosu 1992, 83-116). Elaborating on the same geometrical problem both by synthesis (in the Euclidean model) and by analysis (with the aid of algebra) may be very enlightening for understanding the roles of discovery procedures and of proof in geometry (Bos and Reich 1990; §7.2.b).

Shortly after the Cartesian method had been published, Desargues (and, to a certain extent, Pascal) gave a serious impetus to the synthetic approach with the creation of projective geometry (Kline 1963, Chs.10, 11; Gray 1987a, 16-21). The two points of view are equally acceptable, logically sound and mathematically fruitful but they are methodologically different. Comparing them leads to issues such as a unified perspective on all conic sections (circles, ellipses, parabolas and hyperbolas), whether by the analysis of Descartes (in which a conic is an equation of the second degree in two unknowns), or by the synthesis of Desargues (in which a conic is the projection of a circle); cf. §7.2, b2.

Moving on to a later period, it may also be interesting to observe how the fruitful co-operation of the two approaches, visible in the works of Monge and Carnot, was followed by a confrontation of extreme points of view in the geometers of the next generations. Brianchon, Poncelet, Chasles and Steiner supporting the exclusive use of synthetical methods, while Gergonne, Servois, Möbius and Plücker defended the supremacy of the analytical point of view (Boyer 1985, 572-585). For an account of the history of geometry in this period, see Gray 1987b.

(iv) Revision of implicit assumptions

Our example here is at university level: the efforts of the Irish mathematician William Rowan Hamilton that led him to the quaternion concept (for details and further references see Tzanakis 1995). In the early 19th century it was realised that the product z_1z_2 of two complex numbers is geometrically given by the plane rotation and multiplication of the one by the argument and norm of the other. Hamilton's motivation was to find an extension of complex numbers, so that a similar relation exists between the sought numbers and rotations and similarities in space. That his efforts were for a long time unsuccessful was partly due to his geometric perception of z_1z_2 as a rule for multiplying vectors in the plane, rather than as a (linear) operator defined by (say) z_1 on complex numbers, a useful concept that dominated mathematics much later (cf. §7.2, b). These are mathematically, but not conceptually, equivalent geometric representations. So for several years he confined himself to the study of this question for vectors in R^3. A change of perspective for z_1z_2 in the above sense readily leads to the appreciation that one has to move to R^4: a three-dimensional rotation followed by a similarity requires 4 parameters, a fact already known at that time from Euler's work on mechanics. This is a good starting point for arriving naturally at the quaternion concept, by considering the problem of determining analytically (e.g. in terms of Euler's parameters) the composition of two space rotations (cf. §7.2, a1, a2). This makes it necessary to reject the commutativity of the product of two numbers, an assumption implicit to mathematics before Hamilton's quaternions (cf. 7.2.a1). In this example, history of mathematics inspires the presentation, which appears implicitly (§7.3.2(iii)). This could also serve as an example to highlight for students some aspects of generalising from one set of number-like objects to another; for example, that some properties are lost in generalisation.

(v) Intuitive arguments

Apart from well-known examples, like the intuitive perception of the derivative of a function as a rate of change or as a slope to a curve and the integral as an area, many other examples can be given (Friedelmeyer 1996, 118-119). For instance, Bernoulli's intuitive argument of using Fermat's derivation of the law of refraction in geometrical optics in order to solve the brachistochrone problem (Dugas 1988, 254-256) can be used as a natural first step to introduce the calculus of variations (university level; cf. §7.2, a1, a2); or at the high school level, as a physically interesting problem that needs some computational skill and an understanding of elementary differential calculus (Simmons 1974 §1.6; cf. §7.2, a3). Similar comments hold for the introduction of basic concepts and theorems of vector analysis through fluid dynamics and electrodynamics (university level), by reproducing the non-rigorous, but intuitive proofs of Stokes' and Gauss' theorems given by Maxwell (1873/1954, §§ 21-24; cf. §7.2, a1, a3, a2).

Finally in this sub-section we illustrate how two of these features may come together in an example.

(ii) Alternative conceptions and (iv) revision of implicit assumptions and paradoxes

The long history of the mathematisation of infinite sets is very rich in ideas, paradoxes, alternative conceptions and revision of implicit assumptions, from Zeno's paradoxes (5th century BC), up to Zermelo's, Gödel's and Cohen's work in the 20th century (for the latter see Van Dalen and Monna 1972, 26-62). It is often intermingled with the history of related topics, like the concepts of the continuum, measure, and dimension (e.g. Stillwell 1989, Ch.20; cf. §7.2, a3). Here we comment only on the different approaches in history to the conceptually difficult subject of the cardinality of infinite sets, which may have interesting didactical implications for undergraduate students (§7.2, c2).

There are two historical ways of approaching the question of comparing the cardinal numbers of two infinite sets: inclusion and bijection. Already in 1638, Galileo fully understood and pointed out their mutual incompatibility, by referring to the bijection between the set of positive integers and its proper subset of perfect squares and by drawing the conclusion that, when dealing with the infinite, it is not possible to use words like 'bigger', 'smaller' or 'equal' (Galilei 1638/1954, 31-33). These criteria reappeared in the 19th century in the work of Bolzano, Cantor and Dedekind.

In 1851, Bolzano proposed the inclusion criterion: a set A has less elements than a set B, if A is a proper subset of B, that is if A is contained in B but not equal to B (Bolzano 1851/1950, §§ 19-21). Thus, in his own example, the set of all real numbers between 0 and 5 (0,5) has less elements than (0,12) despite the existence of a bijection between them (e.g. $5y=12x$); (Lombardo Radice 1981, III.2; Moreno and Waldegg 1991, 213-216). Later, Cantor proposed the modern criterion based on bijections: two sets A and B have the same cardinal number, if and only if there exists a bijection between A and B (Cantor 1878, 1895; cf. Fauvel and Gray 1987, 577-580). This conception opposed that of Bolzano, given Dedekind's definition (1888) of an infinite set as one possessing a proper subset with the same cardinal number as the whole set (Dedekind 1888/1963, 63; Lombardo Radice 1981, IV.1; Moreno and Waldegg 1991, 216-219). For a comprehensive account of Cantor's work in this connection, see Dauben 1979.

7.4.7   Historical problems

The history of mathematics provides a vast reservoir of problems that can be stimulating and productive for both students and teachers.   From a didactic perspective, the problems are of various kinds.
(i)   problems with no solution,
(ii)   famous problems still unsolved, or solved with great difficulty,
(iii) problems having clever, alternative, or exemplary solutions,
(iv)   problems that motivated and/or anticipated the development of a whole (mathematical) domain, or simply
(v)   problems presented for recreational purposes (distinct from the previous cases (i)-(iv) which are more closely related to the main mathematical curriculum).
Below we give a small sample that can be used in a variety of ways at various levels of instruction, and some relevant references.

(i) Problems with no solution

(a) The three famous problems of antiquity: doubling the cube, trisecting an angle, squaring the circle (and the construction of the regular heptagon); Bunt et al. 1976, Ch.4. These can also be considered as recreational problems (e.g. Dörrie 1965, §§ 35-37), or as problems which motivated developments in algebra through the algebraicisation of Euclidean constructions (e.g. Courant and Robbins 1941, Ch.III; Bunt et al. 1976, §4-8; Stillwell 1989, §§ 2.3, 5.4, 11.7).

(b) The problem of solving by radicals the general n-th degree algebraic equation for $n = 2, 3, 4, 5$ and higher (e.g. Stillwell 1989, Chs.5, 18). This also falls in (iv), if an introduction to Galois' ideas and group theory is given (cf. Bourbaki 1984, 72-73, Klein 1926-7/1979, 81-84, 99-106).

(c) The impossibility of expressing the arc length of an ellipse and a hyperbola in terms of elementary functions. This destroyed Leibniz's programme of integration in closed form. The problem may serve as an introduction to elliptic integrals and functions (and also falls in category (iv)). The same holds for the physical problem of finding in closed form the period of oscillation of a simple pendulum (e.g. Stillwell 1989, Ch.11).

(d) The impossibility of proving Euclid's parallel postulate from the other axioms, a problem that motivated the development of both Euclidean and non-Euclidean geometry (e.g. Bonola 1955).

(ii) Famous problems still unsolved, or solved with great difficulty

(a) Fermat's Last Theorem—that $x^n + y^n = z^n$ has no solutions for n greater than 2—both as an elementary problem (with proofs for specific exponents, e.g. Rademacher and Toeplitz 1990, §14; Laubenbacher and Pengelley 1999, 156-203) and questions related to its general form (e.g. Edwards 1977). For a semi-popular account, see Singh 1998.

(b) The innocent-looking Goldbach's conjecture (every even natural number is the sum of two primes), as well as, the existence of infinitely many pairs of twin primes (primes with a difference equal to 2); see e.g. Courant and Robbins 1941, 30-31.

(c) Riemann's conjecture (the zeros of the *zeta* function have all real part equal to 1/2) in several contexts; for instance, in connection with the distribution of primes (e.g. Davis and Hersh 1980, Ch.8, and in detail in Edwards 1974).

(iii) Problems having clever, alternative, or, exemplary solutions

(a) Many simple proofs of the Pythagorean theorem, which appeared in different cultures (Loomis 1972; Eves 1983, Ch.4; Nelsen 1993).

(b) Dandelin's proof of the characterisation of conic sections, considered as the intersection of a cone and a plane, as loci of points, by using two spheres tangent to the cone and the plane (e.g. Apostol 1967, §13.18).

(c) Various proofs of the fundamental theorem of algebra (e.g. Stilwell 1989, §§13.6-13.7, Dörrie 1965, §23), and more compact proofs based on the elementary theory of analytic functions (e.g. Knopp 1945, §28).

(iv) Problems that motivated and/or anticipated the development of a whole domain

As well as the examples alluded to above in (i), (iia), and (iic), we may mention

(a) The 'prime number theorem': the number of primes less than n, approaches asymptotically $n/\ln(n)$ (Courant and Robbins 1941, 25-30; Davis and Hersh 1980, Ch.5; Apostol 1976, Ch.13; Hardy and Wright 1975, Ch.XXII.) This motivated developments in number theory (Apostol 1976, 8-9).

(b) The Weierstrass polynomial approximation theorem for real continuous functions (Hairer and Wanner 1996, §III.9), which stimulated developments in approximation theory and functional analysis (see e.g. Bourbaki 1984, 257-258).

(c) The problem of stakes mentioned below in §7.4.9(d), which stimulated the development of probability theory.

(d) The problem of small vibrations of a string, and the debate in the second half of the 18th century among d'Alembert, Euler and Daniel Bernoulli concerning the determination of the general solution of the wave equation. The investigations concerning this problem were recapitulated in Lagrange's report to the Academy of Turin in 1759, *Recherches sur la nature et la propagation du son*, but the problem remained poorly understood for some time and significant progress was made after

Fourier's work (1822) on the solution of partial differential equations (especially the heat equation) with the aid of trigonometric series (Davies and Hersh 1980, Ch.5; Dieudonné 1981, §I.2; Fourier 1822/1955). This problem, apart from stimulating the development of the classical theory of Fourier analysis, also led to a discussion and gradual clarification of the concept of a function, a key development with far reaching consequences to the whole of mathematics (e.g. Struik 1948, 147-148; cf. §7.4.6 (ii)).

(e) Johann Bernoulli's brachistochrone problem: to find the trajectory of a point mass moving on a vertical plane between two fixed points, under its weight only, in least time (Courant and Robbins 1941, 383-384; Simmons 1974, §1.6). This problem and its solution—the cycloid—anticipated and motivated developments in both the calculus of variations and analytical mechanics (Dugas 1988, Ch.III.V).

(f) Closely related to (e), is the study of the cycloid in the 17th century. It constituted a source of problems that motivated developments in calculus. Roberval, Fermat and Descartes proposed ingenious methods to find the tangent through a point of the curve and they showed (together with Pascal, Torricelli) a great virtuosity in the manipulation of indivisibles in order to compute the area under the curve (Clero & Le Rest 1980, Chs. 2, 3). Its length was calculated by Wren and by Pascal, and the latter also determined the centres of gravity of several plane regions and solids associated with the cycloid (Clero & Le Rest 1980, Ch.4, 5; Dugas 1988, 186; Hairer and Wanner, 103). Huygens proved that it is the solution of the problem of the isochrone pendulum (Clero & Le Rest 1980, Ch.6; Dugas 1988, §II.V.6; cf. Sommerfeld 1964, §17) and the Bernoullis showed that it is the solution of the brachistochrone problem (Hairer and Wanner 1996, 136-137). It was a curve also studied by Newton, Wallis and Leibniz (Clero & Le Rest 1980, Ch. 8). The variety of ideas, concepts, points of view, methods and results connected with the history of this curve make it a privileged vehicle for a historical approach to the calculus.

(v) Recreational problems

Many historical examples can be found in the references below, in the *Journal of Recreational Mathematics*, or from Singmaster 1993. Some of them are:

(a) Euler's problem of finding the number of ways in which a plane convex polygon with n sides can be divided into triangles by its diagonals (Dörrie 1965, §7).

(b) Lagrange's problem of proving that any natural number is the sum of 4 squares (Rademacher and Toeplitz 1990, §9).

(c) Steiner's problem of finding the maximum of $x^{1/x}$ for real positive x (Dörrie 1965, §89).

(d) The 'five-colour problem', a much simpler version of the four colour problem (Rademacher and Toeplitz 1990, §12),

(e) 'Napoleon's problem' (and solution presented to Laplace) of specifying the centre of a given circle with the aid only of a compass (Carrega 1981, Ch.7, page 115).

7.4.8   Mechanical instruments

The introduction of mechanical instruments in the mathematics classroom is related to two interconnected problems in mathematics education: the socio-cultural development of mathematical awareness (§7.2, e, §7.3.3(b)), and building up an empirical basis for mathematical proofs (Bartolini Bussi 1998, 5; cf. §7.2, a3). It is possible to illustrate many mathematical concepts and proofs using instruments that have been devised for this purpose, for instance, drawing conic sections, or solving the ancient Greek geometrical problems.   A list of such didactically appropriate mechanical   devices   can   be   found   at   the   WWW-address http://museo.umino.it/labmat/ and further reference to related teaching experiments can be found in (Bartolini Bussi and Pergola 1996) and in §10.2.2 of this volume. Here we mention briefly a few examples.

(a) Descartes in his *Géométrie* of 1637 (Descartes 1954, 153-156), shows how to find *n* mean proportionals (geometric means) between two given lengths *a* and *b*. He gives an actual geometrical construction to do this, which can be used to build a mechanical device in order to perform the construction.   It can easily be simulated using a dynamic geometry software (Dennis and Confrey, 1997, 147-156).   The construction of this machine is not the only example we can find in Descartes.   His method to solve geometrically second degree equations (Descartes 1954, 12-17) can also be done on paper, or simulated on a computer screen.

(b) D'Alembert, in the *Encyclopédie méthodique* (1751/1987, 659-660) describes an apparatus for finding the roots of equations, which can also be simulated.

(c)   Projects   which   involve   more   sophisticated   mathematics,   namely, transformations in the plane, show to the students apparently surprising results and get them involved in the exploration of the underlying mathematics (cf. §7.2, b1, d2, c3).   For instance, Peaucellier's conversion of circular motion into linear motion is a striking example of inversion.   Hart's converter solves the same problem.   These dynamic constructions can be done both with 'real' materials and on the computer screen (see, for example, Courant and Robbins, 155-158).

(d) Finally, devices for testing experimentally the 'brachistochrone' property of the cycloid may be constructed, similar to that constructed in 18th century Paris— which still exists today, together with a bigger, modern one in La Cité des Sciences et de l' Industrie de la Villette in Paris.   Two billiard balls are left to move downwards along a straight line and a cycloidal trajectory, respectively, with the same end points.   It may be seen that the billiard ball along the cycloid arrives first to the lowest point, contrary to what may be naively guessed on intuitive grounds. A similar device for comparing the time along a straight line and a parabolic trajectory, constructed by Galileo, is now in the Science Museum of Florence, Italy (cf. §7.4.6(v), 7.4.7(iv), and Chabert 1993).

7.4.9   Experiential mathematical activities

An experiential mathematical activity would consist of re-living arguments, notations, methods, games and other ways of doing mathematics in the past. Several kinds of activity are possible, of which we mention four classes here.

(a) arguments

The teacher sets a specific question, or problem, taken from the history of mathematics and explains its importance to the scientific community in the past. Then, he encourages the students to think about it, discuss it in the classroom under his supervision, work at home alone, or in groups and re-discuss their findings and opinions (7.2.d2, a4).

For example, students might be asked to consider justifications, or indeed proofs, of Euclid's 5th ('parallel') postulate. In an experiment (Patronis 1997), the interest of 16 year old high school students was stimulated by the teacher's elaborating historical comments on the foundations of geometry. She asked them to think about the following question (cf. §7.2, a2). The existence of a line, passing through a given point and parallel to a given line, is easily proved: what about its uniqueness, which seems to be intuitively evident? This served as an intellectual challenge for some students (cf. §7.2, e1), who proceeded to re-invent arguments put forward in well known 'proofs' of the 5th postulate (cf.. §7.2, b1). Moreover, there was a lively discussion on more general meta-mathematical themes; on what is meant by proving a proposition, or what is it to make a correct mathematical assertion. The answer given by one student that "a correct proposition is one accepted by the majority of people and which does not violate certain rules", may give the teacher an opportunity to discuss further examples at this meta-mathematical level and may give hints to the students about the evolutionary nature of mathematical knowledge (cf. §7.2, b1).

(b) notations

Students can be introduced to ancient numeration systems through their notations and be given the opportunity to practice writing different kinds of numbers in these systems. By implication, they are exposed in an experiential way to the re-appreciation of the (decimal) positional numeration system, whose characteristics are taken so much for granted. By comparing and contrasting, they can analyse the hidden assumptions of their known system and its efficiency (§7.2, a2). A detailed discussion of an example, in which student teachers work on a reproduction of an old Babylonian clay tablet, is discussed in §8.3.1 (see also van der Waerden 1961, 37-45; Smith 1958, 36-39 for a description of the subject in a way adaptable to a didactic approach; Eves 1990, 19-21).

(c) methods

Students can be asked to make use of old finger reckoning methods to make simple calculations as people did in the past. For example, students can experience and practice a simple multiplication method for numbers between 6 and 10, which could make it unnecessary to learn the multiplication table for numbers greater than 5 (Smith 1958, 201):

To multiply 7 by 8, say, raise two fingers on one hand and three on the other, since 5+2=7 and 5+3=8. Then, add the numbers denoted by the raised fingers, 2+3=5, and multiply those denoted by the others, 3 × 2 = 6, and the former result is the tens, 50, and the latter is the units, the product being 56.

High school students with basic knowledge of elementary algebra can be asked why the method works in general. They can also be asked to invent and justify a similar method for numbers between 10-15 (see Smith 1958, 201-202; §7.2, b1, c3). Baumgart (1989, 120-123) provides material that can be used to introduce children to finger reckoning and operations performed using fingers. The classical Greek problems can also provide mathematical activities that could be experiential: Baumgart (1989, 199-200) describes several constructions to trisect an angle (Archimedes' *neusis*, or using the conchoid, etc) which can be reproduced by the students, using simple classroom materials for the tomahawk or hatchet construction with cardboard (see also Eves 1990, 114-115; Aaboe 1964, 108-109). Another example would consist of solving quadratic equations graphically following the methods of Al-Khwarizmi, or Descartes, either with paper and pencil or with dynamic geometry software (Jones 1969, 260-263; Descartes 1954, 12-17).

(d) games

It is often claimed (see e.g. Boyer 1968, 397) that the starting point for the modern theory of probability can be found in correspondence between Fermat and Pascal about the following problem: two gamblers are playing for a stake, which is to go to the one, who first wins n points, but the play is interrupted, when the first has made p points and the second q points. It is required to know how to divide the stakes. Students can be asked to replay a dice game, interrupt it and discuss the ways in which the stakes should be divided between the players. Students can also be asked to play ancient games (for example, games taken from Bell and Cornelius 1988) and analyse their strategies, possible implicit mathematical ideas and the socio-cultural context in which the games appeared (7.2.a4, 7.2.e).

7.4.10   Plays

Plays are usually integrated in education in general as a way to enact human situations, perhaps to illuminate moral or ethical or social quandaries; thus they are usually not associated with mathematical classrooms. History of mathematics nevertheless provides an opportunity to incorporate the use of plays in at least two different ways. Fuller discussion of this area will be found in §10.2.1.

(a) Plays can be designed to re-experience the life of mathematicians in the past, as a way to appreciate the human side of mathematical activity (§7.2, d1). Ponza (1998) carried out such an experiment with her high school students to encourage them in their mathematical studies by researching and reviving episodes of the turbulent and short life of Galois (cf. Ponza 1996). T. Limnaiou, (as reported in the Hellenic Society's Report, 1998) described theatre plays, which were performed in the evenings at school with the participation of students, teachers and parents. The plays were based on ancient Greek texts and the historical comments included in the mathematics textbooks.   She reports mixed results: on the one hand, it was stimulating for the students, yet many teachers argued that "this is not mathematics".

(b) Plays can be designed to re-enact famous arguments in history, to let students revive not only the human aspects of the history of mathematics, but also mathematical issues, as if they were their own (§7.2, b1, d2).   Such plays may be constructed by the class, or the work of other teachers utilised in this context (cf Hitchcock 1992).   Boero and Tizzani 1997, Garuti 1997, Boero *et al.* (1998), describe a teaching experiment and make a theoretical analysis in which it is suggested that by echoing historical voices, students may identify their own and other's conceptions.

7.4.11   Films and other visual means

Films related to the history of mathematics can highlight the human, cultural and social context of mathematics and mathematicians, and/or mathematical ideas, developments and arguments (§7.2, e).   There are only a few movies which are played commercially in theatres.   One of them (the Swedish film *The hill on the dark side of the moon*), was on the life of Sofia Kovalevskaya.   There are some TV programs about mathematics and mathematicians, which were aired in public channels.   For example, the Public Television Broadcasting net in the USA (through the Public station WQED of Pittsburgh) produced in 1998 a collection of seven one-hour videotapes, under the name 'Life by the Numbers'. Devlin (1998, p. vii), in the accompanying book to the video collection, describes the series as being   about 'everyday life and the role played in everyday life by mathematics'. The series presents a diverse group of individuals (scientists, artists, athletes, medical researchers and others), describing their creative and surprising ways of using mathematics to explore the world and improve life.

Films have also been developed with clearly didactical intentions for classroom use, with a strong focus on history.   A notable example is *The Tunnel of Samos*. It is notable because it merges the historical and the mathematical aspects of the tunnel construction, shares with the viewers the consideration of historical hypotheses based on mathematical arguments, and makes use of the graphical and visual power of the media to illustrate the mathematical principles involved (for some details on the mathematical aspects of the subject, see §9.5.1 in this volume; for the historical problem, see Van der Waerden 1961, 102-104. The film has a duration of 30', produced within "Project Mathematics" in 1995, in the California Institute of Technology, Caltech 1-70, Pasadena, CA 91125). In the UK the Open University has also made a number of films about the history of mathematics, which appear regularly on the BBC.

Visual means, other than movies, include posters displaying portraits of mathematicians, facsimiles of famous works, time charts with chronological, or thematic historical developments.   Among the posters, 'Magic Maths' is a series of mathematical stereograms developed by J. Shanks and C. Daniel (Available from the authors at: Otago Maths Education Centre, Dep. of Maths and Stats, University of Otago, P.O. Box 56, Dunedin, New Zealand). These stereograms offer visual appeal and interest to students of all abilities and ages, hiding images which students can interpret and interconnect.

7.4.12   Outdoor experiences

The mathematics of outdoors experiences refers, among other things, to the identification of forms and shapes, patterns in nature, in architecture (past and present) and in art (§7.2, a3).   Exploring historical outdoor instruments with students, such as navigational and surveying equipment in order to learn trigonometry, is another such set of experiences (Kiely 1947).   The term could also refer to visits to museums of science which display mathematical exhibits of different kinds, some of which may include historical background.   In the following we describe a unique example of an outdoors experience taken from Japanese culture.

The Chinese mathematical tradition inspired the development of mathematics in Japan.   This gradually led to its regeneration in the early 17th century.   Wa-San, the original Japanese mathematics, is to be distinguished from European or Western mathematics (Smith and Mikami 1914; Mikami 1913/1974; Ogura 1993; Rothman and Fukagawa 1998): the name *Wa-San* is composed of two Chinese characters: *wa* meaning 'Japan' and *san* meaning 'arithmetic' or 'calculation'.   During the 250 years of the Edo Age, many professional mathematicians established their own schools and developed original numerical mathematics, while the school system for laymen, called 'Terakoya', spread all over Japan and it was through this type of school that laymen learned Wa-San. Since laymen could not afford to publish their problems and solutions, as was common practice among mathematicians of the period, they posted

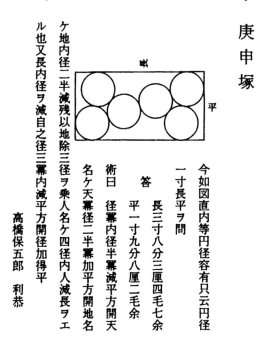

Figure 7.2: If the radius of each circle is 1, what is the length of the rectangle's sides? (Mitiwaki et al 1987, Ohtake 1974)

them in 'San-Gaku', a kind of bulletin board in the temples or shrines of several regions, until about 80 years ago.

Many of these problems can be solved by modern college, or even secondary school, mathematics, but at that time they were solved numerically. Figure 7.2 shows such a problem on a 'San-Gaku'.

Some of these materials are still preserved, and published also in English (Fukagawa and Pedoe, 1989, Fukagawa and Sokolowsky in press). In the Gunma prefecture in Japan there are 74 such bulletin boards. High school mathematics clubs tour these sites during summer vacations, or holidays and report on their investigations at their school cultural festival (Okada 1957, 1975, Ohtake 1974). The problems and their answers are usually provided by the board, but the solution process usually is missing, or it is very sketchy. Thus, student activities consist of understanding the problem, re-creating a solution procedure and checking their answers with those provided by the bulletin board (§7.2, a1). The student methods are very different from those used by the problem authors, since they apply modern western mathematics to solve them. Nevertheless students still have to decipher the statement and the solution of the problem given in the board (§7.2, c5). By having in mind that these problems and their solutions were the creation of ordinary people, students not only exercise problem solving, but they also may have an opportunity to demystify the subject, and to connect it to popular cultural practices (§7.2, e3).

7.4.13 The WWW

The World Wide Web (Internet) can help the integration of history in mathematics education in at least two ways: as a resource, and as a means of communication. The resource aspect is dealt with quite fully in §10.3 below; here we focus on the communication dimension, with an example from Israel of how the Internet can be used to deliver and support entire courses.

Zehavi (1999) has implemented a course for in-service teachers on the history of negative numbers, based on the materials developed in the Weizmann Institute (for a description of the materials see Arcavi *et al.* 1982), taking advantage of the hypertext facilities to explore links according to the user's decision. The course is restricted (by means of a personal password) to those teachers enrolled. Access to an outline of the 10 chapters of the course is possible at

```
http://www.weizmann.ac.il/sci-tea/math/open.htm.
```

After a first face to face meeting, participants work from their homes, they answer the questions from the materials, receive comments and after a certain date, full answers are provided by the course administrators and made available to the participants through the web. As well as being part of an electronic forum for discussion of issues, raised by the learning materials and reflecting on the answers by their peers, teachers are provided with an email address for technical support. Such courses demand a heavy administration, but at the same time, allow the regular updating of materials online and attention to many individual needs.

Bibliography and references for chapter 7

The following abbreviations are used in this bibliography:

Proc. HEM Braga for: *História e Educação Matemática, Proceedings of the 2ⁿᵈ European summer university on history and epistemology in mathematics education and the ICME-8 satellite meeting of the International study group on the relations between history and pedagogy of mathematics,* Braga, Portugal: 24-30 July 1996, 2 volumes.

LR for: preprint, available from the author, submitted to the 'ICMI Study Conference', 19-26 April 1998, CIRM, Luminy, Marseilles, France and included in *The Luminy Reader: Discussion texts for the ICMI Study Conference on 'The role of the History of Mathematics in the teaching and Learning of Mathematics'* (two sets of papers, Spring 1998).

Aaboe, A. 1964. *Episodes from the early history of mathematics*, New York: Random House

Apostol, T. 1967. *Calculus* I, Waltham MA: Xerox

Apostol, T. 1976. *Introduction to analytic number theory*, New York: Springer

Arcavi, A. 1985. *History of mathematics as a component of mathematics teachers background*, Ph.D. Thesis, Rehovot: Weizmann Institute of Science

Arcavi, A. 1987. 'Using historical materials in the mathematics classroom', *Arithmetic teacher* 35 (4), 13-16

Arcavi, A., Bruckheimer, M. 1991. 'Reading Bombelli's x-purgated algebra', *College mathematics journal*, 22 (3), 212-219

Arcavi, A., Bruckheimer, M., Ben-Zvi, R. 1982. 'Maybe a mathematics teacher can profit from the study of the history of mathematics', *For the learning of mathematics* 3 (1), 30-37

Arcavi, A., Bruckheimer, M., Ben-Zvi, R. 1987. 'History of mathematics for teachers: the case of irrational numbers', *For the learning of mathematics* 7 (2), 18-23.

Bai, S-S. (ed.) 1990. *Nine chapters on the mathematical art: a translation in modern Chinese* (in Chinese), Shandong: Shandong Educational Press

Bakar, M., Tall, D. 1991. 'Students' mental prototypes for functions and graphs', *Proc. of the 15th PME Conference* i, Assisi, Italy, 104-111.

Barbin, E. 1996. 'Can proof be taught?', in E. Barbin, R. Douady (eds.), *Teaching mathematics: the relationship between knowledge, curriculum and practice*, Pont-a-Mousson: Topiques Éditions, 195-210.

Barbin, E. 1997. 'Histoire et enseignement des mathématiques: pourquoi? comment?', *Bulletin AMQ*, Montréal, 37 (1), 20-25.

Bartolini Bussi, M. G. & M Pergola 1996. 'History in the mathematics classroom: linkages in kinematic geometry', in H. N. Jahnke, N. Knoche & M. Otte (eds), *History of mathematics and education: ideas and experiences*, Göttingen: Vandenhoeck and Ruprecht, 39-67

Bartolini Bussi, M. G. 1998. 'Drawing instruments: historical and didactical issues', in *LR*

Barwell, M. 1913. 'The advisability of including some instruction in the school course on the history of mathematics', *The mathematical gazette*, 7, 72-79

Bell, R., Cornelius, M. 1988. *Board games around the world: a resource book for mathematical investigations*, Cambridge: University Press

Biggs, N.L., Lloyd, E.K., Wilson, R.J. 1976. *Graph Theory: 1736-1936*, Oxford: University Press

Boero, P., Garuti, R., Pedemonte, B., Robotti, E., 1998. 'Approaching theoretical knowledge through voices from history and echoes in the classroom', in *LR*.

Boero, P., Tizzani, P., 1997. 'La chute des corps de Aristote à Galilée: voix de l'histoire et echos dans la classe', *Proc. of CIEAEM-49*, Setubal (in press).

Bolzano, Bernard 1851/1950. *Paradoxes of the infinite*, D.A. Steele (tr.), London: Routledge and Kegan Paul

Bonola, R., 1955. *Noneuclidean geometry: a critical and historical study of its development*, New York: Dover

Bos, H. J. M., Reich, K. 1990. 'Der doppelte Auftakt zur frühneuzeitlichen Algebra: Viète und Descartes', in: E. Scholz (ed), *Geschichte der Algebra*, Mannheim: Bibliographisches Institut, 183-234.

Bourbaki, Nicolas 1984. *Éléments d'histoire des mathématiques*, Paris: Masson

Boyer C. B. 1956. *History of analytic geometry*, Princeton: University Press

Boyer, C.B. 1959. *The history of the calculus and its conceptual development*, New York: Dover

Boyer C. B. 1968. *A history of mathematics*, New York: Scripta Mathematica, 1985 and Princeton: University Press 1968.

Brin, P., Gregoire, M., Hallez, M., 1993. 'Approche pluridisciplinaire de la naissance de la perspective' in: Lalande *et al.* 1993, 195-214.

British Ministry of Education 1958. *Pamphlet* **36**, London: H.M.S.O., 134-154.

Brousseau, G. 1983. 'Les obstacles épistémologiques et les problèmes en mathématiques' *Recherche en didactique des mathématiques* **4**, 165-198.

Brown, H.I. 1977. *Perception, theory and commitment: the new philosophy of science*, Chicago: University Press

Bruckheimer, M., Ofir, R., Arcavi, A. 1995. 'The case for and against casting out nines', *For the learning of mathematics* **15** (2), 24-35.

Bruckheimer, M., Arcavi, A. 2000. in V. Katz (ed.), *Using history to teach mathematics: an international perspective*, Washington, D.C: Mathematical Association of America.

Buhler, M. 1990. 'Reading Archimedes' measurement of a circle', in J. Fauvel (ed), *History in the mathematics classroom: the IREM papers*, Leicester: Mathematical Association, 43-58.

Bunt, L.N.H., Jones, P.S., Bedient, J.D. 1976. *The historical roots of elementary mathematics*, Englewood Cliffs: Prentice Hall

Cantor, Georg 1878. 'Ein Beitrag zur Mannigfaltigkeitslehre', *Journal für die reine und angewandte Mathematik* **84** (1878), 242-258, reprinted in Cantor's *Gesammelte Abhandlungen*, Berlin: Springer 1930, 119-133.

Cantor, Georg 1895. 'Beiträge zur Begründung der transfiniten Mengenlehre', part I, *Mathematische Annalen* **46** (1895), 481-512, reprinted in Cantor's *Gesammelte. Abhandlungen*, Berlin: Springer 1930, 282-351.

Carrega, J.C. 1981. *Théorie des corps, la règle et le compas*, Paris: Herman

Chabert, Jean-Luc 1993. 'Le problème brachistochrone', in *Histoire des problèmes, histoire des mathématiques*, Commission Inter-IREM, Paris: Ellipses, 179-198; tr. (by C. Weeks) as *History of mathematics histories of problems*, Inter-IREM Commission, Paris, Ellipses 1997, 183-202

Chandrasekhar, S. 1987. *Truth and beauty: aesthetics and motivations in science*, Chicago: University Press

Clero, J.P., Le Rest, E. 1980. *La naissance du calcul infinitesimal au XVIIème siècle*, Paris: Éditions du C.N.R.S.

Courant, R., Robbins, H., 1941. *What is mathematics? an elementary approach to ideas and methods*, Oxford: University Press

Cousquer, Éliane 1998 *La fabuleuse histoire des nombres*, Paris: Diderot

D' Alembert, J. 1751/1987. *Encyclopédie méthodique: mathématiques*, Paris: ACL Éditions

Dalen, D. van, Monna A. F., 1972. *Sets and integration: an outline of the development*, Groningen: Wolters-Noordhoff

Dauben, J.W., 1979. *Georg Cantor: his mathematics and philosophy of the infinite*, Cambridge: Harvard University Press

Davis, P.J., Hersh, R., 1980. *The mathematical experience*, Boston: Birkhäuser

Dedekind R., 1888/1963. *Essays in the theory of numbers*, New York: Dover

De Morgan, A. 1865. 'Speech of Professor De Morgan', *Proc. London Math. Soc.* **1**, 1-9.

Dennis, David & Jere Confrey 1997. 'Drawing logarithmic curves with Geometer's sketchpad: a method inspired by historical sources', in J R King and D Schattschneider (eds), *Geometry turned on! Dynamic software in learning, teaching and research*, Washington: MAA, 147-156

Descartes, René 1637/1954. *The geometry of René Descartes*, (tr.) D.E.Smith, M.L. Latham, New York: Dover

Devlin, K. 1998. *Life by the numbers: the companion to the PBS series*, New York: J. Wiley and Sons

Dieudonné Jean 1981. *History of functional analysis*, Amsterdam: North Holland

Dörrie, H. 1965. *100 great problems of elementary mathematics: their history and solution*, New York: Dover

Du, R-Z., Wang, Q-J., Sun, H-A., Shao, M-H., Qi, Z-P. 1991. *A concise dictionary of history of mathematics* (in Chinese), Shandong: Shandong Educational Press.

Dugas, R. 1988. *A history of mechanics*, New York: Dover

Edwards, H.M. 1974. *Riemann's zeta function*, New York: Academic Press

Edwards, H.M. 1977. *Fermat's last theorem: a genetic introduction to algebraic number theory*, New York: Springer

Ernest, Paul (ed) 1994. *Constructing mathematical knowledge: epistemology and mathematics education*, London: Falmer Press

Eves, Howard 1990. *An introduction to the history of mathematics*, 6th edn, Philadelphia: Saunders

Eves, Howard 1983. *Great moments in mathematics: before 1650*, Washington: MAA

Fauvel, John 1991. 'Using history in mathematics education', *For the learning of mathematics* **11** (2), 3-6

Fauvel, John, Gray, Jeremy (eds) 1987. *The history of mathematics: a reader*, London: Macmillan

Fauvel, John & Van Maanen, Jan, 1997. 'The role of the history of mathematics in the teaching and learning of mathematics: Discussion document for an ICMI Study (1997-2000)', *Educational studies in mathematics* **34**, 255-259

FitzSimons, Gail 1996. 'Is there a place for the history and pedagogy of mathematics in adult education under economic rationalism?', *Proc. HEM Braga* ii, 128-135.

Flashman, M.E., 1996. 'Historical motivation for a calculus course: Barrow's theorem' in R. Calinger (ed.) *Vita mathematica: historical research and integration with teaching*, Washington: MAA, 309-315.

Fourier, J.B. 1822/1955. *The analytic theory of heat*, tr. A. Freeman, New York: Dover

Freudenthal, H. 1983. *Didactical phenomenology of mathematical structures*, Dordrecht: Reidel

Friedelmeyer, J-P. 1990. 'Teaching 6th form mathematics with a historical perspective' in J. Fauvel (ed.), *History in the mathematics classroom: the IREM papers*, Leicester: Mathematical Association, 1-16.

Friedelmeyer, J-P. 1996. 'What history has to say to us about the teaching of analysis', in E. Barbin, R. Douady (eds), *Teaching mathematics: the relationship between knowledge, curriculum and practice*, Pont-a-Mousson: Topiques Éditions, 109-122.

Fukagawa. H. & Pedoe, D. 1989. *Japanese temple geometry problems*, Winnipeg: Charles Babbage Research Foundation

Fukagawa, H., Sokolowsky, D. in press. *Traditional Japanese mathematics problems from the 18th and 19th centuries*, Singapore: Science Culture Technology Publishing.

Galilei G. 1638/1954. *Dialogues concerning two new sciences*, New York: Dover

Garuti, R. 1997. 'A classroom discussion and a historical dialogue: a case study', *Proc. of PME-XXI* ii, Lahti, 297-304

Gispert, H. 1997. 'Les mathématiques dans la culture scientifique et la formation des enseignants', in J. Rosmorduc (ed.), *Histoire des sciences et des techniques*, Centre Régionale de Documentation Pédagogique de Bretagne, 347-355.

Glaisher, J.W.L. 1890. 'Presidential Address to the British Association for the Advancement of Science' in R.E. Moritz (ed.) *On mathematics and mathematicians*, New York: Dover 1942, 96.

Gottlib, O. 1998. 'The development of the concept of function: integrating historical and psychological perspectives', in R. Even et al (eds.), *Functions resource file* (in Hebrew), Department of Science Teaching, Weizmann Institute of Science, Rehovot

Gray, Jeremy 1987a. *European mathematics in the early seventeenth century*, Milton Keynes: The Open University

Gray, Jeremy 1987b. *Projective geometry and the axiomatization of mathematics*, Milton Keynes: The Open University

Grugnetti, Lucia 1994. 'Difficulties and misconceptions of the concept of function', L. Bazzini (ed.), *Proc. 5th international conference on systematic cooperation between theory and practice in mathematics education*, ISDAF, Pavia, Italy, 101-109

Grugnetti, L. 1998. 'The history of mathematics and its influence on pedagogical problems', in *LR*

Hadamard, J. 1954. *The psychology of invention in the mathematical field*, New York: Dover

Hairer, E., Wanner, G. 1996. *Analysis by its history*, New York: Springer

Hallez, M. 1990. 'Teaching Huygens in the rue Huygens' in J. Fauvel (ed), *History in the mathematics classroom: the IREM papers*, Leicester: Mathematical Association, 97-112

Hardy, G.H., Wright, E.M. 1975. *An introduction to the theory of numbers*, 4th edn, Oxford: University Press

Heath Thomas L. 1981. *A history of Greek mathematics* **ii**, New York: Dover

Hellenic Society for the History of Science and Technology 1998. 'Report on the meeting on History of mathematics and mathematics education: a theoretical approach', in *LR*; *Diastasi* **3**, 1998, 123-130 (in Greek).

Hiebert, J., Carpenter, T.P., 1992, 'Learning and teaching mathematics with understanding', in D.A. Grouws (ed.), *Handbook of research on mathematics teaching and learning*, New York: MacMillan.

Hitchcock, Gavin 1992, 'Dramatizing the birth and adventures of mathematical concepts: two dialogues', in R. Calinger (ed.), *Vita mathematica: historical research and integration with teaching*, Washington: Mathematical Association of America 1996, 27-41

Horng, Wann-Sheng 1996. 'Euclid versus Liu Hui: a pedagogical reflection', *Proc. HEM Braga* ii, 404-411.

Horng , Wann-Sheng 1998. 'History of mathematics and learning algebra: A brief report', in *LR*

Jones, Phillip S., 1969. 'The history of mathematics as a teaching tool' in *Historical topics for the mathematics classroom*, Reston, Va: NCTM (31st Yearbook), 1-17

Jozeau, M-F., 1990. 'A historical approach to maximum and minimum problems', in J. Fauvel (ed), *History in the mathematics classroom: the IREM papers*, Leicester: Mathematical Association, 25-42.

Keily, Edmond R. 1947. *Surveying instruments: their history and classroom use*, New York: Columbia University (NCTM 19thYearbook)

Klein, F., 1914/1945. *Elementary mathematics from an advanced stand point*, New York: Dover.

Klein, F. 1926-7/1979. *Development of mathematics in the 19th century*, Brookline MA: Mathematics and Science Press.

Kleiner, Israel 1996. 'A history-of-mathematics course for teachers, based on great quotations', in R. Calinger (ed.) *Vita mathematica: historical research and integration with teaching*, Washington: MAA, 261-268

Kline M. 1963. *Mathematics: a cultural approach*, Massachusetts: Addison-Wesley

Kline M. 1972. *Mathematical thought from ancient to modern times*, New York: Oxford University Press

Kline, M. 1973. *Why Johnny can't add: the failure of the new mathematics*, New York: St. Martin's Press

Knopp, K. 1945. *Theory of functions, part I*, New York: Dover

Kragh, H. 1990. *Dirac: a scientific biography*, Cambridge: University Press

Kronfellner, Manfred 1996 'The history of the concept of function and some implications for classroom teaching', in R. Calinger (ed.) *Vita mathematica: historical research and integration with teaching*, Washington: MAA, 317-320

Lakatos, I. 1976. *Proofs and refutations*, Cambridge: University Press

Lalande, F., Jaboeuf, F., Nouazé, Y., (eds.) 1993. *Proceedings of the First European Summer University on the History and Epistemology in Mathematics Education*, Montpellier: IREM de Montpellier 1993

Lam, T.Y. 1998. 'Representations of finite groups: a hundred years, part I', *Notices of AMS* **45**, 361-372.

Laubenbacher, Reinhard & Pengelley, David 1999. *Mathematical expeditions: chronicles by the explorers*, New York: Springer

Leake, L. 1983. 'What every mathematics teacher should read: twenty four opinions', *The mathematics teacher* **76**, 128-133.

Lefort, Xavier 1990. 'History of mathematics in adult continuing education' in J. Fauvel (ed), *History in the mathematics classroom: the IREM papers*, Leicester: Mathematical Association, 85-96

Le Goff, J-P. 1996. 'Cubic equations at secondary school level: following in Euler's footsteps', in E. Barbin, R. Douady (eds), *Teaching mathematics: the relationship between knowledge, curriculum and practice*, Pont-a-Mousson: Topiques Éditions, 11-34.

Lit, Chi Kai 1999. *Using history of mathematics in junior secondary school classroom: a curriculum perspective* (in Chinese), MPhil Thesis, Hong Kong: Chinese University of Hong Kong

Lombardo Radice, L. 1981. *L'infinito*, Roma: Editori Riuniti.

Loomis, E. S. 1972. *The Pythagorean proposition*, Washington, DC: NCTM, orig. pub. 1927

Lycée group, IREM de Clerrmont-Ferrand 1996. 'Introducing the concept of a function' in E. Barbin, R. Douady (eds), *Teaching mathematics: the relationship between knowledge, curriculum and practice*, Pont-a-Mousson: Topiques Éditions, 1996, 159-170

Maanen, Jan van 1991. 'L' Hôpital's weight problem', *For the learning of mathematics* **11** (2), 44-47

Mahoney, M. S. 1994. *The mathematical career of Pierre de Fermat 1601-1665*, Princeton: University Press

Mancosu Paolo 1992. 'Descartes's *Géométrie* and revolutions in mathematics', in D. Gillies (ed), *Revolutions in mathematics*, Oxford: Clarendon Press, 83-116

Martín, C.M. 1996. 'Géométrie et mouvement: un exemple d'autoformation' in *Proc. HEM Braga* ii, 73-79.

MAA 1935. 'Report on the training of teachers of mathematics', *American math. monthly* 8, 57-61.

Maxwell, J.C. 1873/1954. *A treatise on electricity and magnetism* i, New York: Dover

Menghini, M. 1998. 'Possible co-operation between researchers in history of mathematics and researchers in didactics of mathematics: some reflections' in *LR*

Mikami, Yoshio 1913/1974. *The development of mathematics in China and Japan*, 2nd edn, New York: Chelsea (1st edition 1913).

Miller, G.A. 1916. *Historical introduction to mathematical literature*, New York: MacMillan.

Mitiwaki, Y., Ohtake, S., Ohyama, M., Tanaka, M., Hamada, T. 1987. *The San-Gaku of Gunma*, Takasaki: Jobu-Insatu (in Japanese)

Montesinos Sirera, José, 1996. 'The 'Seminario Orotava de historia de la ciencia': an interdisciplinary work group', in *Proc. HEM Braga*, ii, 261-264

Moreno, L.E., Waldegg, G. 1991. 'The conceptual evolution of actual mathematical infinity', *Educational studies in mathematics* 22, 211-231

NCTM 1969. *Historical topics for the mathematics classroom*, Reston, Va: National Council of Teachers of Mathematics (31st NCTM Yearbook, reprinted 1989)

Nelsen, R.B. 1993. *Proofs without words*, Washington: MAA

Nouet, Monique 1996. 'Using historical texts in the Lyceé' in E. Barbin, R. Douady (eds), *Teaching mathematics: the relationship between knowledge, curriculum and practice*, Pont-a-Mousson: Topiques Éditions, 125-138

Ofir, Ron 1991. 'Historical happenings in the mathematics classroom', *For the learning of mathematics* 11 (2), 21-23

Ofir, R., Arcavi, A. 1992. 'Word problems and equations: an historical activity for the algebra classroom', *Mathematical gazette* 76 (475), 69-84.

Ogura, K. 1993. *Wasan*, English tr. by Norio Ise, Tokyo: Kodansya Ltd, 1993 (Japanese edition by Iwanami Shoten, Tokyo, 1940).

Ohtake, S. 1974. 'San-Gaku of Kou-Shin-Tho' (in Japanese), *News of Annaka*.185, 12

Okada, Y. 1957. 'A research into San-Gaku by the mathematics club' (in Japanese), *Bulletin of Tatebayashi Girls' High School* 4, 36-40

Okada, Y. 1975. 'The activity of the mathematics club in Tatebayashi High School' (in Japanese), *Newsletter of Japanese History of Mathematics in Gunma*, 10, 1

Patronis, A. 1997. 'Rethinking the role of context in mathematics education', *Nordic studies in math. educ.* 5 (3), 33-46

Pérez, Maria, 1996. 'The Proyecto Helena', in *Proc. HEM Braga*, ii, 326-330.

Poincaré, H. 1908. *Science et méthode*, Paris: Flammarion. English tr. by F. Maitland (1914), reprinted by Thoemmes Press, Bristol 1996.

Polya, G. 1954. *Induction and analogy in mathematics*, Princeton: University Press

Polya, G. 1968. *Patterns of plausible reasoning*, Princeton: University Press

Ponza, M.V. 1996. 'La experiencia interdisciplinaria en la realidad educativa de hoy', Revista SUMA (Spain) 21, 97-101.

Ponza, M.V. 1998. 'A role of the history of mathematics in the teaching and learning of mathematics: an Argentinian experience', *Mathematics in school* 27 (4), 10-13.

Rademacher, H., Toeplitz, O. 1990. *The enjoyment of mathematics*, New York: Dover

Radford, L; Guérette, G 1996. 'Quadratic equations: reinventing the formula. A teaching sequence based on the historical development of algebra' in *Proc. HEM Braga* ii, 301-308.

Ransom, P., Arcavi, A., Barbin, E., Fowler, D. 1991. 'The experience of history in mathematics education', *For the learning of mathematics* **11** (2), 7-16.

Rickey, V.F. 1995. 'My favourite ways of using history in teaching calculus', in F. Swetz et al (eds), *Learn From the Masters!* Washington: MAA, 123-134.

Rickey, V.F. 1996. 'The necessity of History in teaching Mathematics' in R. Calinger (ed.) *Vita mathematica: historical research and integration with teaching*, Washington: MAA, 251-268.

Rodriguez, Michel 1998. 'L'histoire des mathématiques, pourquoi je m' y intéresse... et ce que j' en fais', in *LR*

Rogers, Leo 1998. 'Ontogeny, phylogeny and evolutionary epistemology', in *LR*

Rothman, T., Fukagawa, H. 1998. 'Japanese Temple Geometry', *Scientific American*, May, 62-69.

Schoenfeld, A.H. 1992, 'Learning to think mathematically: problem solving, metacognition and sense making in mathematics', in D.A. Grouws (ed.), *Handbook of research on mathematics teaching and learning*, New York: MacMillan

Schoenfeld, A.H., Smith, J. and Arcavi, A. 1993. 'Learning: the microgenetic analysis of one student's evolving understanding of a complex subject matter domain', in R. Glaser (ed.) *Advances in instructional psychology* **4**, New Jersey: Erlbaum, 55-175

Schubring, G. 1978. *Das genetische Prinzip in der Mathematik-Didaktik*, Stuttgart: Klett

Schubring, G. 1988. 'Historische, Begriffsentwicklung und Lernprozess aus der Sicht neuerer mathematikdidaktischer Konzeptionen (Fehler, 'Obstacles', Transposition)', *Zentr. Didaktik für der Mathematik* **20**, 138-148.

Sierpinska, A. 1991. 'Quelques idées sur la méthodologie de la recherche en didactique des mathématiques, liée a la notion d'obstacle épistémologique', *Cahiers de didactique des mathématiques* (Thessaloniki, Greece) **7**, 11-28.

Sierpinska, A. 1994. *Understanding in mathematics*, London: The Falmer Press

Simmons, G.F. 1974. *Differential equations with applications and historical notes*, New Delhi: McGraw Hill

Singmaster, David, 1993, *Sources in recreational mathematics*, (sixth preliminary edition), London: South Bank University

Siu, Man-Keung 1990. *Mathematical proofs* (in Chinese), Nanjing: Jiangsu Ed. Press

Siu, Man-Keung 1993. 'Proof and pedagogy in ancient China: examples from Liu Hui's commentary on Jiu Zhang Suan Shu', *Educ. studies in math.*, **24**, 345-357.

Siu, Man-Keung 1995. 'Concept of function: its history and teaching', in F.J.Swetz et al. (eds.), *Learn from the masters!*, Washington: MAA, 105-121.

Siu, Man-Keung 1997. 'The ABCD of using history of mathematics in the (undergraduate) classroom', *Bull. Hong Kong Math. Soc.* **1**, 143-154.

Siu, Man-Keung 1998. 'The (in)complete quadrangle: historians of mathematics, mathematicians, mathematics educators and teachers of mathematics', Plenary talk at the ICMI Study Conference, 19-26 April 1998, CIRM, Luminy, Marseilles, France.

Smith, D. E., Mikami, Y. 1914. *A history of Japanese mathematics*, Chicago: Open Court Publishing Company

Smith, D.E. 1958. History of mathematics ii: special topics of elementary mathematics, New York: Dover Publications

Sommerfeld, A. 1964. *Mechanics*, New York: Academic Press

Stewart, Ian 1989. *Les mathématiques*, Paris: Pour la Science

Stillwell, John 1989. *Mathematics and its history*, New York: Springer

Struik, D.J. 1948. *A concise history of mathematics*, New York: Dover

Swetz, F.J., Kao, T.I. 1977. *Was Pythagoras Chinese? an examination of right triangle theory in ancient China*, Pennsylvania State University Press

Székely, G.J. 1986. *Paradoxes in probability theory and mathematical statistics*, Dordrecht: Reidel

Thomas, Ivor 1941. *Selections illustrating the history of greek mathematics* ii, Loeb Classical Library, London: Heinemann and Cambridge: Harvard University Press.

Toeplitz, O. 1927. 'Das Problem der Universitätsvorlesungen über Infinitesimalrechnung und ihre Abgrenzung gegenüber der Infinitesimalrechnung an der höheren Schulen', *Jahrberichte der Deutschen Mathematiker Vereinigung* **36**, 88-100.

Toeplitz, O. 1963. *Calculus: a genetic approach*, Chicago: University Press

Tzanakis, C. 1995. 'Rotations, complex numbers and quaternions', *Int. J. Math. Educ. Sci. Technol.* **26**, 45-60.

Tzanakis, C. 1996. 'The history of the relation between mathematics and physics as an essential ingredient of their presentation', *Proc. HEM Braga* ii, 96-104

Tzanakis, C. 1997. 'The quest of beauty in research and teaching of mathematics and physics: an historical approach', *Nonlinear analysis, theory, methods and applications* **30**, 2097-2105.

Tzanakis, C. 1998. 'Discovering by analogy: the case of Schrödinger's equation', *European J. of physics* **19**, 69-75.

Tzanakis, C. 1999. 'Unfolding interrelations between mathematics and physics, motivated by history: two examples', *Int. J. Math. Educ. Sci. Technol.* **30**, 103-118.

Tzanakis, C., 2000. 'Presenting the relation between mathematics and physics on the basis of their history: a genetic approach', in V. Katz (ed.), *Using history to teach mathematics: an international perspective*, Washington, D.C: Mathematical Association of America

Tzanakis, C., Thomaidis, Y., to appear. 'Integrating the close historical development of mathematics and physics in mathematics education: Some methodological and epistemological remarks', *For the learning of mathematics.*

Vasco, C. E. 1995. 'History of mathematics as a tool for teaching mathematics for understanding' in D.N. Perkins et al. (eds.) *Software goes to school*, New York: Oxford University Press, 56-69

Ver Eecke, P. 1982. *Pappus d' Alexandrie, la collection mathématique* ii, Paris: Albert Blanchard

Vinner, S., Dreyfus, T. 1989. 'Images and definitions for the concept of function', *Journal for research in mathematics education* **20** (4), 356-366.

Van der Waerden, B.L. 1961. *Science awakening* i, New York: Oxford University Press

Wagner, D.B. 1978. 'Liu Hui and Tsu Keng-Chih on the volume of a sphere', *Chinese science* **3**, 59-79.

Youschkevitch, A.P. 1976. 'The concept of function up to the middle of the 19th century' *Archive for history of exact sciences* **16**, 37-85.

Zehavi, N. 1999. 'History of mathematics on the web: from flow of information to significant learning'. Preprint, Rehovot: Weizmann Institute of Science.

Chapter 8

Historical support for particular subjects

Man-Keung Siu

with Giorgio T. Bagni, Carlos Correia de Sá, Gail FitzSimons, Chun Ip Fung, Hélène Gispert, Torkil Heiede, Wann-Sheng Horng, Victor Katz, Manfred Kronfellner, Marysa Krysinska, Ewa Lakoma, David Lingard, João Pitombeira de Carvalho, Michel Rodriguez, Maggy Schneider, Constantinos Tzanakis, Dian Zhou Zhang

Abstract: *This chapter provides further specific examples of using historical mathematics in the classroom, both to support and illustrate the arguments in chapter 7, and to indicate the ways in which the teaching of particular subjects may be supported by the integration of historical resources.*

8.1 Introduction

Some of the ways in which history of mathematics can help mathematics students, teachers and researchers were examined in the previous chapter. Reasons were put forward for concluding that history can help us to

(i) grasp more profoundly the meaning of concepts, theories, methods and proofs in mathematics;
(ii) identify crucial steps, difficulties and obstacles in the evolution of a subject;
(iii) organise teaching better and provide motivation for the study of a subject;
(iv) build up a reservoir of examples, problems and alternative viewpoints about a subject;
(v) appreciate mathematics better as a creative process;
(vi) see mathematics as a human endeavour which is related to other human activities;
(vii) maintain an open and humane attitude towards the study of mathematics.

Many examples were given to illustrate various ways of implementing the integration of history of mathematics with mathematics teaching in mathematics

John Fauvel, Jan van Maanen (eds.), *History in mathematics education: the ICMI study*, Dordrecht: Kluwer 2000, pp. 241-290

education. While these many examples are certainly illustrative and useful, they are by design, in order to illustrate a wide range of implementations, a potpourri with only sketchy descriptions. In this chapter we will offer a further list of selected examples from classroom teaching experience, to be accompanied by discussion more detailed than could be afforded in chapter 7.

To help readers better orientate their attention and interests we discuss these examples against the background of a three-dimensional framework:
1. the level of the curriculum, which in most countries has a tripartite layering, from primary or elementary school (from age 6 to 12) to secondary or high school (from age 12 to 18) up to university or college (from age 18 and beyond);
2. the mathematical topic within the curriculum, such as algebra, geometry, analysis, probability theory, etc.;
3. the ways by which history of mathematics is integrated with mathematics teaching in mathematics education.

Clearly, a historical example will not often have a set of clear-cut coordinates in this three-dimensional framework. The same topic may be presented at different levels (often to different depth) or with the historical content integrated in different ways, and the same example may involve different areas of mathematics. In fact, as we witness time and again in history, many instances of mathematical development arose from or resulted in the fruitful marriage of different areas of mathematics. Thus, readers are requested to regard this framework only as a rough schematic tool in a broad sense rather than as a strict compartmentalisation. One example (§8.4.7) is even selected to display how the same piece of historical material can be used at different levels in different subject areas for different purposes.

Of course, there are as many different ways to integrate history of mathematics into classroom teaching as there are teachers. Different teachers have different styles, hold different beliefs and place emphasis on different aspects, despite the fact that they all agree on the value of history of mathematics—and even on this point teachers may differ in their conception of what history means, not to mention the different views a historian of mathematics, a mathematician and a mathematics teacher may adopt on this issue! This is natural and not a bad thing: variety implies richness, which when gathered under combined effort will yield a fuller vista. Hence, instead of attempting to tailor the variety of examples contributed by different authors into one uniform mould, we prefer to retain the individual style and emphasis, while grouping the examples into a more structured whole in a format closely related to the general directions and emphases given in section 7.2 and 7.3 of chapter 7. Name(s) of contributing author(s) are attached to each section.

Section 8.2 consists of examples of teaching specific topics in which history of mathematics inspires the whole structure of the teaching. Section 8.3 includes examples that unfold the evolving nature of mathematics, both in content and in form, as well as to present (some small piece of) mathematics in the context of different cultures. Section 8.4 treats some specific examples from various areas of mathematics and levels of the curriculum. Finally, section 8.5 highlights the social and cultural aspects of mathematics in a broad sense. Readers are requested not to interpret this structure too inflexibly—sometimes examples in one section can equally be placed in another. Certainly such a small list of examples can hardly do justice to the wide variety of possible ways of integrating history in teaching and learning mathematics. But we hope to exhibit a wide coverage so as to stimulate other teachers across the world to think of more examples and to make available further didactical source material.

This chapter, then, is both supplementary and complementary to chapter 7: supplementary in the sense that it provides further specific examples to support the arguments presented there; complementary in that the historical dimension of teaching, as illustrated mainly through the content of these examples, complements the practical implementations described in section 7.4 of chapter 7. With this in mind, we present the examples for the convenience of the reader by giving clearly marked references to relevant sections in chapter 7 in bold face such as **7.2.c2**.

8.2   Teaching projects inspired by history

8.2.1   Examples from algebra and analysis

Marysa Krysinska, with the collaboration of Christiane Hauchart

Where history inspires the presentation of mathematics, there can be a global reorganisation of the conventional, deductively organised teaching approach (**7.3.2**). The two examples which follow are taken from the Belgian teaching project *De Question en Question* (DQQ), which led to a series of textbooks with the same title for the first 4 years of high school (Thomas-Van Dieren and Rouche 1993; GEM 1996). In these textbooks the teaching approach is heuristic, in this sense: a sequence of problems and problem-situations is given, on the basis of which new concepts are progressively constructed, in order to solve a problem or to provide a proof. The foundational questions and their answers appear at the end, in contrast with conventional textbooks. In elaborating these textbooks, history played a dominant role.

Negative numbers

Negative numbers appear in the project in the context of several models, like gain and loss, debt and credit, stairs up and down. Above all, they are used for locating the points on a straight line. Their addition is introduced by means of

John Fauvel, Jan van Maanen (eds.), *History in mathematics education: the ICMI study*, Dordrecht: Kluwer 2000, pp. 243-245

- two graded rulers, sliding along each other, thus providing a mechanical device for addition;
- successive movements of the rulers forward and backward on the number line, when the numbers to be added exceed the gradation of the rulers.

In history, the multiplication of negative numbers constituted a very important epistemological obstacle **(7.2.c2)**. Today's students share in experiencing such obstacles. In the DQQ textbook, after it is noticed that the intuitive models which work for the addition of negative numbers do not work for their multiplication, the multiplication of negatives is introduced by trying to extend the multiplication table of positive numbers to negative ones and conserving the observed regularities. This is done by introducing, first the multiplication of a positive number by a negative one, and then the multiplication of two negative numbers. In the first case, multiplication of a positive number by a negative one is expressed by the succession of two geometric transformations. For instance, multiplication by (-2) means, to take the opposite and to multiply by 2, or the reverse. In the second case, the product of two negative numbers is found by extending geometrically a linear function table, e.g. (3,-6), (2,-4), (1,-2), to negative values of x; that is, in our example, by computing $y = -2x$ for negative values of x.

History suggests that the conceptual extension from positive to negative numbers is facilitated **(7.2.a1)** in the context of analytical geometry. Freudenthal (1983) observed that mathematicians who applied Descartes' method could no longer avoid allowing the letters to take negative values. If straight lines are to be described algebraically in their totality, or curves described algebraically in all cases, negative values of the variables must necessarily be admitted.

Functions

Functions appear as a means to give a model of phenomena in various contexts. For instance, in the DQQ textbook a model of the dependence of the stopping distance of a car on its velocity v is given. From a data table, we observe first that the stopping distance is a sum of the "thinking distance" d_T and the "braking distance" d_B. We represent the data graphically by vertical sticks, in the manner of Nicole Oresme (Calinger 1995, 253-260; Clagett 1959, Ch.6]. This representation suggests the laws of dependence: the d_T-graph is a straight line, while the d_B-graph looks like a parabola; after this, we verify in the tables that d_T/v and d_B/v^2 are constant.

This teaching approach takes into account the following historical facts:

- Proportionality, ratio conservation and linearity are fundamental concepts which can lead to the discovery of non-linearity (Freudenthal 1983)
- Historically, the study of motion has been closely related to the emergence of the function concept, and more generally, to the development of analysis (Boyer 1959, Chaps 4,5). In the DQQ project this favourable relation is preserved **(7.2.a1)**.
- The representation of functions by the use of vertical sticks (cf. Oresme) makes algebraic operations with functions easier. The representation of any magnitude $f(x)$ (length, area, volume, time, etc) by a line segment seems to support the function concept (Souffrin and Weiss) **(7.2.a1)**.

- In contrast with the usual teaching approach to analysis (in which the function concept, in its abstract form as a relation, is introduced right from the beginning), distinguishing the independent from the dependent variable leads to the less general, but more intuitive, conception of a function as a rule by which the second is expressed in terms of the first (**7.4.6**(ii) and references therein).

References for §8.2.1

Boyer, C.B. 1959. *The history of the calculus and its conceptual development*, New York: Dover

Calinger, R. (ed.) 1995. *Classics of mathematics*, Englewood Cliffs: Prentice-Hall

Clagett, M. 1959. *The science of mechanics in the middle ages*, Madison: University of Wisconsin Press

Freudenthal, Hans 1983. *Didactical phenomenology of mathematical structures*, Dordrecht: Reidel

GEM 1996. *De question en question* 3 (Groupe d'Enseignement Mathématique), Bruxelles: Didier Hatier

Souffrin, P., Weiss, J.P., 'Remarques à propos de la traduction du traité des configurations des qualitiés et des mouvements', *Cahiers du Seminaire d'Epistemologie et d'Histoire des Sciences de Nice*, No. 19

Thomas-Van Dieren, F., Rouche, N., 1993. *De question en question* 1, Bruxelles: Didier Hatier

8.2.2   A heuristic introduction to analysis implicitly inspired by its historical development

Maggy Schneider

The following is an outline of examples taken from the Belgian teaching project *Approche Heuristique de l'Analyse* done by the Groupe AHA (consisting of P. Bolly, A. Chevalier, M. Citta, C. Hauchart, M. Krysinska, D. Legrand, N. Rouche, M. Schneider) (Groupe AHA 1999), concerning the last two years of high school and in which history inspires the presentation (**7.3.2**).

An introduction to the concept of the instantaneous rate of change of a quantity

This introduction is based on the study of problems like this:

A pump is filling up a conical vase with an angle 90° at the vertex, in such a way that the level h of the water increases by 1 cm per minute. At what stage does the flow of the pump reach 100 cm^3/min?

This problem motivates students to test the applicability of the concept of a steady flow, and to lead them gradually to understand that the flow is continuously increasing. Hence, they begin to understand that it is necessary to cut the time into smaller and smaller intervals. With more working they may come to see that this problem leads to the idea of a new calculus (putting $\Delta t = 0$ in the expression for the

John Fauvel, Jan van Maanen (eds.), *History in mathematics education: the ICMI study*, Dordrecht: Kluwer 2000, pp. 245-248

average flow of water in the interval $[t, t + \Delta t]$) and the physically intuitive assertion (correct if numerically $h = t$) that the flow is 100cm^3/min when the cross-section of the vase is 100cm^2. Finally, in order to help the students overcome their uneasiness about the boldness of this new calculus, and their hesitation to accept the concept of instantaneous flow, a thought experiment is proposed (based on a reformulation of the intuitive assertion above, adaptable to the case h is different from t), which may convince them that the result obtained in this way is exact. It consists of a comparison of the flow in a conical vase with a steady flow in a cylinder, in particular cases. Epistemological and didactic aspects of this problem are analysed in Schneider 1992.

The concept of a tangent line

The concept of a tangent line is originally studied separately from the concept of velocity. Calculations of lines tangent to polynomial functions, obtained by linear approximations, challenge the intuitive geometric idea that students have for a tangent line, namely a straight line intersecting the curve as a whole only at this point. Subsequently, the connection between linear approximations and differential quotients is established by the formal similarity of the results obtained from two problems, one about velocities and another about tangents. The instantaneous velocity and the slope of a tangent, though *a priori* conceptually different, appear henceforth as two aspects of the same concept. Then, the calculation of instantaneous rates of change helps the determination of the slope of tangents to the graph of non-polynomial functions, without using a linear approximation (for details see Grand'Henry-Krysinska and Hauchart 1993).

The limit concept

One gets closer to the concept of the limit in order to prove that an infinite filling yields an exact result for a curvilinear area or a volume. This can be seen, for example, when one fills up the area under $y = x^3$ from $x = 0$ to $x = 1$ using rectangles. The sum of the area of n circumscribed rectangles gives the sequence $(1 + 2/n + 1/n^2)/4$. The sum of the area of $n - 1$ inscribed rectangles gives the sequence $(1 - 2/n + 1/n^2)/4$. These two sequences have the same limit 1/4. However, not all students are convinced that this limit is the exact area sought. Hence, it is interesting to propose a proof, on the basis of which aspects of the abstract concept of a limit are built up. Suppose that this area is bounded above and below by the two given sums above. Therefore, the area under $y = x^3$ cannot equal $1/4 + \varepsilon$ with ε as small as we like, for by subdividing the interval into a sufficiently large number of segments, we may make the sum of the area of circumscribed rectangles to be between 1/4 and $1/4 + \varepsilon$. Hence the area sought is larger than one of its approximations from above, which is a contradiction. Similarly, the area cannot be equal to $1/4 - \varepsilon$.

As shown in Schneider 1988, such a proof prepares the way to the (ε, N) formulation of the concept of the limit of a sequence. For example, the quantifiers $\forall$ and $\exists$ and the traditional way in which they are put forward ($\forall$... $\exists$....), appear as a kind of 'watermark' in this proof. Namely, on the one hand, one has to verify inequalities *whatever* the value of ε. On the other hand, the contradiction appearing in this proof follows already from the *existence* of a value of n fulfilling a given condition. Of course, the limit concept is constructed here only in an implicit and sketchy way. But actually, this proof is based on the possibility of finally getting the term of each sequence above as close as one wishes to 1/4. Thus, it leads the students to appreciate the technical role to be played later by the abstract rigorous formulation of the limit concept.

In several respects, this approach is inspired by history (**7.3.2**):

— *Order and choice of topics* (**7.2.a1**). The concepts (velocities, tangents, areas and volumes) are first introduced in physical and geometrical contexts without any *a priori* connection between each other. This connection is gradually made evident: slopes of tangents and velocities are two aspects of the derivative concept; area and volume calculations appear later as the reverse procedure of the computation of derivatives. Finally, the general relationship between these problems is established through the rigorous formulation of the limit concept. Studying the properties of this concept, one is led to sharpen the nature of the numbers used.

— *The evolutionary nature of the form of mathematics* (**7.2.b2**). In the project, limits are first defined as results obtained by cancelling terms with Δt or $1/n$ as was done in the 17th century. At the end of the project, limits are defined in terms of "ε, δ".

— *The evolutionary nature of mathematical research activity* (**7.2.b1**). The initial motivation is the solution of problems. In this way, the new calculus, though not a rigorously founded method, is nevertheless a powerful tool, which produces new results. And then the new method is validated by thought experiments for velocities and linear approximations for tangents. Later, concepts of the final (rigorous) theory are proof-generated concepts (in the sense of Lakatos 1976, Appendices 1, 2), that is, concepts mathematically sharpened in order to meet the requirements of a rigorous proof.

— *Epistemological obstacles* (**7.2.c2**). This approach takes into account difficulties encountered by mathematicians in the past, and by students today (even after a first course in analysis). For example, some students seem unwilling to accept the concept of an instantaneous rate of change of a quantity, feeling rather that to obtain a flow, it is necessary that a small volume remains. Others are sceptical about calculating the exact value of a curvilinear area by cancelling terms in the sum of areas of rectangles, because their conception of a limit rests on their visual perception of magnitudes, where rectangles have to be gradually narrowed until they become real line segments (Schneider 1988).

— *Intellectual style of classical works* (**7.2.c4**). Newton's kinematic arguments have inspired this approach: the idea of variation, hence the idea of a differential quotient, was originally introduced more easily in terms of velocity than in terms of the tangent concept. Other problems leading to the Fundamental Theorem of Calculus are also inspired by Newton's kinematic arguments. The proof by *reductio ad absurdum* described above is inspired by the ancient Greek exhaustion method.

References for §8.2.2

Grand'Henry-Krysinska, M., Hauchart C. 1993. 'Réflexions épistémologiques à propos du concept de tangente à une courbe', in: IREM de Montpellier, *Actes de la 1ère Université d'été européenne d'histoire et d'épistémologie dans l'éducation mathématique*, 431-442
Groupe AHA 1999. *Une approche heuristique de l'analyse*, Louvain-la-Neuve: De Boeck
Lakatos, I. 1976. *Proofs and refutations: the logic of mathematical discovery*, Cambridge: University Press
Schneider, M. 1988. *Des objets mentaux « aire » et « volume » au calcul des primitives*, thèse de doctorat, Université de Louvain-la-Neuve
Schneider, M. 1992. 'A propos de l'apprentissage du taux de variation instantané', *Educational studies in mathematics* **23**, 317-350

8.2.3   How may history help the teaching of probabilistic concepts?

Eva Lakoma

In the past 20 years, there has been a continuous evolution of ideas on the nature of mathematics and its teaching and learning. Instead of simply transmitting the definitions of basic concepts, presenting the formal structure of mathematical theories and giving some straightforward applications, it is now accepted that mathematics teaching at all educational levels should also stimulate students' interest and promote their abilities to use mathematics as a language for communicating and describing mental, physical, or social phenomena. This point of view, however, requires that mathematics teaching must take into account the actual cognitive development of the students (cf. Freudenthal 1983; Sierpinska 1994, 1996). Just to supply a simpler version of already-made mathematics is insufficient to provide a didactical approach compatible with the above-mentioned point of view (7.1, 7.2). This point is especially important in the domain of stochastics (as we call probability theory and statistics). Probabilistic concepts cannot be understood in depth by simply giving their logical connections to other concepts and their place in modern probability theory, founded for instance on Kolmogorov's axioms, which are too abstract to be understood by students. Didactically, a heuristic (non-axiomatic) approach is needed, which presents stochastics as a live part of mathematics, making possible the solution of real problems by describing real situations on the basis of simple models which have a great explanatory value (Lakoma 1990).

John Fauvel, Jan van Maanen (eds.), *History in mathematics education: the ICMI study*, Dordrecht: Kluwer 2000, pp. 248-252

The
cup of probability

fortuna   probare

casus   probus

haphazard   probatus

veritas   probabilis

truth probabilitas

fair share

fair stake

expectatio

probability

verisimilis

verisimilitudo

frequency

by chance

value

possible

probable

equipossible

equiprobable

event frequency

independence probability

Figure 8.1 The 'cup of probability' points at the historical development of the dual probability concept. On the left side are the mental objects that correspond with the aleatory aspect, on the right side the epistemological nature of probability is represented. Before 1660 the two aspects existed independently, from Pascal's time on they join, leading to the 20th century foundation through the notion of independence and Kolmogorov's axioms.

A knowledge of the ways by which probabilistic thinking naturally appears and of the peculiarities of understanding probabilistic concepts is extremely desirable. In this kind of educational research, it is very helpful, and turns out to be fruitful, to take into account the historical development of this domain (Lakoma 1999a). Giving a historical perspective to the teaching of stochastics is helpful, not only for exploring and understanding the student's ways of probabilistic thinking (**7.2.a1**), but also for inspiring the design of a teaching approach to stochastics, the Local Models Approach (LMA), which takes into account the student's actual cognitive abilities (Lakoma 1990; 1996; 1998; 1999b; 1999c; **7.2.c2** and **7.3.2**).

An analysis of the historical phenomenology of probabilistic concepts led to the following conclusions that have been taken into account in the LMA (Lakoma 1990).

(a) The concept of probability has a dual nature

(i) an *epistemological* aspect, implied by the general state of our knowledge of a given phenomenon. It is related to the degree of our belief, conviction or confidence on an argument concerning this phenomenon, and which is supported by this argument.

(ii) an *aleatory* aspect, related to the physical structure of the random mechanisms under consideration and with their tendency to produce stable relative frequencies of events (for the historical analysis see Hacking 1975).

The "chance calculus" is based on (i) and the "frequency calculus" on (ii). History shows that a necessary condition for understanding probability is to make explicit its dual nature. The analysis of original or reconstructed probabilistic reasonings shows (Hacking 1975; Lakoma 1992) that from Pascal's era onwards, both these aspects are inseparable, affecting each other deeply. This analysis suggests that emphasising only one of these aspects, or treating them separately in teaching, prevents students' understanding of the fundamentally dual nature of the probability concept. Also see figure 8.1 for a schematic representation of the historical development of the concept of probability.

(b) There is an interplay between the concepts of probability and of expectation

Development of the dual nature of probability goes in parallel with the emergence and establishment of the concept of expectation (expected value). This appears already in the pre-Pascal period in both the probabilistic reasonings used and the content of the problems studied, which were focused on the estimation of chances for winning in a game or on the distribution of a stake, naturally anticipating the notion of expectation. In a more sophisticated form, expectation appears around 1660 in the reasoning of Pascal and Huygens, who also appreciated the dual nature of the probability concept (Freudenthal 1980; Hacking 1975; Todhunter 1865). Apparently, the emergence of the concept of expectation and its careful distinction from probability made probability calculus more understandable and clear for many people in the past, thus enhancing its development. Therefore, didactically, probability and expectation can and should be introduced and developed in parallel, while always keeping in mind that it is necessary to distinguish and contrast them; e.g. one may consider probabilistic problems in connection with answering not only the question "How often?" but also the question "Is it worthwhile"

(c) Local models are useful

Historically, the probability concept emerged through the solution of concrete problems, arising from the needs of everyday life. In order to solve them, people tried to observe the random phenomenon from which the problem arose, to discover some (if any) of its regularities and to develop arguments sufficient for providing conclusions related to the answer to the problem. These activities were connected with a theoretical modelling of the random phenomenon, which emphasised some of its features, neglecting others as less important. The conclusions drawn from such models were not considered as absolute, but were tested in practice, confirming or questioning the validity of the model. Hence, the natural activities on which learning probabilistic concepts could be based is the determination, formulation and search for solution of concrete problems, according to the scheme (1) discovery of the problem; (2) formulation of the problem; (3) construction of a model representing the real phenomenon under consideration; (4) analysis of the model; (5) comparison of the results obtained with the real situation. Originally, students build models, adequate only for concrete random phenomena and having an explanatory value, which may be called local models. At a more advanced level, these models become mathematically more sophisticated and general, appropriate for the description of a whole class of phenomena.

(d) Problems with a finite and infinite probability space should be presented

All the old probabilistic problems considered in the pre-Pascal era could be described in modern terminology in terms of a finite probability space. The problem of 'waiting for the first success' (one of the simplest problems described by an infinite probability space) appeared originally in Huygens's *De ratiociniis in ludo aleae* (1657) (Hacking 1975, Ch.11). Why was such a natural problem posed so late? The answer seems to be connected with the establishment of the dual nature of the probability concept, and the emergence of the concept of expectation. It is worthwhile to notice that the first attempt to solve this problem went back to Cardano's *De ludo aleae* (ca. 1550) (Hacking 1975, Ch. 6), who anticipated this dual nature and the concept of expectation. By means of such examples involving infinite probability spaces, it may be possible to help students realise this dual nature of the probability concept. Accordingly, in teaching probability, it would be good not only to present problems formulated in finite probability spaces, but also problems in infinite probability spaces, which are naturally stimulating for the students and which can be solved by methods that take into account the student's actual level of cognitive development.

The conclusions (a)-(d) have been taken into account at secondary school level in several cases which have been carefully analysed (Lakoma 1990; 1998), suggesting that the LMA teaching strategy has a positive effect at this level.

References for §8.2.3

Freudenthal, Hans 1980. 'Huygens' foundations of probability', *Historia mathematica* **7**, 113-117

Freudenthal, Hans 1983. *Didactical phenomenology of mathematical structures*, Dordrecht: Reidel

Hacking, Ian 1975. *The emergence of probability*, Cambridge: University Press

Lakoma, E. 1990. *Local models in probability teaching* (in Polish), doctoral thesis, Department of Mathematics, Informatics and Mechanics, Warsaw University

Lakoma, E. 1992. *The historical development of the probability concept* (in Polish), Warsaw: CODN-SNM

Lakoma, E. 1996. 'Local models in stochastics teaching: an example of post-modern approach to mathematics teaching', in *Proceedings of the 8th European Seminar on Mathematics in Engineering Education SEFI-8*, Prague: Technical University, 123-127

Lakoma, E. 1998. 'On the interactive nature of probability learning', in *Proceedings of the 49th International Commission for the Study and Improvement of Mathematics Education CIEAEM-49*, Setubal: University, 144-149

Lakoma, E. 1999a. 'On the historical phenomenology of probabilistic concepts: from the didactical point of view', *Proceedings of 7è Université d'Été d'Histoire des Mathématiques, July 1997*, Nantes: IREM des Pays de la Loire, 439-448.

Lakoma, E. 1999b. 'The diachronic view in research on probability learning and its impact on the practice of stochastics teaching', in *Proceedings of the 50th International Commission for the Study and Improvement of Mathematics Education CIEAEM-50*, Neuchatel: University, 116-120

Lakoma, E. 1999c. 'Del cálculo probabilístico al razonamiento estocástico: un punto de vista diacrónico' in *Uno: revista de didactica de las matematicas* 22, 55-61

Sierpinska, A., 1994. *Understanding in mathematics*, Falmer Press, London

Sierpinska, A., 1996. 'The diachronic dimension in research on understanding in mathematics: usefulness and limitations of the concept of epistemological obstacle', in: H.N. Jahnke *et al.* (eds.), *History of mathematics and education: ideas and experiences*, Göttingen: Vandenhoeck Ruprecht

Todhunter, I. 1865. *A history of the mathematical theory of probability*; repr New York: Chelsea, 1965

8.2.4   Trigonometry in the historical order

Victor Katz

Most modern trigonometry texts begin the subject by defining the basic trigonometric ratios, calculating these ratios using some elementary geometry for 30, 45, and 60 degrees, and then assume that students will use calculators to find the trigonometric ratios for any other value. Students are thus led to believe that their calculators are "magic boxes" with little people inside measuring sides of triangles. Furthermore, when the half angle, sum, and difference formulas are derived, students wonder what their purpose is.

It is much more natural to adapt the original order of treatment of Ptolemy (or even Copernicus) and develop the subject in a manner inspired by history (Aaboe 1964, Ch.4; **7.3.2**). That is not to say that one should work with chords. There are

John Fauvel, Jan van Maanen (eds.), *History in mathematics education: the ICMI study*, Dordrecht: Kluwer 2000, pp. 252-253

good reasons for using sines (and cosines and tangents). But given the definition of a trigonometric ratio, one can use geometry to determine these values, not only for 30, 45, and 60 degrees, but also for angles of 18, 36, 54, and 72 degrees. The initial goal of the course is, then, to calculate, at least in principle, the values of sine, cosine, and tangent for every integral-valued angle from 1 to 90 degrees. So one derives the half angle and difference formulas and uses them to calculate. Students will soon realise that it is not possible to calculate the sine of 1 degree exactly, however. (Teachers may well want to relate this impossibility to the question of trisecting an angle (**7.2.a3**).) However, if students notice that the sine function is essentially linear for small values, one can then approximate the sine of 1 degree to a reasonable level of accuracy and then use the sum formulas to calculate in principle the sine (and cosine and tangent) of any angle of an integral number of degrees. (Again, the linearity of the sine function for small values is an important idea for later use.) With the trigonometric tables now calculated, one can use them to solve triangles of various types. In particular, another goal of the trigonometry course, again one based on history, should be to solve spherical triangles as well as plane triangles. After all, the major use of trigonometry, from the time of its invention, was to solve spherical triangles related to astronomy. Such questions still prove to be of great interest to students. (See relevant chapters in (Katz 1998) for a more detailed account.)

References for §8.2.4

Aaboe, A. 1964. *Episodes from the early history of mathematics*, New York: Random House
Katz, V. J. 1998. *A history of mathematics: an introduction*, 2nd ed., New York: HarperCollins

8.3   Cultural aspects of mathematics in a historical perspective

8.3.1   Number systems and their representations

8.3.1.1   Counting and symbol systems

Gail FitzSimons

Historical support is valuable for students of all ages; not least, adults seeking to develop their numeracy skills. For instance, it is usually fascinating for mature-age students to reflect on the number system currently used. Where did the digits come from? How have they evolved? What about zero? What brought about the change from Roman numerals? What is the history of the evolution of symbols indicating the decimal point, percentage, basic operations, index numbers etc? What is the

John Fauvel, Jan van Maanen (eds.), *History in mathematics education: the ICMI study*, Dordrecht: Kluwer 2000, pp. 253-254

history of vulgar (common) fractions? What about the term "vinculum"? What about counting systems in non-European cultures: finger reckoning (e.g. Smith 1958), body counting systems (e.g. Bishop 1995), etc? Such questions (generated by teachers or students) may lead to the realisation that mathematics as we know it was not always there, and that there is a historical necessity across different cultures in the creation of mathematical solutions to mankind's challenges and problems (7.2.e). Smith 1958 provides an excellent resource for teachers, both in text and in illustrations.

From counting systems it is but a short step to representations in the form of calendars: "probably the idea of applying the syntax of causal and temporal

Figure 8.2: A simplified Aboriginal seasonal calendar
(source: Davis, Harris & Traynor 1980, 3)

chaining to arithmetic and geometry was the origin of mathematics as we know it" (Schweiger 1994, 300). Much work has been done on the history of calendars, yet inspection of an Australian Aboriginal calendar of the seasons (figure 8.2), for example, will illustrate a totally different world view (FitzSimons 1992).

References to §8.3.1.1

Bishop, A. J. 1995. 'What we can learn from the counting systems research of Dr. Glendon Lean', keynote address to the *International Study Group for the Relations of History and Pedagogy of Mathematics*, Cairns, June/July 1995

David, S., Harris, J., Traynor, S. 1980. 'Community based science programs for Aboriginal schools', *Developing Education* 7(4), 2-10

FitzSimons, Gail E. 1992. 'Mathematics as a way of knowing', *Vinculum* 29(1), 4-12

Schweiger, F. 1994. 'Mathematics is a language', in D.F. Robitaille, D. H. Wheeler, C. Kieran (eds) *Selected lectures from the 7th International Congress on Mathematical Education*, Université Laval, Québéc City, 297-309

Smith, D. E., 1958. *History of mathematics* ii, Dover, New York

8.3.1.2   A Babylonian tablet

Torkil Heiede

Based on an Old Babylonian tablet of around 1700 BC, the following is an example
of how it is possible to help students—at all levels of education—to discover for

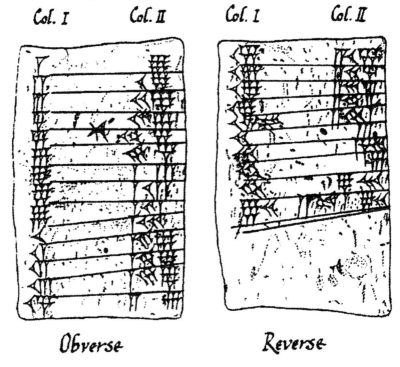

Figure 8.3: A Babylonian tablet, object of study for the mathematics class

themselves with minimal guidance how to read some ancient numerals, understand
an ancient number system and find traces of this number system in their everyday
experience.

Figure 8.3 (reproduced from and discussed in Aaboe 1964, 6-10) shows the
obverse (O) and reverse (R) of the tablet, with writing consisting of combinations of
just two symbols, a vertical wedge (vw) and a corner wedge (cw). The signs in the
left-hand column (Col. I) of the first nine lines on O may be read as the numbers
from 1 to 9, since they consist of one to nine vw (grouped in threes for easy
reading). In the next line they see a cw, naturally read as 10, especially since the
entries in the last four lines on the O and the first four lines on R can then be read as
the numbers from 11 to 18. The students can be told that, to read R, the tablet is
turned 180° around its lower edge. The next line in Col. I on R ought to show 19,

John Fauvel, Jan van Maanen (eds.), *History in mathematics education: the ICMI study,*
Dordrecht: Kluwer 2000, pp. 255-257

and it does, but with a special sign (the usual sign is the one expected by the students). This is supported by the fact that the signs on the next four lines can easily be interpreted as 20, 30, 40, 50, respectively. Only multiples of 10 are shown, but it is easy to guess that e.g. 27 would be written by combining the signs for 20 and 7.

With all this in mind, the students can now turn to the right-hand column (Col. II). The first six lines on O can easily be read as 9, 18, 27, 36, 45, 54, and students guess that the tablet is a multiplication table for 9. Hence, the following lines on O should say: 63, 72, 81, 90, 99, 108, 117, 126; but what the students find is something looking like 1,3; 1,12; 1,21; 1,30; 1,39; 1,48; 1,57; 2,6. This makes sense if the digits to the left of the commas are interpreted as 1×60 and 2×60 respectively (e.g. 1,48 means $1\times60+48=108 = 12\times9$ which stands in Col. II against 12 in Col. I). Then all (except the last) lines on R can be similarly interpreted (e.g. 2,33 corresponds to $2\times60+33=153=17\times9$ standing against 17 in Col. I) and the tablet gives all multiples of 9 from 1×9 to 59×9 by combining different entries (e.g. $27\times9=20\times9+7\times9$). But what about 60×9? A bright student might point out that, as a vw can mean 1×60 as well as 1, we can go back to the first line of the O and find 60×9 in Col. II as 9 which can be understood as $9\times60=540$, standing against 1 in Col. I which is now read as 1×60. Moreover, the entry in the second line of Col. I can be read as $2\times60=120$, and the entry in Col. II as $18\times60=1080$ which is precisely 120×9, and so on throughout the whole tablet. Then by returning to the beginning once more and interpreting 1 as 1×60^2, 2 as 2×60^2, and so forth, the tablet can be used eventually to find all multiples of 9! The students will observe that everything is expressed in a position system just like ours, based on 60 rather than 10. In our system we must memorise ten different symbols for the numbers from 0 to 9, but it would be much harder to memorise fifty-nine different symbols for the numbers from 1 to 59. It is remarkable that the Babylonians managed with just two symbols, the vw and the cw, and using the second for 10 as a sort of auxiliary base (probably they had inherited this from an earlier repetitive system based on 10).

But there is even more to be said: just as we can express common fractions as (finite or infinite) decimal fractions in base 10, the Babylonians could also understand their vw as $1/60$, $1/60^2$, ... and thereby express common fractions as sexagesimal fractions (e.g. 1,15,20 could mean $1+15/60+20/60^2$). However, the system was complicated by the absence of a sexagesimal 'point' and the meaning of the text had to be deduced from the context (in our transcription we put a semicolon to distinguish e.g. 1;15,20 = $1+23/90$ from 1,15;20 = $75+1/3$). There was no symbol for zero (until very late in their history) so that 1,15;20 could also mean 1,0,15;20 = $3615+1/3$. They tried to compensate for both weaknesses by appropriate spacing (e.g. in Col. II, seventh line on the O, four vw should be read, not as 4 but as 1,3 = $1\times60+3 = 63$). It is also clear that Babylonian scribes must have had tablets with other multiplication tables. This explains the mysterious last line on R of our tablet; it is simply a sort of heading for the next tablet in the set.

What can be gained by exposing students to such a discovery procedure?

(i) By coming to understand a number system completely different from ours, they may be able to appreciate that the same thoughts and insights can be expressed in very different but equally valid ways (**7.2.c5, 7.2.d2, 7.2.e3**).

(ii)   They may thereby come to understand our own number system better and wonder where it comes from (**7.2.b2**).

(iii) They may get a sense of the history of mathematics by being led to understand that our subdivisions of the time and angle units are reminiscences of this 4000 year-old number system, passed on to us through ancient Greek astronomers.

(iv)   Last but not least, one should mention the pure joy of discovery (**7.2.e1**).

Reference to §8.3.1.2

Aaboe, A. 1964. *Episodes from the early history of mathematics*, New York: Random House

8.3.1.3   Abacus in mind

Dian Zhou Zhang

Making history an integral part of mathematics classes for students, young and old, has the possibility of stimulating further research on their part.   Real personal interest stimulates continuous questioning and non-routine branching of inquiry (**7.2.b2; 7.2.c1**). The following is an example.

Although the calculator is very common in many primary schools in China, mental arithmetic is still a popular tradition.   In recent times, many teachers and students are using 'abacus in mind', that is, imaginary manipulations of the abacus, to do the basic operations on whole numbers.   This raises the question: "The procedure of operation with the abacus is from left to right (i.e. from higher to lower digits), but in the normal pencil-paper operation it is usually from right to left. Which way is better?"

Some ten years ago, in the German curriculum the convention for order of calculating changed from 'right to left' to 'left to right', the author has been informed.   Many Chinese mathematics educators want to unify the two systems.   If 'abacus in mind' is really a good way of operating, China intends to change the curriculum just like in Germany.   In this connection, it may be helpful to clarify the historical aspects of the subject.   The *Sun zi suan jing* (c. 5th century) was the earliest mathematical text in China in which an explicit description of the method of multiplication and division appeared.   Multiplication was done from left to right (Lam and Ang 1992, §3.3).   The same method was later employed in the Islamic mathematical world.   But in the earliest printed book on arithmetic in Europe, the *Treviso Arithmetic* (1478), multiplication was done from right to left (Swetz 1987). When and why were the rules changed? The clarification of this situation would be of interest in mathematics education research.   These questions are within the scope of many schools and pre-service trainee teachers or in-service teacher, in schools of education.

John Fauvel, Jan van Maanen (eds.), *History in mathematics education: the ICMI study*, Dordrecht: Kluwer 2000, pp. 257-258

References to §8.3.1.3

Lam, L.Y., Ang, T.S. 1992. *Fleeting footsteps : tracing the conception of arithmetic and algebra in ancient China*, Singapore: World Scientific
Swetz, F.J. 1987. *Capitalism and arithmetic: the new math of the 15th century*, La Salle: Open Court

8.3.2   The Pythagorean theorem in different cultures

Wann Sheng Horng

The so-called Pythagorean theorem has witnessed multiple discoveries over the course of history **(7.2.e3)**. It has been demonstrated in different civilisations—the word 'demonstrated' rather than 'proved' is used here because the traditional meaning of a 'proof' (an English term equivalent to Greek *apodeixis*) is a specific deductive procedure leading to what is to be concluded. This can be traced back to Greek primary concern about methodology in order to secure the certainty of mathematics. For example, the Pythagorean theorem (Proposition 47 in the first book of Euclid's *Elements*) is proved as a logical consequence of earlier propositions together with the five postulates and five common notions at the beginning of the book. It is interesting to note that the 17th century English philosopher Thomas Hobbes was convinced, by reading this proof, of the claim that certainty of knowledge is attainable through mathematics **(7.2.e)** (Siu 1990).

However, the teacher should not regard the Euclidean demonstration as the only legitimate approach to this proposition in the classroom, especially since multicultural concerns are now an issue of mathematics education. In fact, ancient mathematicians in both China and India usually used some other approaches to explain why their formulas or algorithms work (see Siu 1993). For example, the term *upapatti*, which appears often in ancient Indian mathematical texts, bears a meaning close to 'convincing demonstration'. Two arguments were associated with the *upapatti* in Bhaskaracharya's (b.1114 AD) *Bijaganita* (Joseph 1994). Modern versions of these two demonstrations for the Indian 'Pythagorean theorem' are as follows:

1. Since   triangles   CDB, ADC   and   ACB   are similar, $a/c = d/a$ implies

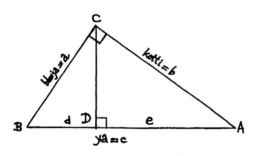

Figure 8.4

John Fauvel, Jan van Maanen (eds.), *History in mathematics education: the ICMI study*, Dordrecht: Kluwer 2000, pp. 258-262

$d = a^2 / c$ and $b/c = e/b$, implies $e = b^2 / c$ respectively. Therefore, $c = d+e = (a^2 +b^2)/c$, and so $c^2 = a^2 +b^2$ (see figure 8.4).

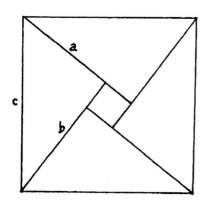

Figure 8.5

2. Let $ya = c$, *bhuja* $= a$ and *kotti* $= b$, then $c^2 = (b-a)^2 +4 \cdot \frac{1}{2}ab = a^2 +b^2$ (see figure 8.5).

According to Saraswati Amma (1979, 3), the Indian *upapatti* is different from Greek *apodeixis*:

There was an important difference between the Indian proofs and their Greek counterparts. The Indian's aim was not to build up an edifice of geometry on a few self-evident axioms, but to convince the intelligent student of the validity of the theorem so that visual demonstration was quite an accepted form of proof. This leads us to another characteristic of Indian mathematics which makes it differ profoundly from Greek mathematics. Knowledge for its own sake did not appeal to the Indian mind. Every discipline (*sastra*) must have a purpose.

To teachers sharing multicultural concerns the first of these Indian approaches to the Pythagorean theorem is of particular interest, since a similar method is also found in ancient Chinese mathematics. The third century Chinese mathematicians Zhao Shuang and Liu Hui gave their commentaries to the *Zhou Bi Suan Jing* (The mathematical canon of the gnomon of Zhou) and the *Jiu Zhang Suan Shu* (Nine chapters on the mathematical art) respectively. In showing how the 'Pythagorean theorem' works, both of them present visual demonstrations similar to that of Bhaskaracharya. The Chinese 'Pythagorean theorem' was related to the treatment of the so-called *Gou Gu* problem, namely, given two sides of a right-angled triangle, to find the third side. Note that literally *Gou* and *Gu* denote the least and the medium side respectively. It is due to this fact that the Pythagorean theorem is also called the *Gou Gu* theorem in Chinese mathematics textbooks (see Swetz and Kao 1977).

Let us first see how Zhao Shuang commented on the *Gou Gu* problem and its solution in his commentary to the *Zhou Bi Suan Jing* (Cullen 1995, 83):

The base and altitude are each multiplied by themselves. Add to make the hypotenuse area. Divide this to open the square, and this is the hypotenuse. In accordance with the hypotenuse diagram ['Xian Tu', see figure 8.6]. You may further multiply the base and altitude together two of the red areas. Double this to make four of the red areas. Multiply the difference of the base and altitude by itself to make central yellow area. If [one such] difference area is added [to the four red areas], the hypotenuse area is completed.

Figure 8.6: The 'Xian Tu' diagram

We leave to the reader to translate Zhao Shuang's demonstration into modern algebraic notation. Essentially it is very similar to that of Bhaskaracharya. We can add one similar example, namely, Liu Hui's explanation of how the algorithm for the *Gou Gu* problem works. In his commentary to Chapter Nine (entitled 'Gou Gu') of *Jiu Zhang Suan Shu*, Liu Hui explained as follows (Martzloff 1997, 296):

Base-squared makes the red square, leg-squared makes the azure square. Let the Out-In mutual patching [technique] [be] applied according to the categories to which [these pieces] belong by taking advantage of the fact that what remains does not move and form the surface of the hypotenuse.

Since Liu Hui's original diagram had been lost by the thirteenth century, perhaps the rational reconstruction by the late Qing mathematician Gu Guanguang (1799-1862) (Figure 8.7, from (Martzloff 1997, 297) would help us understand how the visual demonstration was actually carried out.

As to whether such explanation is related to the Greek sense of proof, Cullen comments: "It may be misleading to call Liu Hui's "suasive explanations" by the same name as the rather differently directed and structured rhetorical machinery provided by writers such as Euclid, for which we may reasonably use the modern term 'proof'." (Cullen 1995, 92). Martzloff expresses a similar point: "[T]he explanation of Pythagoras' theorem may only suggest how to set about it and since the commentator's excessively laconic text is clearly, on its own, not sufficient to reconstitute the details of the process, it follows that it is not only what the student will read or heard that is important but the manipulation which he will have seen the master undertake. The fact that these two- or three-dimensional figures of Chinese

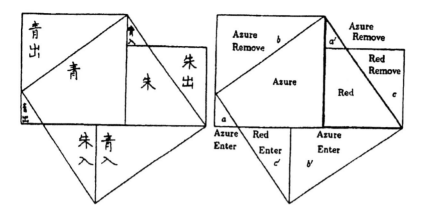

Figure 8.7: out-in patching

geometry often refer to actual concrete objects reinforces this interpretation."
(Martzloff 1997, 72). In using these different explanations in the classroom,
teachers can use them to make contrasts, emphasising not only the methodology but
the epistemology as well (**7.2.b1, 7.2.b2, 7.2.c5**). In other words, teachers should
try to stress that to prove is not only to convince but also to enhance understanding
(**7.2.b1**). After explaining what these proofs or explanations are about, the teacher
can go on to urge students to explore their socio-cultural meaning. In this
connection, the teacher is encouraged to introduce a critical re-evaluation of
mathematics in different civilisations and thereby share with the students a sense of
multiculturalism in mathematics (Nelson *et al.* 1993; Joseph 1991; Gerdes 1994;
7.2.e3).

For general information on the Pythagorean Theorem, teachers may like to refer
to Loomis 1968, in which over three hundred proofs have been collected. Teachers
who want to introduce to the class some related ethnomathematics will find useful
material in Gerdes 1994. Those who are critical of Eurocentrism and
Hellenocentrism in the history of mathematics should keep in mind that
multicultural concerns help to promote in students a flexible and open mind to
mathematical culture of any origin.

References for §8.3.2

Amma, T. A. Saraswati 1979. *Geometry in ancient and medieval India*, Varanasi: Motilal
    Banarisidass
Cullen, C. 1995. 'How can we do the comparative history of mathematics? Proof in Liu Hui
    and the Zhou Bi', *Philosophy and the history of science* 4 (1), 59-94
Euclid 1925. *The thirteen books of the elements* ed. T. L. Heath, 2nd edn, Cambridge:
    University Press 1925; reprint New York: Dover 1956
Gerdes, Paulus 1994. *African Pythagoras: a study in culture and mathematics*, Maputo:
    Instituto Superior Pedagogico

Joseph, G.G. 1991. *The crest of the peacock: non-european roots of mathematics*, London: Tauris

Joseph, G.G. 1994. 'Different ways of knowing: Contrasting styles of argument in Indian and Greek mathematical traditions', in P. Ernest (ed), *Mathematics, education and philosophy: an international perspective*, London: Falmer Press, 194-204.

Loomis, E.S. 1968. *The Pythagorean proposition*, Reston, Va: National Council of Teachers of Mathematics

Martzloff, J-C. 1997. *A history of Chinese mathematics*, Springer-Verlag, Heidelberg-New York; originally published in French, Mason, Paris, 1987

Nelson, D., Joseph, G.G., Williams, J. 1993. *Multicultural mathematics: teaching mathematics from a global perspective*, Oxford: University Press

Siu, Man-Keung 1990. *Mathematical proofs* (in Chinese), Nanjing: Jiangsu Educational Press

Siu, Man-Keung 1993. 'Proof and pedagogy in ancient China: examples from Liu Hui's Commentary on Jiu Zhang Suan Shu', *Educational studies in mathematics* **24**, 345-357

Swetz, Frank J., Kao, T.I. 1977. *Was Pythagoras Chinese? an examination of right triangle theory in ancient China*, Pennsylvania State University Press

8.3.3   Measuring distances: Heron vs. Liu Hui

Chun Ip Fung, João B. Pitombeira de Carvalho

With the introduction of trigonometric ratios in the middle school or early secondary mathematics curriculum, students are often confronted with problems which rely on the notion of angle. Some students accept without question the use and availability of angles of elevation or depression. The activity described below helps to re-instate for students the centrality of similar triangles in simple surveying situations.

The Source

According to the Chinese classic *Hai Dao Suan Jing* written by Liu Hui, the surveying of the height of distant objects could be done by the method of double difference. Using this method, the Chinese achieved complicated surveying of remote objects without the notion of angle (**7.2.a2; 7.4.3**). Problem 1 of the nine problems in the book reads (Swetz 1992, p.20: *zhang* and *bu* are ancient Chinese length units):

Now for [the purpose of] looking at a sea island, erect two poles of the same height, 3 *zhang* [on the ground], the distance between the front and the rear [pole] being a thousand *bu*. Assume that the rear pole is aligned with the front pole. Move away 123 *bu* from the front pole and observe the peak of the island from the ground level; it is seen that the tip of the front pole coincides with the peak. Move backward 127 *bu* from the rear pole and observe the peak of the island from the ground level again; the tip of the back pole also coincides with the peak. What is the height of the island and how far is it from the pole?

John Fauvel, Jan van Maanen (eds.), *History in mathematics education: the ICMI study*, Dordrecht: Kluwer 2000, pp. 262-264

The activity

Ask students to estimate the height of an island without actually going onto the island (**7.3.2**).

Target student group

Junior secondary school students with knowledge of similar triangles, congruent triangles, Pythagorean theorem, and trigonometric ratios.

The organization of the activity

Step 1: Let students discuss how to do it with simple apparatus.

Step 2: Show how the Chinese did it by calculating the area of suitably chosen rectangles, as explained in Yang Hui's commentary (1275) on Liu Hui's *Hai Dao Suan Jing* (Wu 1982). (See Example 8.4.4 for a related discussion in Greek mathematics.)

Step 3: Ask students to go on a field trip to estimate (in groups) the height of an island in a familiar community.

Step 4: Ask students to compare their results with the data available in relevant agents/authorities.

Objectives of the activity:

1. To let students appreciate that as far as solution of right triangles is concerned, the trigonometric ratio technique is simply a tool derived from the properties of similar triangles;
2. To let students see that the development of mathematics does not follow a unique path, independent of the civilisations in which it has developed (**7.2.e3, 7.2.a1**);
3. To let students see how mathematics intertwined with surveying in ancient time (**7.2.a3, 7.2.d**).

The problem described above, was tackled in a different culture (namely, that of Hellenistic Greece) by Heron of Alexandria (1st century AD). (Heath 1921, ii, 345; for his life see Drachmann 1972). It is interesting to present the problems from that perspective, thus implicitly stressing the cultural dimension of mathematical activity (**7.2.e**). To avoid repetition, we consider a Greek solution to the somewhat different problem of measuring the distance *AB*, when *B* is inaccessible from *A.*, e.g. because there is a river between *A* and *B* (as in figure 8.8).

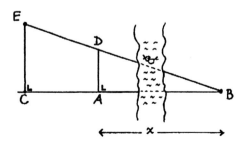

Figure 8.8: Determine the distance AB, when B is inaccessible from A

On the straight line *BA produced*, choose a point *C* and erect perpendiculars *AD* and *CE* to *BC*. The point *D* is chosen to lie on *BE*. The points *C* and *E* are chosen in such a way that *AC*, *AD* and *CE* can be measured. Using similar triangles, we have that *CE/AD* = *BC/BA*. Let *AB* = *x*. Then $(AC+x)/x$ = *CE/AD*. If we call *CE/AD* *k*, which is known, we have $AC+x = kx$, and thus $x(k-1) = AC$, and so $x = AC/(k-1)$ (Katz 1998; 4.3.1).

This solution requires a surveyor's measuring chain (or tape) and a sighting instrument that can measure angles (in particular, right angles).

References for §8.3.3

Drachmann, A. G. 1972. 'Hero of Alexandria', in C. C. Gillispie (ed), *Dictionary of scientific biography*, New York: Charles Scribner's Sons, vi, 310-315

Heath, T.L. 1921. *A history of greek mathematics*, 2 vols, Oxford: Clarendon Press; repr New York: Dover 1981.

Katz, V. J. 1998. *A history of mathematics: an introduction,* 2nd ed., New York: HarperCollins

Swetz, F. J. 1992. *The sea island mathematical manual: surveying and mathematics in ancient China*, Pennsylvania State University Press

Wu, W.J. 1982. 'Investigation on the source of the ancient methods' in W. J. Wu (ed), *Hai Dao Suan Jing, Jiu Zhan Suan Shu and Liu Hui*, (in Chinese), Beijing: Normal University Press, 162-180

8.4   Detailed treatment of particular examples

8.4.1   Introducing complex numbers: an experiment

Giorgio T. Bagni

The introduction of imaginary numbers is an important step in the high school mathematics curriculum (students 15-19 years old). High school students of 11 to 14 years old are often reminded about the impossibility of calculating the square root of negative numbers. However, at a later stage, they are asked to accept the presence of '$\sqrt{-1}$', named *i*. This inconsistency can be a source of confusion.

On the other hand, we may consider the solution of cubic equations following the work of Niccolo Fontana (Tartaglia, 1500-1557), Girolamo Cardano (1501-1576), and Rafael Bombelli (1526-1573): imaginary numbers were not introduced via *quadratic* equations, but via *cubic* equations, an approach having a basic advantage. Their solution does not take place entirely in the set of real numbers, but one of the final results is always real. A recent study was motivated by this fact (**7.3.2**). In this research 97 high school students (age 16-18), who did not know

John Fauvel, Jan van Maanen (eds.), *History in mathematics education: the ICMI study*, Dordrecht: Kluwer 2000, pp. 264-265

complex numbers, were interviewed (Bagni 1997). For the equation $x^2 + 1 = 0$, hence $x=\pm i$, only 2% accepted the solution, 92% rejected it and 6% did not answer. Afterwards, the solution of the cubic equation $x^3 = 15x + 4$, namely $x = (2 + 11i)^{\frac{1}{3}} + (2 - 11i)^{\frac{1}{3}}$ so that $x = (2+i) + (2-i) = 4$, was accepted by 54%; 35% rejected it and 11% did not answer.

Under the same conditions, a similar test was then proposed to 52 students of the same age group, where the equations were presented in the *reverse order*: 41% accepted the solution of the cubic equation (25% rejected and 34% did not answer). Immediately after that, the solution of the quadratic equation was accepted by 18% of the students, with only 66% rejecting it (16% did not answer). These experimental results suggest that teaching a subject by taking account of some basic facts in its historical development may help students to acquire a better understanding of it (Weil 1978; Fauvel 1990; Swetz 1995).

References for §8.4.1

Bagni, G.T. 1997. ' "Ma un passaggio non è il risultato..." L'introduzione dei numeri immaginari nella scuola superiore', *La matematica e la sua didattica* 2, 187-201

Fauvel, John (ed.) 1990. *History in the mathematics classroom: the IREM papers,* Leicester: The Mathematical Association

Swetz, F.J. 1995. 'To know and to teach: mathematical pedagogy from a historical context', *Educational studies in mathematics* 29, 73-88.

Weil, A. 1978. 'History of mathematics: why and how', in *Proc. of International Congress of Mathematicians*, Vol. I, O. Letho ed., Helsinki: Academia Scientiarum Fennica, 227-236.

8.4.2 Intertwining a mathematical topic with other (non-) mathematical topics

a Duplication of the cube

Manfred Kronfellner

Tasks often play an important role in mathematics teaching, as well as in preparing for teaching and in assessment. One strategy for introducing history would be to offer suitable tasks in which a traditional curriculum topic is connected with history. Such tasks might act as 'kernels of crystallisation' for some further historical information in order to connect these kernels, by and by, to a network-like overview of some steps in the historical development of mathematics (**7.2.a3**).

An example of such a network may be based on the problem of the duplication of a cube, one of the 'three classical problems' which have been stimulating mathematicians for more than two thousand years. The question of how to duplicate a cube—that is, the geometric construction of $\sqrt[3]{2}$ —was expressed by Menaechmus

John Fauvel, Jan van Maanen (eds.), *History in mathematics education: the ICMI study,* Dordrecht: Kluwer 2000, pp. 265-269

using the proportion $1 : x = x : y = y : 2$. By transforming this proportion, we get $y = x^2$, $xy=2$ so that $x^3 = 2$. This implies that the solution x can be obtained from the intersection of the parabola $y=x^2$ with the hyperbola $xy=2$.

To find the solution, the ancient Greeks invented devices such as Plato's

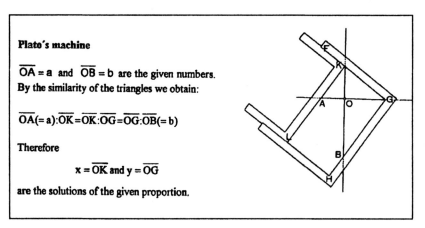

Plato's machine

$\overline{OA} = a$ and $\overline{OB} = b$ are the given numbers.
By the similarity of the triangles we obtain:

$\overline{OA}(=a):\overline{OK}=\overline{OK}:\overline{OG}=\overline{OG}:\overline{OB}(=b)$

Therefore

$$x = \overline{OK} \text{ and } y = \overline{OG}$$

are the solutions of the given proportion.

Figure 8.9: Plato's machine for constructing OK and OG such that a:OK=OK:OG=OG:2a, in which case OK is the side of the doubled cube.

machine and Eratosthenes' plates, see figures 8.9, 8.10 (Heath 1963; Eves 1976; Kaiser 1996), or they created new curves like Diocles' cissoid, see figure 8.11.

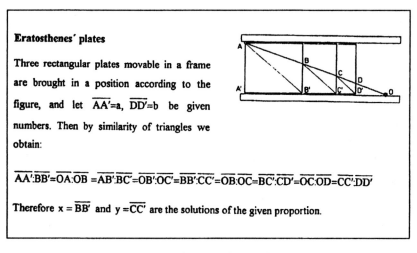

Eratosthenes' plates

Three rectangular plates movable in a frame are brought in a position according to the figure, and let $\overline{AA'}=a$, $\overline{DD'}=b$ be given numbers. Then by similarity of triangles we obtain:

$\overline{AA'}:\overline{BB'}=\overline{OA}:\overline{OB}=\overline{AB'}:\overline{BC'}=\overline{OB'}:\overline{OC'}=\overline{BB'}:\overline{CC'}=\overline{OB}:\overline{OC}=\overline{BC'}:\overline{CD'}=\overline{OC}:\overline{OD}=\overline{CC'}:\overline{DD'}$

Therefore $x = \overline{BB'}$ and $y =\overline{CC'}$ are the solutions of the given proportion.

Figure 8.10: Eratosthenes' plates

In high school, the explanation of each of these machines or curves can be posed, independently of each other, as tasks concerning applications of similar triangles and proportions.

Let C be a circle with diameter OA and radius r. Let D be the tangent at A. Through O a line g is drawn, which intersects the circle in O and P_1 and the tangent in P_2. The cissoid consists of all point that fulfil the condition:

$$\overline{OP} = \overline{P_1P_2} \, .$$

The cissoid can be used to duplicate the cube, since it allows to construct two mean proportionals between two given lengths $\frac{a}{2}$ and a.

Let M be a point with $\overline{ZM} = \frac{1}{2}r$ and $\overline{ZM} \perp \overline{OA}$. Then for each point of the cissoid: $\dfrac{\frac{a}{2}}{\overline{LP}} = \dfrac{r}{\frac{1}{2}r} = 2 \Rightarrow \overline{LP} = \dfrac{a}{2}$.

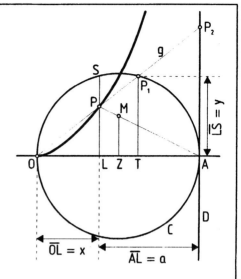

We have to show now that $x = \overline{OL}$ and $y = \overline{LS}$ satisfy the proportion $\dfrac{\frac{a}{2}}{x} = \dfrac{x}{y} = \dfrac{y}{a}$.

Because of the definition of the cissoid $\overline{OP} = \overline{P_1P_2}$ and therefore $\overline{AT} = \overline{OL}$, resp. $\overline{AL} = \overline{OT}$. From $\triangle OLP \sim \triangle OTP_1 \cong \triangle ALS \sim \triangle SLO$ we find $\overline{LP} : \overline{OL} = \overline{OL} : \overline{LS} =$ $= \overline{LS} = \overline{AL}$. In other words: $\dfrac{a}{2} : x = x : y = y : a$, and so x is the side of the cube which has twice the volume of the cube with side $\dfrac{a}{2}$. An arbitrary cube with side s can now be doubled by determining t such that $\dfrac{a}{2} : x = s : t$.

Figure 8.11: Doubling the cube with the cissoid of Diocles

Extensions to a historical network (see the table below):

One possible extension is to deal with the other two classical problems as well (trisection of an angle, quadrature of a circle) and the ancient Greek methods to solve them (using further new curves such as conchoid, quadratrix, Archimedean spiral), or to reveal connections of this problem with musical scales (see Example 8.4.2b following).

Another possibility is to elaborate on the scientific/philosophical background of Greek mathematics. The Greeks were not satisfied with the methods described above because they wanted solutions using only compasses and the straightedge. Why? The restriction of geometric constructions to compasses and straightedge, a tradition that went back to Plato and possibly earlier (Boyer 1959, 27; Wussing 1965, 75), is reflected in the postulates of Euclid's *Elements,* which formed a secure basis for mathematics. The need for such a basis seems to be connected not only with the discovery of the incommensurable magnitudes, but also with Zeno's paradoxes. Among other things (in particular, criticism of the concept of motion and time (Boyer 1959, 24; Whitrow 1980, section 4.4)), Zeno (at least implicitly) tried to criticise the mathematics of his period. (For Zeno's intention, see Boyer 1959, 23-24; Kirk et al 1983, sections 327-329.) An analysis of his famous paradoxes shows that application of discrete methods to infinity may cause problems: does a line consist of (indivisible) points (atoms)? Do points exist? Do we get these points when we bisect the line infinitely often? Can we make up a line out of points? (Struik 1967, 44) Such questions could not be answered at the time. Therefore, in addition and in parallel to the fact that the discovery of the irrationals produced a deep crisis of mathematics by showing the incompleteness of mathematical argumentation based exclusively on rational numbers (a view not universally supported by recent historians: for a new interpretation of Greek work on incommensurable magnitudes, see Fowler 1987; Knorr 1975), the difficulties revealed by Zeno's paradoxes concerning the relation between the discrete and the continuous led Greek mathematicians to try to consolidate the basis of mathematics and to develop a secure method.

The axiomatic method in Euclid's *Elements*—based on Plato and Aristotle— fulfilled this need (Eves and Newsom 1958, §2.2; Kaiser and Nobauer 1998, 18). The starting point of this work, the postulates, grounded it implicitly (though not explicitly) on constructions by compasses and straightedge: the first postulate (to draw a straight line through two given points) and the second postulate (to continue a straight line in either direction) allow the use of a straightedge; the third postulate (to draw a circle with a given centre and a distance) allows the use of a pair of compasses. This implies that all one can construct with compasses and straightedge is also deducible from the postulates. With the axiomatic method the mathematicians possessed a tool which produces indubitable results as long as the postulates are indubitable. Zeno's paradoxes can already be found in school books as tasks, but mostly they remain isolated as an oddity; their role in the history outlined above is rarely explained in textbooks.

There are also possible connections to more recent developments: besides Kepler's use of conic sections in astronomy, we can cope in some sense with Galois' theory which leads to the impossibility of a solution of the problem using only compasses and straightedge. This proof can be explained heuristically in nonlinear analytical geometry, when teaching the intersection of circles (compasses) and lines (straightedge); these intersections always lead to equations of degree 2^n, but never to equations of degree 3; therefore cubic roots cannot be constructed with these tools.

A historical network based on the duplication of the cube

starting point / topic	historical topic	prerequisites (+ repetition and application of)	further goals / connections
irrational numbers	Menaechmus' proportion, Eratosthenes' plates Plato's machine	similar triangles, proportions, construction of square roots	trisection of an angle squaring a circle proportions in (Pythagorean) mathematics, esp. music
series and / or axiomatic method	Zeno's paradoxes Euclid's axiomatic method		modern mathematics (Hilbert, Bourbaki, ...)
nonlinear analytical geometry	construction of $\sqrt[3]{2}$ by intersection of a parabola and a hyperbola (Menaechmus); $\sqrt[3]{2}$ cannot be constructed with straightedge and compasses		Kepler's use of conic sections in astronomy Galois, modern algebra
curves	cissoid	similar triangles, proportions	conchoid, quadratrix, Archimedian spiral, ...

References for §8.4.2 a

Boyer, C.B. 1959. *The history of the calculus and its conceptual development*, New York: Dover

Eves, H., Newsom, C.V, 1958. *An introduction to the foundations and fundamental concepts of mathematics*, New York: Holt, Reinhart & Winston

Eves, H. 1976. *An introduction to the history of mathematics*, 4th edn, New York: Holt, Rinehart & Winston; originally published in 1953.

Fowler, David 1987. *The mathematics of Plato's Academy: a new reconstruction*, Oxford: University Press (2nd edn 1999)

Heath, T.L. 1963. *A manual of greek mathematics*, Dover, New York, 1963; orig publ 1931

Kaiser, H. K. 1996. 'The problem of the duplication of a cube', in A. S. Posamentier (ed), *The art of problem solving*, Thousand Oaks: Corwin Press, 371-382.

Kaiser, H., Nöbauer, W. 1998. *Geschichte der Mathematik*, 2nd edn, Munich: Oldenbourg

Kirk, G.S., Raven, J.E., Schofield, M. 1983. *The presocratic philosophers : a critical history with a selection of texts*, 2nd edn, Cambridge: University Press

Knorr, W.R. 1975. *The evolution of the euclidean elements*, Dordrecht:.Reidel

Struik, D.J. 1967. *A concise history of mathematics*, revd edition, New York: Dover

Whitrow, G.J. 1980. *The natural philosophy of time*, 2nd edn, Oxford: Clarendon Press

Wussing, H. 1965. *Mathematik in der Antike*, Leipzig: Teubner

b Musical scales

Michel Rodriguez

All mathematics teachers in the world know that music was an integral part of mathematics in Greek civilisation, but few know why it was so. Though many of them have learned music theories and participate in musical activities, yet they have no idea of the close relationship between the two disciplines, except for the famous name of Pythagoras. The history of mathematics can help to clarify this point. Below we outline an activity which touches upon some of the principal characteristics of Greek mathematical culture, like the constructions with straightedge and compasses, and the theory of proportion (8.4.2a above, 8.4.4 below; **7.2.a3; 7.2.d**).

The activity was carried out in two 3-hour sessions with 15 students (half the class) of a French high school 2nd class (15-16 years old). It was devised as a series of activity modules so that teachers can have more freedom when using them, in contrast to the rigidity and apparent constraints of the official curriculum. First we revisit the duplication of the cube.

First Part: Problem of Delos - duplication of the cube (7.2.a2)

1. A voyage from Eudoxus to Descartes (constructions with straightedge and compasses): (a) construction of line segments, areas and volumes, (b) emergence of unit segment, construction of the product, quotient, square root (duplication of square) of numbers, their geometric mean, the golden ratio etc.

2. The central problem, the duplication of the cube: (a) research with straightedge and compasses, until the conjecture of the impossibility of the problem comes to the mind of the students ("What is the number that we want to construct?" Emergence of $2^{1/3}$).

3. Mechanical solutions: (a) setting up in parallel the problem of finding the double mean proportional, (b) presentation of Eratosthenes' *mesolabe*; students are supplied with identical rectangular tiles on each of which the diagonal is already marked, and they have to find out why these tiles can be used to find $2^{1/3}$, (c) with the aid of handouts, a rapid presentation of Plato's machine (see figure 8.9).

Second Part: Musical scales, Pythagorean and equal-tempered

1. What is a musical note?: (a) presentation of the inverse proportional relation between the frequency of the wave emitted and the length of the vibrating string (we measure the lengths of the strings of a guitar); (b) the notion of a resonance interval: octave and fifth (respective ratios 2 and 3/2); (c) principles of the Pythagorean scales, algorithm and computation of the first 11 ratios by this method; (d) "Is the Pythagorean scale constructible with straightedge and compasses?" Yes, because all ratios are rational; (e) "Does the stem of the guitar represent a Pythagorean scale?"

John Fauvel, Jan van Maanen (eds.), *History in mathematics education: the ICMI study*, Dordrecht: Kluwer 2000, pp. 270-272

No, there are noticeable differences, (particularly at the level of the 4th space), but the reading of the measurements help to observe a geometric regularity of this scale.

2. The equal-tempered scale, which appeared in Europe in the end of the 17th century: (a) the scale of Werckmeister in 1691, temperament of 12 notes; (b) the ratio of frequencies will form a geometric sequence; find the ratio r, say. One finds $r^{12} = 2$, thus $r = 2^{1/12}$; (c) particular case, the 4th space corresponds to the ratio of frequency $(2^{1/12})^4 = 2^{1/3}$. Here we come back to $2^{1/3}$. (See also Chapter 9 of Land 1975, whose figure on p.132 is reproduced here as figure 8.12.)

3. Return to Delos, by posing a simple problem: "Is the equal-tempered scale constructible with straightedge and compasses?" For concluding this activity, two different possibilities have been envisaged:

(a) A classical one is to point out that this problem is equivalent to the problem of the duplication of the cube: (i) To find a construction of the tempered scale with straightedge and compasses means to settle the problem on the 4th space, which at the same time solves the Delian problem. (ii) To find the solution of the problem of Delos means that the 4th space of a guitar is constructible. Now the sixth is already constructible ($2^{6/12} = 2^{1/2}$, which we have already come across). Thus, we find the 5th space as the geometric mean, and the ratio of the 4th to the 5th (or of the 5th to the 6th) will be the ratio which enables us to construct the whole scale using proportionalities!

(b) A more exotic possibility is to claim that we can find an acoustic solution to the problem of Delos (**7.4.8, 7.4.9**). Suppose that we want to find the edge of a cube twice in volume of a given cube. We start by transferring the length of the edge of the given cube to a stretched monochord and observe the resonant interval that this vibration gives with a string of a guitar when a finger lies on the 4th space (one can even adjust the tension of the monochord or of the guitar for tuning the two vibrations, and play them at unison). Then, lengthen the monochord without

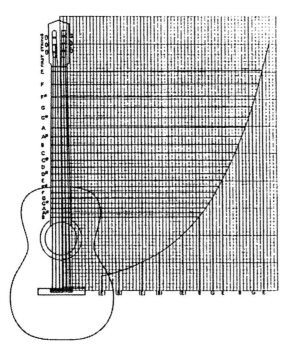

Figure 8.12: Guitar and scale (from Land 1975, 132)

modifying its tension and compare now with the same string of the guitar vibrating empty (space 0). When we find out the same resonance interval (or the unison), the lengths will be in the ratio, and the only thing left is to construct the double cube!

The activity in this section takes a lot of time, but it is worthwhile, since it touches on a variety of domains (**7.2.a3**). However, one may remark that at the end of the activity there are still doubts about the constructibility of the duplication of the cube and of the scale of Werckmeister. Isn't this a good illustration of an essential aspect of science? In science, there is always something left to look for.

Reference for §8.4.2 b

Land, Frank 1975. *The language of mathematics*, revised edition, London: John Murray; originally published in 1960.

Rodriguez, Michel 2000. 'Sure le manche de ma guitare', in *Les nombres, actes des journées académiques de l'IREM de Lille*, Paris: Ellipses

c Leonardo's geometric sketches

Chun Ip Fung

Fascinating geometric sketches are found in the notebooks of Leonardo da Vinci. Most of these figures are directly related to the squaring of curvilinear regions. The following cat's eye diagram (figure 8.13) is one among them (Wills 1985, 11).

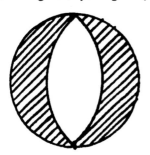

Figure 8.13: The cat's eye, or what part of the circle is the shaded area?

The following activity is designed for junior secondary school students (age 12 to 14) with knowledge of the Pythagorean theorem, the area ratio of similar figures, and knowing formulae for area computation of simple figures including circle, sector, rectangle. The task is to calculate the area of the shaded part in the above diagram in terms of the radius of the circle (**7.3.3.b; 7.3.1**). .

Design of the activity:

Step 1: Ask students to compute the area, making assumptions where necessary (**7.2b1**).

Step 2: Show how Leonardo did it, using a simple cut-and-paste method , as displayed in figure 8.14 for a different area.

Purposes of the activity:
1. To let students appreciate the.aesthetic nature of mathematics (**7.2.e1**);

John Fauvel, Jan van Maanen (eds.), *History in mathematics education: the ICMI study*, Dordrecht: Kluwer 2000, pp. 272-273

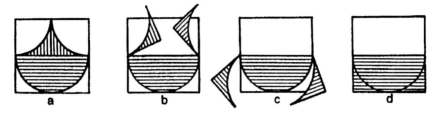

Figure 8.14: How Leonardo solved this type of problems, i.e. by simple cut-and-paste (from Wills 1985).

2. To warn students that the existence of mathematical products, such as formulae for area computation, does not automatically downplay the importance of having an alert and active mind (**7.2.d**).

Reference for §8.4.2 c

Wills, H., 1985. *Leonardo's dessert: no pi*, Reston, Va: National Council of Teachers of Mathematics

8.4.3   Surveyors' problems

João B. Pitombeira de Carvalho

The following are two examples whose purpose is to show how concepts of elementary Euclidean geometry were used to solve surveyors' problems in times gone by (**7.2.a2**). The tools used are simple and easily constructed, to enable high school students to actually solve similar problems (**7.4.8**). An ordnance map, a compass, measuring chains or tapes can be easily procured and offer the opportunity of letting the students practise their skills (**7.4.12**).

a   The tunnel of Eupalinos on the island of Samos

What is striking about this example is that the tunnel was constructed, around 530 BC, starting simultaneously from *both* sides of the mountain, as would be done today. However, today we have very sophisticated instruments that enable us to dig both segments of the tunnel in such a way that both working crews meet as planned. How did ancient builders proceed, without our sophisticated surveying instruments? The answer is fairly simple: they used plane Euclidean geometry (particularly the similarity of triangles) and had sighting instruments, called a dioptra, that enabled surveyors to measure angles with good accuracy. (For further discussion of this celebrated tunnel, in a classroom context, see §9.5.1.)

John Fauvel, Jan van Maanen (eds.), *History in mathematics education: the ICMI study*, Dordrecht: Kluwer 2000, pp. 273-276

As a matter of fact, students can easily build themselves such sighting devices and practice measuring angles, as in ancient times. Suppose we want to dig a tunnel with ends at A and B (see figure 8.15), which are initially assumed to be level.

Draw an arbitrary straight line segment *BC*. From *C*, draw the perpendicular *CD* to *BC*, then *ED* perpendicular to *CD*, then *EF* perpendicular to *DE*, and so on, till we are close to *A*. These straight line segments are all drawn in such a way that from *C* you can see *D*, from *D* you can see *E*, and so on. We now have a polygonal line *BCDEFGH*. Note that the angle at each vertex *C*, *D*, *E*, *G* is a right angle. All these right angles can be drawn using a dioptra.

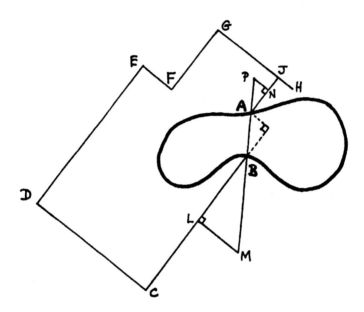

Figure 8.15: Making a straight tunnel, starting from both sides of the mountain (530 BC)

Choose *J* on *GH* such that *JA* is perpendicular to *GH*. Let *AK* be the perpendicular from *A* to *BC*. Since the lengths *DC*, *EF* and *GJ* are known, and using the fact that our polygonal line is made up of adjoining perpendicular segments, it is very easy to find the length *AK*. Similarly, it is possible to find the length *KB* and thus the ratio *BK/AK*. Let this ratio be called *k*. Construct now the right triangles *BLM* and *ANB* such that the ratios *BL/LM* and *NA/PN* are both equal to *k*. The similarity of the triangles *BLM*, *BKA* and *ANP* assure us that the points *P*, *A*, *B* and *M* are collinear. It is now very simple to dig the tunnel; just make sure that the crews working at *Q* and *R*, inside the mountain, can be sighted from *P* and *M* respectively (van der Waerden 1974, 102-104).

This description assumes that all points are in the same plane. But it can easily be modified to take account of differences in height between the points considered.

We have a description of a dioptra given by Hero, and thus it was possible to reconstruct this very ancient and useful instrument (van der Waerden 1974, 104; see Drachmann 1972 for more on Hero's work). The *alidade* (from the Arabian word al'Dad), a very simple sighting instrument, is still in use today by armies in the field or prospectors. It is simply a pocket compass placed on a horizontally held board.

b Heron's formula for the area of a triangle

It is known that the Greeks did not use trigonometry to solve surveying problems (Katz 1998, 158-162). Instead, they relied on plane geometry. We have just seen how they could solve surveying problems using elementary facts of plane Euclidean geometry. Heron's name is also attached to a formula, albeit one that is probably due to Archimedes (Fauvel and Gray 1987, 205-206; Thomas 1941, 470-477): let *ABC* be a triangle with sides *a*, *b* and *c*. If $p = (a+b+c)/2$, then Hero's formula states that the area *S* of the triangle is given by $S = \sqrt{p(p-a)(p-b)(p-c)}$.

Using this formula, it is easy to find the area of any plot of land bounded by a polygonal line, if we are able to measure the distances between its vertices. Thus, to find the area of *ABCDEFG*, we can decompose the polygon as shown (see figure 8.16). If we can measure the distances *AB, BC, CD, DE, EF, FG, AG, AC, CG, GC* and *GE*, we can find the required area using only a very simple instrument, a measuring chain or tape, without having to worry about measuring angles and 'solving' triangles using trigonometry.

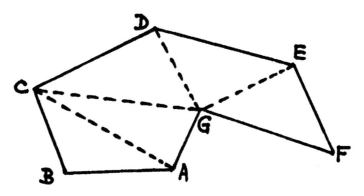

Figure 8.16: Dissecting a polygon in order to determine its area

These examples illustrate the way knowledge of its history may help teachers and students to appreciate the importance of mathematics for the solution of real problems of vital importance, by elementary means (7.2.e3; 7.2.a3). More generally, vocational contexts offer opportunities for the use of history, e.g. in the history of quality control, a vital aspect of most modern industries (COMAP 1990). In trade and technician areas, it should be possible to briefly trace the evolution of formulas and techniques, generally presented to students as a *fait accompli*, to enhance the depth of understanding and to stimulate interest. Adopting such an historical perspective will offer the possibility of stimulating discussion on the social uses to which mathematics is put; a competency not commonly found, if at all, in current adult or vocational mathematics curricula.

References for §8.4.3

COMAP 1990. *Against all odds: the sample mean and control charts*, Video No. 18, Lexington: Consortium for Mathematics and Its Applications

Drachmann, A. G. 1972. 'Hero of Alexandria', in C. C. Gillispie (ed), *Dictionary of scientific biography*, New York: Charles Scribner's Sons, vi, 310-315

Fauvel, John, Gray, Jeremy. (eds.) 1987. *The history of mathematics: a reader*, London: Macmillan

Katz, V. J. 1998. *A history of mathematics: an introduction*, 2nd ed., New York: HarperCollins

Thomas, I. 1941. *Selections illustrating the history of greek mathematics* ii, Cambridge: Harvard University Press

Van der Waerden, B. L. 1974. *Science awakening*, 3rd edition, A. Dresden tr., Groningen: Wolters Noordhoff; orig Dutch publ 1950

8.4.4   Theory of proportion and the geometry of areas

Carlos Correia de Sá

The theory of proportion played a central role in Greek mathematics. However, the early Pythagorean approach, which took into account only the positive integers, proved insufficient when incommensurable magnitudes were discovered. Eudoxus eventually created a new theory of proportions (exposed in book v of Euclid's *Elements* (Euclid 1925)) that worked both in the commensurable and in the incommensurable cases. Meanwhile, the need for another method of proof was certainly felt. Although there is no historical evidence that the geometry of areas was created as an alternative method of proof, it allowed formulations and proofs of old results without appealing to the concept of proportion.

The interrelations between the theory of proportions and the geometry of areas constitute a considerable wealth of resources that can be put to use in the mathematics classroom: the concepts of ratio and proportion, the ideas of number and area, several geometric constructions with straightedge and compasses, the Pythagorean theorem, the geometric solution of 2nd degree equations and (perhaps most importantly) many opportunities to practise the translation from the geometrical to the numerical context and vice versa (**7.2.a3; 7.2.a2**).

A significant part of the geometry of magnitudes that Euclid exposed in the *Elements* admits an arithmetic-algebraic interpretation that may be explored in the high school mathematics classroom, in order to reveal the interrelations between arithmetic operations, algebraic procedures and geometric constructions.

John Fauvel, Jan van Maanen (eds.), *History in mathematics education: the ICMI study*, Dordrecht: Kluwer 2000, pp. 276-279

The geometric analogues of addition and subtraction are obvious. In Greek geometry, it is the construction of a rectangle with given sides that is usually interpreted as a multiplication of line segments. Propositions 1 and 4 of Book ii of the *Elements* may then be regarded as the geometric versions of the distributive law of multiplication with respect to addition and of the formula for the square of a sum, respectively. This part of the geometry of areas has no relation to the theory of proportions; it may have been created independently, probably long before the discovery of incommensurability. But the notions of the fourth proportional of three line segments, and of the mean proportional of two line segments, admit alternative formulations in terms of the geometry of areas.

The standard constructions of the *fourth proportional* (*Elements* vi, 2) and of the *mean proportional* (*Elements* vi, 13), may have been the first ones to be used. However, they require a theory of proportions for the incommensurable case. Alternative formulations of these concepts, that avoid any reference to proportionality, are the following.

– Let the line segments a, b, c be given; their fourth proportional is a line segment x such that $a{:}b{=}c{:}x$ or, equivalently, such that $ax{=}bc$. Thus, x is the side of a rectangle that admits a as a side and has the area of the rectangle with sides b and c.

– In an analogous way, let line segments a, b be given; their mean proportional is a line segment y such that $a{:}y{=}y{:}b$, or, equivalently, such that $y^2{=}ab$. Thus, y is the side of a square with the same area as the rectangle with sides a and b.

These are examples of problems (I) of 'application' of an area to a line segment and (II) of the 'quadrature' of an area, respectively:

(I) *To apply a figure F to a line segment s* means to construct a rectangle with the same area as F and having s as one of its sides (it is enough to construct its other side x); written as $sx{=}F$, where x is the line sought, this geometrical construction clearly admits an arithmetical interpretation as a division.

(II) *To find the quadrature of a figure F* is to construct a square with the area of F (it is enough to construct its side y); written as $y^2{=}F$, where y is the line sought, this geometrical construction corresponds to the extraction of a square root.

Euclid presented these constructions, in the context of the geometry of areas, using only straightedge and compasses; the application of a rectangle to a line segment (*Elements* i, 43) and the quadrature of a rectangle (*Elements* ii, 5 and 6). These propositions constitute alternative constructions of the fourth and mean proportionals:

(a) Implicit to proposition *Elements* i, 43 is the notion of the diagonal decomposition of a parallelogram; in the classroom, however, one may prefer to use only the case of the rectangle, which is the only one needed in this context, although the general case is not harder to prove. The proof uses only that a parallelogram is bisected by any of its diagonals, and that if one subtracts equals from equals, then one obtains equals.

(b) The idea behind propositions *Elements* ii, 5 and ii, 6 is the same: to transform a given rectangle into a gnomon which is the difference of two squares. The proofs use only the equality of the areas of the two complements of a diagonal decomposition (for the particular case of the squares).   Once the rectangle is expressed as the difference of two squares, the Pythagorean theorem immediately yields its quadrature.

The drawing of parallels and perpendiculars to given lines, passing through given points, is the only technical skill that is needed in order to be able to manipulate areas by means of the above mentioned propositions. Of course, this is a topic where appropriate tasks provide many opportunities to practice these elementary but important constructions with straightedge and compasses.

It is natural to deal with applications and quadratures of rectangles before considering other figures: the geometric constructions are the simplest ones and the links to the theory of proportions are very close.   However, both operations are easily generalised to an arbitrary polygonal figure, by means of its decomposition into (a finite number of) triangles.   One need only construct, for each triangle of the decomposition, a rectangle with the same area; the application of all the rectangles to the same line segment, as in layers, yields a rectangle equal in area to the initial polygon; by squaring each of the rectangles and by adding all the resulting squares (by means of the Pythagorean theorem) one always obtains new squares as sums. A typical task may be, for example, to give the areas A and B (A bigger than B) and the line segment s and to ask for (1) a rectangle equal to $A+B$ and a side equal to s, (2) a square equal to $A-B$ (see figure 8.17).

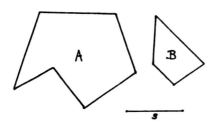

Figure 8.17: Application of areas

Mastering the procedures used in the case of polygons leads to an understanding of the difficulties met in the case of most curvilinear figures and in particular, of the reason why the quadrature of the circle was such an important problem for so long (**7.2.c1**).

A generalisation of the concept of application of areas, considering 'deficient' and 'exceeding' applications, can also be found in Euclid's *Elements* (although his search for generality forces him to postpone it to book vi, where, after the Eudoxan theory of proportions has been exposed, he deals with the similarity of plane figures).   The Greek names for the concepts of *deficiency* and *excess* were used by Apollonius in order to classify the conic sections, and are still in use today in the words 'ellipse' and 'hyperbola' respectively.   These generalisations of the concept of the application of areas are particularly relevant for the mathematics classroom, since they allow for the solution of second degree equations by methods based on the geometry of areas (**7.2.a3**); there are historical texts of medieval Arab mathematicians containing such solutions.   In this context, it is of course very interesting to compare this approach with the radically different one proposed by Descartes in his *Géométrie*.

The Greeks also considered an important generalisation of the notion of the mean proportional. If one inserts any (finite) number of line segments between two given line segments, in such a way that the ratio of any consecutive two segments is constant, then one obtains magnitudes in continuous proportion; this concept corresponds to that of a geometrical progression. An important illustration comes from Hippocrates' reduction of the problem of the duplication of the cube, to that of inserting two mean proportionals between the edge of the given cube and the double of that edge (cf. example 8.4.2a above).

This topic also extends, in a natural way, to consideration of the ratios of other types of magnitudes. The ratio of two areas reduces to the ratio of two lengths by means of the application of both areas to the same line segment. Finally, if one is willing to incorporate the use of curves into the presentation, then the quadratrix and the spiral are most easily introduced, as curves that transform ratios between angles into ratios between lengths. In particular, one obtains easy solutions of the problem of the trisection of the angle and one may also obtain less trivial solutions of the problem of the quadrature of the circle (via the rectification of the circumference).

References for §8.4.4

Aaboe, A. 1964. *Episodes from the early history of mathematics*, New York: Random House
Euclid 1925. *The thirteen books of the elements* ed. T. L. Heath, 2nd edn, Cambridge:
   University Press 1925; reprint New York: Dover 1956

8.4.5   Deductive vs intuitive thinking: an example from the calculus

Dian Zhou Zhang

Mathematics is an exact science, hence in its context deductive thinking is indispensable. However, mathematics is not equivalent to logic. In China, most school mathematics teachers believe that the sole core of mathematics teaching and learning is the development of pupils' logical thinking ability. Any test problem is almost exclusively designed as a logical process, deductively organised. Therefore, in China's 'examination kingdom', *mathematics = logic* is a very popular ideal. Even at the university level, including teacher training courses, every professor emphasises in analysis the importance of the 'epsilon-delta' language. There is even a well-known motto: "Everything is inferior, only epsilon-delta is superior!"

As a reaction to this, the work of Fermat (1638) may be presented, to show that a great mathematical work in analysis might be non-rigorous, without any use of the epsilon-delta formulation, but based more on mathematical intuition than on deductive reasoning (7.2.b).

In his study entitled *Methodus ad disquirendam maximam et minimam* ('Method of finding maximum and minimum') (Fermat 1638/1891; Struik 1969, 223-4; Fauvel and Gray 1987, 358), Fermat gave the following example: Given a segment

John Fauvel, Jan van Maanen (eds.), *History in mathematics education: the ICMI study*, Dordrecht: Kluwer 2000, pp. 279-280

OB, it is required to find a point *A* on it such that the area of a rectangle with sides *OA* and *AB* is a maximum. This area is $A(B-A) = AB-A^2$. He replaced *A* by *A+E* where *E* is an infinitesimal quantity. Then the length of the other segment is *B-(A+E)* and the areas of the rectangle becomes $(A+E)(B-A-E)$. By arguing that near a maximum the values of a function (that is, here, the two areas) do not change, he put them equal, obtaining:

$$(AB-A^2- AE) + (EB -AE-E^2) = AB - A^2$$

Dividing by *E*, he got $B = 2A + E$. He proceeded further by discarding the *E*-term and got $B=2A$, i.e. the rectangle is a square with a side half of the initial segment (see e.g. Boyer 1959, 155-156).

Many school teachers suggest that we can use the following deductive argument: by noticing that

$$AB-A^2 = B^2/4 - (A-B/2)^2$$

it becomes clear that for $A=B/2$, we get the maximum value $B^2/4$.

Which argument is preferable? If we want to solve this particular problem only, the second is more compact. However, Fermat's demonstration is more powerful and deeper, giving a non-rigorous but intuitive elementary application of infinitesimal calculus, of a much more general domain of applicability and capable of considerable generalisation at a higher level. Notice that, by essentially the same argument, one may introduce the concept of the variation of a functional in the calculus of variations (**7.2.c1**).

References for §8.4.5

Boyer, C.B. 1959. *The history of the calculus and its conceptual development*, New York: Dover

Fauvel, John, Gray, Jeremy. (eds.) 1987. *The history of mathematics: a reader*, London: Macmillan

Fermat, P. 1638/1891. 'Methodus ad disquirendam maximam et minimam', in *P. Fermat: Oeuvres*, Vol. I, Paris: Gauthier-Villars, 133-179

Struik, D.J. (ed.) 1969. *A source book in mathematics: 1200-1800*, Cambridge: Harvard University Press

8.4.6   Tracing the root of the abstract concept of a set

Dian Zhou Zhang

In past decades more and more books in mathematics begin with very general abstract concepts such as sets, axioms and categories. In particular, the notion of a set has become a basic concept for every mathematics learner. In 1994, at the beginning of a Chinese graduate course on real and complex analysis, the class was asked to say something about the historical background of set theory. The students replied that Cantor was the creator of set theory in the 19th century and in Cantor's view, any collection of things of any kind could be a set. They also said that, because the concept is too abstract, Cantor suffered from mental illness by thinking too hard on it and finally died at a mental hospital! Notwithstanding the inaccuracy of the story, this seems to be all that the class knew about Cantor and set theory!

However, we know that every mathematical concept has its concrete root. In order to understand better the thinking process of a mathematician we must seek its original historical source. This is well illustrated by the example of Cantor.

Cantor was led into investigating infinite sets when he got interested in the uniqueness problem of representing a function by its Fourier series. He extended the uniqueness theorem of Heine, as well as that of himself, to the case when an infinite set of 'exceptional' points (that is, points at which one knows nothing more about the sum of the trigonometric series) exists. The more general question is: "What kind of infinite sets can be admitted as exceptional sets for the uniqueness theorem still to hold?" Cantor considered a point set A in the interval $[a,b]$. The set of limit points of A is called the derived set A' of A. The derived set of A' is called the 2-derived set of A, and so on. An infinite set with a finite derived set is called a set of the 1st kind. Likewise, if the n-derived set of A is finite, then A is called a set of the nth kind. In 1872 Cantor published a paper in which he pointed out that if in the interval $[0, 2\pi]$ a trigonometric series represents zero for all x, except possibly on an exceptional set of the nth kind, then all the coefficients of the trigonometric series must vanish. This means that the uniqueness theorem on Fourier series is valid for an exceptional set of the nth kind. This work opened up the way for his point set theory to follow. (See Kline 1972, Ch. 40, 41; Dauben 1979, Ch.2 for more detail).

This story helps students to understand the real mathematical thinking process, and is beneficial for them to acquire a correct insight into mathematics in general (7.2.a2, 7.2.d1).

References for §8.4.6

Dauben, J. 1979. *Georg Cantor: his mathematics and philosophy of the infinite*, Cambridge: Harvard University Press

Kline M. 1972. *Mathematical thought from ancient to modern times*, New York: OUPress

John Fauvel, Jan van Maanen (eds.), *History in mathematics education: the ICMI study*, Dordrecht: Kluwer 2000, p. 281

8.4.7   Discrete mathematics: an example

Man Keung Siu

The following is an example taken from discrete mathematics, which cuts across
different levels from school to university; there are different purposes for using this
example as well as different ways of using it. The topic is the famous problem of
the seven bridges of Königsberg, which asks for a way to walk across all seven
bridges, each exactly once, and back to the starting point (see e.g. Ball 1974, ch.9).
A solution was presented by Euler to the St. Petersburg Academy on 26 August
1735 (Euler 1736; for an English translation of the original, in full or in part, see
Biggs et al. 1976, Ch.1; Calinger 1995, 503-506; Struik 1969, 183-187; Wolff 1963,
197-206).

   Told in the form of a story, perhaps with some embellishment (**7.3.1**), this
example can serve as a nice starter for a public lecture for school pupils, along with
an exposition on related topics such as mazes, one-stroke line drawing and real-life
applications under the heading of the so-called Chinese Postman Problem, i.e. to
find an optimal way, in terms of cost or length, to cover all edges in a given network
(see e.g. Biggs *et al.* 1976, Chap.1; Chavey 1992; Steen 1988, Chap.1). Besides the
arousal of their interest, and learning some graph theory and its applications, the
audience can watch a problem expressed in a different cultural context (**7.2.e3**;
Ascher 1991, Chap.2) and experience a taste of problem solving (**7.2.a2**).

   Through consulting contemporary works in the 60s and 70s on matching and
routing algorithms (see Edmonds *et al.* 1973; Guan 1962), this example can be used
to enhance understanding of those algorithms in an undergraduate course in
operational research (**7.2.a3**).

   This example can also be used in an introductory undergraduate course in
discrete mathematics or graph theory. It provides excellent material for students to
witness how an important notion (in this case the degree of a vertex in a graph) and
a basic theorem (in this case the so-called Handshaking Lemma) arise from their
original forms and evolve into the familiar forms in modern textbooks (**7.2.b1;
7.2.c1**). Students can see how a good formulation (not necessarily in the form we
know it today initially) facilitates a solution and gives rise to new developments.
Throughout the memoir of Euler there is no mention of the term "graph" or
"degree", and no record of any picture which resembles our modern notion of a
graph. It is even interesting to note that Euler's explanation is different from, yet
related to, the standard exposition given in a modern textbook on discrete
mathematics (**7.2.b2**). "What the first solution lacked in completeness and polish, it
made up for in clarity, wealth of ideas, and revelation of the author's train of
thoughts" (Siu 1995, 281).

John Fauvel, Jan van Maanen (eds.), *History in mathematics education: the ICMI study*,
Dordrecht: Kluwer 2000, pp. 282-283

A comparison of the differences and similarities between Euler's solution and the standard exposition in modern textbooks, makes for a fruitful case-study in a course for in-service school teachers on the methodology of problem solving (see Siu 1995, 280-281) and on the nature of proofs in mathematics (see Siu 1990, Ch.4; **7.2.c3**). It can also be used in a course on mathematics, or its history through the study of original documents (Euler 1736) (**7. 4.3**; chapter 9 below). Papers on the history of this problem, such as Sachs *et al.* 1988 and Wilson 1986 are helpful.

References for §8.4.7

Ascher, Marcia 1991. *Ethnomathematics: a multicultural view of mathematical ideas*, Pacific Grove: Brooks/Cole; repr. Chapman and Hall, New York, 1994

Ball, W.W.Rouse 1974. *Mathematical recreations and essays*, 13th ed. (with H.S.M. Coxeter), New York: Dover; originally publ. 1892Biggs, N.L., Lloyd, E.K., Wilson, R.J. 1976. *Graph theory: 1736-1936*, Oxford: Clarendon Press

Calinger, R. (ed.) 1995. *Classics of mathematics*, Englewood Cliffs: Prentice-Hall, 1995

Chavey, D. 1992. *Drawing pictures with one line: exploring graph theory*, HistoMAP Module 21, Lexington: Consortium for Mathematics and Its Applications

Edmonds, J., Johnson, E.L. 1973. 'Matching, Euler tours and the Chinese postman', *Math. program* **5**, 88-124

Euler, L. 1736. 'Solutio problematis ad geometriam situs pertinentis', *Commentarii Academiae Scientiarum Imperialis Petropolitanae* **8**, 128-140

Guan, M-G. 1962. 'Graphic programming using odd or even points', *Chinese math.* **1**, 273-277

Sachs, H., Stiebitz, M., Wilson, R.J. 1988. 'An historical note: Euler's Königsberg letters', *J. graph theory* **12**, 133-139

Siu, Man-Keung 1990. *Mathematical proofs* (in Chinese), Nanjing: Jiangsu Educational Press

Siu, Man-Keung 1995. 'Mathematical thinking and history of mathematics', in F. Swetz *et al* (eds), *Learn from the masters!*, Washington: Mathematical Association of America, 279-282

Steen, L.A. (ed.) 1988. *For all practical purposes: introduction to contemporary mathematics*, New York: W.H. Freeman

Struik, D.J. (ed.) 1969. *A source book in mathematics: 1200-1800*, Cambridge: Harvard University Press

Wilson, R.J. 1986. 'An Eulerian trail through Königsberg', *J. graph theory* **10**, 265-275

Wolff, P. 1963. *Breakthroughs in mathematics*, New York: American Library

8.4.8   The relation between geometry and physics: an example

Constantinos Tzanakis

The study of the historical evolution of mathematics and physics reveals their continuous fruitful interaction. By following an approach inspired by history (**7.3.2**), this interaction can and should be unfolded in the teaching process, contrary to what usually happens. Many examples can be given (Tzanakis 1996; 1999;

John Fauvel, Jan van Maanen (eds.), *History in mathematics education: the ICMI study*, Dordrecht: Kluwer 2000, pp. 283-286

2000). Here, we give an example at the undergraduate level, by contrasting its conventional presentation to one inspired by history. This example is also indicative of the close relation between differential geometry and physics (general relativity: GR) (7. 2.a3), of the way by which history can motivate the introduction of a new concept (7.2.a1; 7.2.c1) and may suggest a way to present it (7.2.a1). We are concerned with the introduction of the concept of a *connection*, which describes the idea of parallelism on an arbitrary manifold *M*. Conventionally it is introduced *ad hoc* in the following rather mystifying and unintelligible way (see e.g. O' Neill 1983, Ch.3; Bishop *et al* 1980, §5.7; Choquet-Bruhat *et al* 1982, 300-301):

– A connection ∇ is a mapping from pairs of vector fields to vector fields, $\nabla:(X, Y)$ $\rightarrow \nabla_X Y$, which is linear in the first argument (over real valued functions f) and satisfies the Leibniz rule in the second:

$$\nabla_X(fY) = f\nabla_X Y + X(f)Y$$

– *Y* is called parallel along (the integral curves of) *X*, if $\nabla_X Y = 0$.
– A curve with velocity *X* parallel to itself is called a geodesic ("straightest" line).

If $\left(\partial/\partial u^i \equiv \partial_i\right)$ are the basis vector fields induced by coordinates $\left(u^i\right)$, the Christoffel functions are defined by $\nabla_{\partial_i}\partial_j \equiv \Gamma_{ji}^k\partial_k$ (henceforth, repeated indices denote summations over them). Subsequently it is shown that if $\Gamma_{ji}^{\prime k}$ are the corresponding functions for other coordinates $\left(u^{\prime i}\right)$, then

$$\Gamma_{\alpha\beta}^{\gamma} = \frac{\partial u^{\gamma}}{\partial u^{\prime\nu}}\left(\frac{\partial u^{\prime\kappa}}{\partial u^{\alpha}}\frac{\partial u^{\prime\lambda}}{\partial u^{\beta}}\Gamma_{\kappa\lambda}^{\prime\nu} + \frac{\partial^2 u^{\prime\nu}}{\partial u^{\alpha}\partial u^{\beta}}\right) \tag{1}$$

Conversely, functions transforming as above under a change of coordinates define a connection uniquely.

This approach leaves the following natural questions unanswered:

– Why does ∇ illustrate parallelism? Specifically, why does $\nabla_X Y = 0$ give a 'straightest' line?
– What motivates the use of the term connection?

An answer presupposes the proof of the (local) existence of normal coordinates, hence, it is necessarily *a posteriori* (see e.g. O' Neill 1983, 59, 72-73).

In contrast, by taking into account the historical development of the subject (7.3.2) we outline below another possible approach:

1. A general knowledge of the history of the subject

Although the appearance of Riemannian geometry precedes its physical applications, (i) it was motivated by physical intuition (see quotations from Gauss, Riemann and Clifford in (Mehra 1972, 111; Spivak 1979, 152-153; Clifford 1876/1956, 569), (ii) its further development was (and still is) greatly stimulated by its applications to general relativity (see e.g. Levi-Civita 1927/1977, vii-viii). Therefore, some aspects of differential geometry may be better understood on the basis of (i) and (ii).

2. Crucial historical steps

(a) Galileo's remark that all freely falling bodies in earth's (homogeneous) gravitational field have the same acceleration (Galilei 1632/1954, 65). (b) Einstein's ingenious generalisation to a universal principle, on which he founded general relativity: at an arbitrary point of any gravitational field, all freely falling bodies, irrespective of their nature, move with the same acceleration, which is constant in a sufficiently small neighbourhood of that point (Einstein 1901/1950a, 100; Pais 1982, 195, 205). (c) This implies that Newton's law of inertia is locally valid. By choosing a coordinate system moving with this common acceleration, a body on which non-gravitational forces do not act moves rectilinearly and uniformly.

3. Reconstruction

(a) Mathematically, 2(c) says that at every point p and for any direction (vector) V, there exists an appropriate curve γ and local coordinates (x^a) in which γ has constant velocity V, i.e. zero acceleration, $d^2x^a/dt^2 = 0$. In a Euclidean space, this describes a straight line. In general, this is true only in a neighbourhood of each point p, i.e. γ is locally 'straight'. Now, it is a computational exercise (7.2.a2) to show that in arbitrary coordinates (u^a), this equation takes the form

$$\frac{d^2u^a}{dt^2} + \Gamma^a_{kl}\frac{du^k}{dt}\frac{du^l}{dt} = 0 \text{ where } \Gamma^a_{kl} = \frac{\partial u^a}{\partial x^m}\frac{\partial^2 x^m}{\partial u^k \partial u^l} \tag{2}$$

with the functions Γ^a_{kl} transforming by (1) in a change of coordinates. That is, (2) are the well known geodesic equations.

   This approach answers the 'natural' questions left untouched by the conventional presentation: in 3. above, the idea of a straightest curve is expressed, i.e. a curve with velocity parallel to itself. Hence, in arbitrary coordinates, Γ^a_{kl} expresses 'parallelism', i.e. the possibility to decide whether two vectors at different points are 'parallel'. Hence Γ^a_{kl} establish the connection between the two vectors. Finally, since Γ^a_{kl} in (2) transforms by (1) in a coordinate change, we have the equivalence of the (physically motivated) existence of normal coordinates with the previously given abstract definition of a connection (if the latter is assumed to be symmetric).

   In this example, history appears implicitly (7.3.2iii), given that: (i) Originally, Einstein did not arrive at (2) in this way (Einstein 1916/1950b, section 9; Pais 1982, 203, 220; Mehra 1972, 103), although a few years later he outlined this approach qualitatively (Einstein 1922/1956, 76). (ii) The concept of parallelism was introduced geometrically by Levi-Civita in 1917 in a different way (Levi-Civita 1923/1977, viii and Chap..V(b); Eisenhart 1926, section 24 and references therein). The approach in 3(a) appeared a few years later in Weyl's work (Weyl 1918/1950, 206; Weyl 1952, section 14).

References for §8.4.8

Bishop, R.L., Goldberg, S.I. 1980. *Tensor analysis on manifolds*, New York: Dover
Choquet-Bruhat, Y., DeWitt-Morette, C., Dillard-Bleick, M. 1982. *Analysis, manifolds and physics*, Amsterdam: North Holland
Clifford, W.K. 1876/1956. 'On the space theory of matter' (1876), in: J.R. Newman (ed), *The world of mathematics* **i**, New York: Simon and Schuster, 568-569
Einstein, A. 1911/1950a. 'On the influwnce of gravitation on the propagation of light' (1911), in A. Sommerfeld (ed.), *The principle of relativity*, New York: Dover
Einstein, A. 1916/1950b. 'The foundations of the General Theory of Relativity' (1916), in A. Sommerfeld (ed.), *The principle of relativity*, New York: Dover
Einstein, A. 1922/1956. *The meaning of relativity*, 6th edition, London: Chapman and Hall
Eisenhart, L.P. 1926. *Riemannian geometry*, Princeton: University Press
Galilei, Galileo 1632/1954. *Dialogues concerning two new sciences*, New York: Dover
Levi-Civita, T. 1923/1977. *The absolute differential calculus*, M. Long tr., New York: Dover; repr. from translation published by Blackie in 1926; original Italian edition 1923
Mehra, J. 1972. 'Einstein, Hilbert and the theory of gravitation', in: J. Mehra (ed.), *The physicist's concept of nature*, Dordrecht: Reidel
O'Neill, B. 1983. *Semi-Riemannian geometry*, New York; Academic Press
Pais, A. 1982. *Subtle is the Lord: the science and the life of A.Einstein*, Oxford: Univ.Press
Spivak, M. 1979. *Differential geometry* **ii**, Houston: Publish or Perish
Tzanakis, C. 1996. 'The history of the relation between mathematics and physics as an essential ingredient of their presentation', in *Proc. HEM Braga* **ii**, 96-104.
Tzanakis, C. 1999. 'Unfolding interrelations between mathematics and physics on the basis of their history: two examples', *Int. J. Math. Educ. Sci. Technol.* **30** (1), 103-118
Tzanakis, C., 2000. 'Presenting the relation between mathematics and physics on the basis of their history: a genetic approach', in V. Katz (ed.), *Using history to teach mathematics: an international perspective*, Washington, D.C: Mathematical Association of America
Weyl, H. 1918/1950. 'Gravitation and electricity' (1918), in: A. Sommerfeld (ed.), *The principle of relativity*, New York: Dover
Weyl, H. 1952. *Space, time, matter*, 4th edn, New York: Dover (original Gerrman edn 1918.)

8.5   Improving mathematical awareness through the history of mathematics

8.5.1   History of mathematics education

Hélène Gispert, Man Keung Siu

Mathematics education develops alongside mathematics, each exerting its influence over the other, sometimes in a gradual or indirect way. In this very broad sense the study of the history of mathematics education is helpful to the training of a mathematics teacher. Teacher education belongs to a domain which connects school disciplines and society, posing questions concerning the function of schools and what knowledge society needs. Thus, views on the role and nature of mathematics

John Fauvel, Jan van Maanen (eds.), *History in mathematics education: the ICMI study*, Dordrecht: Kluwer 2000, pp. 286-288

interact with the goals assigned to the school system, which is affected strongly by the society and the culture in which the school system is embedded (**7.2.e2**)). The history of mathematics education, of how school knowledge was constructed, of how social issues partly determined and influenced education, is of pedagogical benefit in the development of a mathematics teacher (**7.3.3b**).

In (Siu 1995), the author attempts to illustrate through a preliminary study of the history of mathematics in ancient China the thesis that "[the] development of mathematics education, and with it, the development of mathematics itself, is to a large extent dictated by the general prevalent Anschauung of mathematics of the community at the time at the place", and to discuss what lesson we can learn from the study. By "Anschauung of mathematics" is meant "the conception one holds of the subject called mathematics, which breeds a frame of mind that will mould one's action". In parallel with mathematics education in the narrow sense, which consists in the transmission of mathematical knowledge, in the broad sense mathematics education is the formation of an Anschauung of mathematics. A teacher who has acquired a historical perspective on mathematics education will be in a better position to help students in this respect. On a more down-to-earth level, this historical study can help a teacher to understand not only the way of teaching the syllabus, but also the origin and reason for its content. In Siu and Volkov 1999, the authors probe further into this area and discuss the state examinations in mathematics in the Tang Dynasty (618-907), thereby offering a somewhat rehabilitated view from the one hinted at in Siu 1995 and helping to shed light on the question of possible cultural difference in the learning of mathematics; see, for instance, Biggs 1996 (**7.2.e2**).

In Gispert 1997 the author reports on her work with pre-service primary and secondary school teachers in studying different mathematics syllabi in France of the 19th and 20th centuries. This includes the study of the accompanying commentaries as well as the debates which were aroused at different times in their political, economical, scientific and pedagogical contexts (**7.3.3b**). Such activities help to render prospective teachers less naive, better equipped for the syllabus they are going to teach and better prepared for any change in the syllabus which they will encounter during their teaching career.

References for §8.5.1

Biggs, J.B. 1996. 'Western misperceptions of the Confucian-heritage learning culture', in: D.A.Watkins, J.B.Biggs (eds), *The Chinese learner: cultural, psychological and contextual influences*, Comparative Education Research Centre, Univ. Hong Kong, 45-67

Gispert, Hélène 1997. 'Les mathématiques dans la culture scientifique et la formation des enseignants', in: J. Rosmorduc (ed.), *Histoire des sciences et des techniques*, Centre Régionale de Documentation Pédagogique de Bretagne, 347-355

Siu, Man-Keung 1995. 'Mathematics education in ancient China: What lesson do we learn from it?' *Historia scientiarum* **4**, 223-232

Siu, Man-Keung, and Volkov, Alexei 1999. 'Official curriculum in traditional Chinese mathematics: how did candidates pass the examinations?' *Historia Scientiarum* **9**, 85-99

8.5.2   Teaching secondary mathematics in a historical perspective

Victor Katz

In a recently established private school in the USA, for which the author has been acting in an advisory capacity, the basic philosophy involves centring the curriculum on the cultural history of the world (**7.3.3b**). All aspects of the curriculum, including language arts, visual arts, science, and mathematics are tied into that core cultural history. Thus, ideally, students should be studying the mathematics of a particular time period at the same time they are considering the history, art, and literature of that period. In this way, students will understand the role of mathematics in the development of civilisation (**7.2.e3**). It is, of course, also necessary to structure the mathematics curriculum so that students, by the time they graduate, master all of the mathematics that a typical high school student in the USA will have learned by that time. Although this curriculum is only under development, the following is an indication of how this is working out in grades 5 (10-11 year olds) and grade 9 (14-15 year olds).

In grade 5, the curriculum explores ancient Mesopotamia, Egypt, and India from approximately 3000-1000 BC. Thus, the students study such topics as the development of the base-60 place value system and its connection with the decimal place value system; the extension of these systems to fractions; the basic formulas for perimeter, area, and volume; the notion of a square root; an introduction to algebra using false position; and the Pythagorean theorem. In grade 9, where the students are studying the period from about 1450-1650, the mathematical topics include solid geometry, especially the geometry of the sphere (so that students can understand something of navigation in the age of discovery); similarity and its application to perspective; the basics of the conic sections; the solution of polynomial equations, including the cubic formula in the work of Cardano and the subsequent discovery of complex numbers; trigonometry, through a reading of the first book of Copernicus' *De Revolutionibus (1543)*; the idea of a mathematical model in the work of Galileo; and the beginnings of analytic geometry in Descartes and Fermat. Although it remains to be seen whether the entire secondary curriculum can be dealt with in this manner, this development is an exciting new way of integrating history of mathematics with mathematics teaching and learning.

John Fauvel, Jan van Maanen (eds.), *History in mathematics education: the ICMI study*, Dordrecht: Kluwer 2000, p. 288

8.5.3   Adults' mathematics educational histories

Gail FitzSimons

This section includes a description of some activities utilised by the author in a class of women returning to study mathematics, in an informal setting. Although the subject was part of a recognised credential, curriculum and assessment were at that time negotiable. The intention of the course was to provide the students with the mathematical skills they wished to learn; possibly, but not necessarily, with a view to further study or to gaining employment. For some, the expressed intention was to be able to help their children with mathematics homework throughout the different stages of schooling—an important social and economic consideration according to Faure et al 1972. Accordingly, the aim of integrating history was in the form of a general cultural and social awareness of mathematics and ethnomathematics (**7.2.e**; **7.3.3b**). The goals included assisting the women to overcome mathematics anxiety, to better connect mathematics with the rest of their lives, to view mathematics as a fallibilist discipline, and to enhance their metacognitive skills by reflecting on their previous mathematics learning experiences (for more detail, see §6.2.3, above).

At various times the history of mathematics was used for:

a) teaching *through* history and ethnomathematics (**7.3.2**),

b) teaching *about* history and ethnomathematics (**7.3.1**) and

c) encouraging students' reflection on their personal history of mathematics education to encourage metacognition.

Thus, aspects of mathematics related to philosophy, art, architecture, natural and social sciences, for example, were integrated into classes, as was the cultural heritage of different societies at different periods (**7.2.e3**). These three foci could equally apply in other sectors of education, as will be demonstrated below.

There are many reasons why adults return to study mathematics to pursue further or vocational education (FitzSimons 1994). Along with some teacher education students, adults frequently exhibit signs of anxiety, if not low self-esteem, in mathematics at least. They are likely, at some point, to have experienced mathematics as absolute, cold and unwelcoming, with instruction having been aimed primarily at other more able students in the class. Mathematics may have even been used to classify and position them. Re-entry to the study of mathematics *per se* is not always the choice of the student, but may be a requirement imposed by course regulations or other authorities. The task of the mathematics instructor is not only to teach mathematics, but also in many cases to help the students find new approaches to the subject and how it might be learned; even to overcome difficulties arising from past experiences of learning mathematics.

Asking adult (and teacher education) students to reflect on their past mathematics education experiences serves many purposes. It enables the instructor to know more about the students and to plan more appropriate and meaningful learning experiences. More importantly, it helps the students to articulate their

John Fauvel, Jan van Maanen (eds.), *History in mathematics education: the ICMI study*, Dordrecht: Kluwer 2000, p. 289-290 beliefs and attitudes about the nature of mathematics and of how it is learned. Once these are made explicit it is more likely that they can be addressed, enabling the possibility of a greater breadth of perspective.   It is also a step on the way to developing metacognitive skills.

Both cognitive and affective domains are likely to be invoked in the presentation of personal mathematics education histories. A study (FitzSimons 1995) of women voluntarily returning to study mathematics given an open-ended task of reflection, indicated that the following categories were considered important in their memories of previous schooling:

(a) *content*: lists of topics covered, especially the four basic processes and the emotions evoked by these items;

(b) *pedagogical practices* of their teachers, both positive and negative, and the resulting self-images produced (somewhat different from the majority of mathematics educators!);

(c) *external influences* which affected their mathematics and other education, such as the effects of their parents' and teachers' attitudes towards their gender, the experience of war (common to many immigrants to Australia), and the setbacks associated with moving house, country, or even mathematics groups; and

(d) *thoughts and emotions* about the act of returning to study mathematics.

The study also presented evidence of journal writing reflecting the integral part played by the use of history of mathematics in the classes.

References for §8.5.3

Faure, E., Herrera, F., Kaddoura, A-R., Lopes, H., Petrovsky, A. V., Rahnema, M., Champion Ward, F. 1972. *Learning to be: the world of education today and tomorrow*, Paris: UNESCO

FitzSimons, Gail E. 1994. *Teaching mathematics to adults returning to study*, Geelong: Deakin University Press

FitzSimons, Gail E. 1995. 'The inter-relationship of the history and pedagogy of mathematics for adults returning to study', paper presented to the *International Study Group for the Relations of History and Pedagogy of Mathematics*, Cairns

Chapter 9

The use of original sources in the mathematics classroom

Hans Niels Jahnke

with Abraham Arcavi, Evelyne Barbin, Otto Bekken, Fulvia Furinghetti, Abdellah El Idrissi, Circe Mary Silva da Silva, Chris Weeks

Abstract: *The study of original sources is the most ambitious of ways in which history might be integrated into the teaching of mathematics, but also one of the most rewarding for students both at school and at teacher training institutions.*

9.1   Introduction

Among the various possible activities by which historical aspects might be integrated into the teaching of mathematics, the study of an original source is the most demanding and the most time consuming. In many cases a source requires a detailed and deep understanding of the time when it was written and of the general context of ideas; language becomes important in ways which are completely new compared with usual practices of mathematics teaching. Thus, reading a source is an especially ambitious enterprise, but, as we want to show, rewarding and substantially deepening the mathematical understanding. In this chapter we describe some ideas and international experiences concerning the use of original sources in the mathematics classroom, referring to teaching at schools as well as at teacher education institutions.

In principle, the aims and effects which might be pursued by way of an original source will not be different from those attained by other types of historical activities. However, there are three general ideas which might best be suited for describing the special effects of studying a source. These are the notions of *replacement, reorientation* and *cultural understanding*. By these we mean:

John Fauvel, Jan van Maanen (eds.), *History in   mathematics education: the ICMI study*, Dordrecht: Kluwer 2000, pp. 291-328

(i) replacement

Integrating history in mathematics replaces the usual with something different: it allows mathematics to be seen as an intellectual activity, rather than as just a corpus of knowledge or a set of techniques.

(ii) reorientation

Integrating history in mathematics challenges one's perceptions through making the familiar unfamiliar. Getting to grips with a historical text can cause a reorientation of our views. History of mathematics has the virtue of 'astonishing with what comes of itself' (Veyne 1971). All too often in teaching, what happens is that concepts appear as if already existing. This is true for the concept of a set, for example, but just as true for the concept of a triangle or a function. And concepts are manipulated with no thought for their construction. History reminds us that these concepts were invented and that this did not happen all by itself.

(iii) cultural understanding

Integrating history of mathematics invites us to place the development of mathematics in the scientific and technological context of a particular time and in the history of ideas and societies, and also to consider the history of teaching mathematics from perspectives that lie outside the established disciplinary subject boundaries.

In this chapter we begin with discussing motivations, aims and uses which are especially connected with the study of original sources (section 9.2). Of course, there is some overlap with the general aims underlying the introduction of historical components, but we concentrate on those dimensions specific for our topic. We discuss especially the hermeneutic process of interpreting a source and the special role of language in it (section 9.3). In a further step we investigate four examples, two taken from the context of teacher education (sections 9.4.1 and 9.4.2) and two from school teaching (sections 9.5.1 and 9.5.2). The special reference to teacher education is motivated by our conviction that the reading of original sources should become an obligatory part of mathematics teacher education at all levels. In section 9.6 we deal with didactic strategies, and in section 9.7 discuss some research questions and issues of concern. Section 9.8 is the bibliography for this chapter, and in the appendix, 9.9, the reader will find hints on useful resources.

9.2   Motivations, aims and uses

9.2.1   The specific value and quality of primary sources

The role of primary sources in the integration of history of mathematics into mathematics education should be considered in the light of different possible purposes. Incorporating primary sources is not good or bad in itself. We need to establish the aims, including the target population, the kind of source that might be suitable and the didactical methodology necessary to support its incorporation.

In the following, we describe some objectives and examples of how primary sources help to pursue them. There are almost certainly further ones we do not mention (cf. Arcavi & Bruckheimer 1998; Fauvel 1990 (see especially the papers by Jozeau, Bühler, Hallez, Horain); Furinghetti 1997; IREM de Montpellier 1995; Jahnke 1995; Laubenbacher & Pengelley 1996, 1998; Lefebvre & Charbonneau 1991; LeGoff 1994; Logarto *et al.* 1996; M:ATH 1991; Métin 1997; Nouet 1992).

In contrast to merely relying on secondary literature the reading of primary sources may help to

a) clarify and extend what is found in secondary material,
b) uncover what is not usually found there,
c) discern general trends in the history of a topic (secondary sources are usually all-topic chronological accounts, and some topics are very briefly treated or omitted altogether), and
d) put in perspective some of the interpretations, value judgements or even misrepresentations found in the literature.

Reading historical texts may produce a cultural shock, by which we may experience the *replacement* and *reorientation* referred to above. This will only happen, however, if the reading is not *teleological,* that is, provided we do not attempt to analyse the text uniquely from the point of view of our current knowledge and understanding. Such a reading could carry with it erroneous interpretations, given that the writer may be using an idea according to a conception quite different from ours. If the value of history lies in reorientation, in understanding rather than judging, then texts need to be *contextualised,* that is located in the context of their time. We need to remind ourselves that the writer was addressing not us, but a contemporary audience.

To have our perspectives of knowledge challenged is beneficial. Thus, it is important to read Descartes' *Geometry* (1637) being aware that the text was not understood by his contemporaries. We would then pay more attention to the changes brought about by Cartesian geometry, for example by the introduction of a unit segment, which appears so 'natural' in coordinate geometry that it passes by almost unnoticed. We can also show that the coordinate geometry system works in a way that can be related with the Section Theorem in the geometry of the triangle (Euclid's *Elements* vi.2, sometimes called Thales' Theorem: that a line parallel to one side of a triangle cuts the other sides in the same ratio), something which appears to be quite absent now from the official curriculum in many countries. This example shows that the *replacement* and *reorientation* aspects of history are directly linked to didactical considerations.

Reading historical texts in class introduces history in an explicit way. Nevertheless, this activity has to be integrated into the mathematics lessons and not provided just as an extra. It also presupposes that the teachers have a sense of history and, of course, that they are able to handle the mathematics involved. Thus, reading sources presupposes adequate preparation (see Chapter 4).

9.2.2   Understanding the evolution of ideas

There is a common belief held by many, teachers and students alike, about the static nature of mathematical concepts: once a concept is defined, it remains unchanged. Even those who do not hold this belief may not have had opportunities to experience the evolving nature of ideas. Take for example the concept of *function*. At some early stage, functions were restricted to those which could be expressed by algebraic relationships. Later, the concept was extended beyond correspondences which can be expressed algebraically, and later still to correspondences not involving sets of numbers at all. Thus we have the more general and formal definition today: a subset of the Cartesian product of two sets with certain properties. In another sense, the concept was restricted to univalent relationships. (For a detailed discussion of the history of the function concept see, for example Youschkevitch 1976 and for a brief survey Kleiner 1989.) We suggest that primary sources can offer the experience of a non-mediated contact with the way in which ideas were defined at a certain time, different from that in use today.

Another example is the notion of a *curve*. Curves seem to be considered the same throughout the school programme. The circle, however, can be variously presented: as a static object in geometry, consisting of points at equal distance from its centre; as a dynamic object produced by the rotation of a line segment about one of its (fixed) extremities; as an object in algebra, namely an equation; or as a functional object. History can make us aware of the significance of these different ways of thinking about a curve through letting us understand the problems that led mathematicians to pass from one notion to the other, and also to see the nature of the changes in conception that came about (Barbin 1996). For example, the dynamic notion of a curve in the 17th century is linked to problems about movement that scientists of the time were considering. In particular, we can see from reading *Dialogues on the two new sciences* how Galileo changes the (static) parabola of his study into the (dynamic) trajectory of a cannon ball. Whereas the parabola of Greek geometry is the intersection of a cone and a plane, the Galilean parabola becomes the trajectory of a moving body, subject to a uniform horizontal and a uniformly accelerated vertical movement.

In order to see how the idea of a curve evolved and became refined, it is interesting to read and compare several historical texts, for example to look at the methods for finding tangents found in the works of Euclid, Apollonius, Roberval, Fermat, Descartes, Leibniz and Newton. Similarly, to see how the idea of function or number has evolved and become refined, it is important to read texts related to stages of their history.

Primary sources provide also lively examples of how different *representational systems* were used in the past. These examples may help students to put into perspective our current representational systems as just one of many possible ways of performing operations and handling and communicating concepts. Moreover, by comparing and contrasting our representations with those in the past as they appear in original sources, students might appreciate the crucial role representations play in the inception and evolution of ideas.

In Arcavi 1987 an activity for elementary school students is described, in which a brief extract from the Rhind Papyrus is presented; with the aid of an accompanying 'dictionary', the challenge consists of deciphering the arithmetical operations performed, explaining how they work, and applying them to further examples. This activity serves as the basis for discussion of the characteristics of the Egyptian numeration system as opposed to ours, including advantages and disadvantages of both. Van Maanen (1997) describes similar experiences with primary sources from a later period. His students report that they find it a difficult but very interesting puzzle, first to find out what the handwritten text says and then what it meant and why it worked. Furthermore such a problem makes students aware that methods and standards are changing. When students compare and contrast the representations they know and use at school with those in original sources, they not only learn about the latter, but most importantly, their attention is re-focused on the former, providing an opportunity to re-discover properties taken for granted and which were "clogged with automatisms" (Freudenthal 1983, 469).

9.2.3   Experiencing the relativity of truth and the human dimension of mathematical activity

The fact that the idea of truth is relative can be seen when we consider how the significance of proof has changed in history (Barbin 1994). While the first reasonings in Greek geometry had to do with explaining real problem situations, like the problem of finding inaccessible distances, the purpose of logical proof in Euclid was to convince, or even defeat, the (supposedly sceptical) reader. This idea of proof was denounced in the 17th century by geometers who preferred to enlighten rather than defeat their readers. As for the idea of proof in Hilbert's geometry, it is conceived of as a way of deciding the validity of a proposition, that is to determine whether or not it is consistent with a set of formal axioms. To obtain a feeling for what proof means, it is interesting to read a variety of proofs of the same theorem, for example the different proofs for the sum of the angles of a triangle given over the two thousand years from Euclid to Hilbert (Barbin 1995).

It is also illuminating to study examples of doubts and errors which arose when mathematicians were working on new problems and concepts. This is different from the usual presentation of mathematical activity, described by Kessel in this way (Kessel 1998, 44):

> This detached style of speaking and writing about mathematics suggests to listeners and readers that mathematics is independent of time and place ... ideas that are not tied to specific people, times, and places, but which are abstract and timeless ... and which avoids mentioning concrete doers.

Thus, in many classrooms all over the world, mathematical activity is generally perceived as the production of clean and correct answers to problems. Alternative, recent experiences (e.g. de Abreu 1998; Arcavi *et al.* 1998; Farey & Métin 1993; Lampert 1990; Pirie & Schwarzenberger 1988; Voigt 1985; Wood 1998) are beginning to include the sharing of intuitions, conjectures, the development of heuristics, and the encouragement of reflection and communication. All these legitimise the explicit raising of doubts, committing errors, entering blind alleys, and discussing seemingly non-solvable contradictions.

Thus, primary sources can provide lively documented examples of genuine mathematical activity in the making, and reading them may legitimise and humanise it ("if famous mathematicians went through it, why not I?"). Moreover, these doubts become issues for discussion with the potential of enriching students' formal and informal knowledge of a topic, and their ability to 'talk mathematics'.

For example, one could confront teachers with the doubts mathematicians had in the 16th and 17th century regarding the nature of irrational numbers (Arcavi *et al.* 1987). In the discussion of the source, teachers may dare to express their own discomfort and/or uncertainties about the 'infinite' decimal representation of irrational numbers. They can also share in the struggle between the usefulness of the concept of irrationals when rationals fail (e.g. in geometrical measurements) and their uncertain nature as numbers. For those less troubled by such problems, the discussion serves to develop an awareness that the infinite digits in the decimal expansion of an irrational were regarded as problematic to the point that their status as numbers was questioned. By implication, this leads to a recognition that this can be an issue with students as well, and to reflecting on the crucial importance of the role of representations of a concept, their influence on the way the idea is conceptualised, questioned, and ultimately accepted or rejected.

9.2.4   Relations between mathematics and philosophy

The contribution that the history of mathematics makes to our understanding of the cultural context is an excellent opportunity, or a necessary reason, for relating mathematics to other fields of knowledge (see Furinghetti & Somaglia 1998). Frequently, mathematicians were also philosophers and it is quite artificial to separate their disciplines (Barbin & Caveing 1996). In any case, it is often beneficial to read mathematicians with an awareness of the prevailing philosophy of their time. Consideration of the relationship between mathematics and the real world will benefit enormously when mathematics teachers work collaboratively with teachers of the physical sciences. The example we quoted above concerning Galileo illustrates this point.

Reading a source can be the trigger for establishing a dialogue with the ideas expressed. The source then becomes an interlocutor to be interpreted, to be questioned, to be answered and to be argued with. This applies especially to sources which discuss meta-mathematical issues such as the nature of the mathematical objects we handle, and the essence of mathematical activity. For example, one can use extracts taken from *The principles of algebra* (1796) by William Frend (1757-1841) in which it is proposed that negative numbers should be banned. Frend's arguments against negative numbers raised, and continue to raise for students today, serious discussions on issues such as the use of models, analogies, or metaphors in mathematics (such as debts in accounting); the legitimacy of creating new ideas, provided they are well-defined and internally consistent; the ambivalence of symbols when used in allied but yet different meanings; and the need for formal definitions of concepts such as negative numbers.

9.2.5   Simplicity, motivation and didactics

Occasionally, primary sources can be used because they are simpler and friendlier than their later elaboration.   One notable example is Dedekind's (1831-1916) definition of real numbers, as it appears in his essay *Stetigkeit und Irrationalzahlen* (1872) (see *Essays on the theory of numbers*, 1924).   His style is didactical and clear, first explaining the method to be followed, then using an analogy in order both to engage readers' established knowledge and also to share with readers his sources of insight.   Only after that are the formal definitions carefully developed step by step.   Simplicity and friendliness can also be found in the sense-making explanations proposed in some primary sources for basic but formal mathematical laws, which teachers and curriculum designers struggle to find.   As we progress in history, especially through the 20th century, many texts tend to adopt formal justifications to formal laws, and many students may feel alienated.   However, some older texts often resort to everyday language and reasonable explanations which can enrich the didactical repertoire of teachers by appeal to students' sense making.   Such is the case with Viète's (1540-1603) presentation of simple algebraic laws, in his *In artem analyticem isagoge* (see Bruckheimer & Arcavi 1997).

9.2.6   Perspectives on mathematics education

Primary sources seem to be a most reasonable way to learn about the central topics taught in schools in the past, curricular trends in general and various approaches to learning and teaching.   One activity that Bruckheimer *et al.* (1995) designed for classroom use with 12-13 year old students is based on old arithmetic textbooks, which give the flavour of what and how students studied in the past: methods of calculation were a central topic, and accuracy was a major preoccupation.   There are whole sections devoted to calculation checks, such as 'casting out nines'.   This checking method, as it appears in primary sources, provides an opportunity to deal with many fundamental topics: why does the method work, which kinds of errors can and which cannot be detected, why 9 is preferred to, say, 2, or 7, and so on.

Besides the flavour of past textbooks and dealing with mathematical issues, the sources provide, by implication, the realisation that the goals for mathematics education have changed rather dramatically over the last 100-150 years.   In the past, mathematics instruction for all ('all' in the past was probably more restricted than 'all' is regarded today) may have been mainly devoted to producing good clerks who could calculate accurately.   Today, with the emergence of freely available calculators and the demands of a technological society, the emphasis in arithmetic shifts towards estimation, reasonableness of answers, etc., and other signs of mathematical literacy.

9.2.7   Local Mathematics

Primary sources can also be used in mathematics to rediscover and emphasise the heritage of the culture in which students learn. As most cultures have written mathematical documents (and certainly verbal accounts of everyday mathematical practices), it is not hard to find appropriate sources suitable for classroom use or teachers' workshops.

9.3   Sources, hermeneutics and language

Reading an original source is a specific activity of relating the *synchronous* and the *diachronous* mathematical culture to each other (cf. Jahnke 1994, 154 ff.). The term synchronous culture refers to dialogue and work in the classroom as well as the role of mathematics in public life, in economy, technology, science and culture and the image which is attached to it. The diachronous culture means the development of these elements through history and has to be related to the synchronous culture and the life and thinking of the learners. However, it should not simply affirm the synchronous culture, but should rather widen and deepen the understanding of the learner.

In traditional theories of hermeneutics the relation between the historical meaning of a text (the intention of its author) and its meaning for a modern reader is amply reflected and identified as the essential problem of interpretation. In fact, seen under the aspect of method, history of mathematics, like any history, is essentially an hermeneutic effort. If history of mathematics is not to deteriorate into a dead dogma, teachers should have some ideas about the hermeneutic process and the fruitful tension between the meaning of a text in the eyes of its author and the meaning for a modern reader.

The process of interpreting an original mathematical source may be described by a twofold circle. Texts and their authors (or theories and their creators) are interpreted by a modern reader, and the interpreter should always be aware of the hypothetical and intuitive character of his interpretation. The interpretation takes place in a circular process of forming hypotheses and checking them against the text given. In the case of history of science, the objects of this process of interpretation, the scientific subjects (individuals or groups) are themselves involved in a hermeneutic process of creating theories and checking them against phenomena which they want to explain or against intended aims they want to reach. Thus, the whole process of interpreting a source may be described by a twofold circle where in a primary circle a scientist (or a group of scientists) is acting and in a secondary circle the modern reader tries to understand what is going on. Those concerned with history have to engage with a complex network of relations between their own interpretations of a certain concept or theory and the interpretation of the original author.

Teachers should be aware of this twofold circle and able to move in it. Only this will create a climate in the classroom adequate for encouraging students to generate their own hypotheses about a text and so become ready for thinking themselves into other persons who have lived in another time.

This thinking into other persons and into a different world seems to be the core of an educational philosophy underlying the reading of original sources. She who thinks herself into a scientist doing mathematics at a different time has herself to do mathematics; she moves in a mental game in the primary circle reflecting what the person under study might have had in mind. One has to ask for the theoretical conditions this person is explicitly or implicitly supposing, and one will have to mobilise imagination to generate hypotheses about them.

Thinking themselves into other persons motivates students to reflect about their own views of the subject matter. This reflection, in turn, is made objective by the material (the text) they are studying. Certain aspects of the historical persons and their ideas will be easily accessible, others will remain alien. As a crucial point in hermeneutics, the student's self will unavoidably enter the scene, not as a disturbing factor, but as a decisive prerequisite to insight.

Even if an original source is given in the native language of the students its interpretation presupposes a considerable linguistic competence. This requirement should be accepted by teachers and students. Oral and written language are equally important. The students should have the opportunity of extensive discussions, but they should also be asked to produce their own written texts. The idea of a 'mathematical essay' is old and sounds, since it is never realised, a bit antiquated. Historical subjects would provide natural starting points for such activities.

An important aim should be the elaboration of the individual language of the students. In reading a source they are confronted with at least three different languages: the mathematical language of their usual lessons, the language of the original source, and their own way of speaking about mathematics. These three languages have to be related to each other, and the students should be able to move freely from one language into the other. This should be a general educational aim of mathematics teaching beyond the special occasion of history of mathematics. When in their future lives students practise mathematics, they will need above all to communicate and translate ideas and facts into mathematical language and vice versa. History of mathematics contributes considerably to the development of this ability.

9.4   Integrating original sources in pre-service teacher education

As we said above, the reading of original sources should become an obligatory part of mathematics teacher education at all levels. This will not only contribute substantially to their mathematical competence, but is also a necessary condition if they are expected to include historical components into their future mathematics teaching. In the following we describe two experiences with original sources from teacher education institutions, the first from Morocco, the second from Norway.

9.4.1   Example 1: Egyptian measures of angles

In this section, we present an example of using an original text in the pre-service education of mathematics teachers. The objective was to initiate an analysis of trigonometric notions, in particular the concepts of cotangent, tangent and angle. The example was treated at the École Normale Supérieure in Marrakech, Morocco (see El Idrissi 1998). The text used is an extract of the Rhind Papyrus, written in the 17th century B.C. and now in the British Museum, London. It contains problems together with their solutions. The text was originally in Egyptian hieratic script; we refer to a 20th century English translation (Gillings 1972). The example here concerns reckoning a pyramid, problem 56 from the Rhind Papyrus (*RP 56*) and its solution:

A pyramid has a height of 250 cubits and a base of 360 cubits. What is its sekt?

Solution:

1) Find 1/2 of 360: 180.

2) How many times is 250 in 180: 1/2 1/5 1/50 yard

3) Now a yard is 7 palms.

4) Then multiply 7 by 1/2 1/5 1/50: 5 1/25 palms. This is its sekt.

The above extract was presented to teacher students, and they were confronted with questions and proposals for activities. Actually, an analysis of the problem was done even before the questions were posed. The most important elements emerging from this analysis were:

1. The calculation is given by means of unit fractions.
2. The question is asked about an empirical case, a pyramid.
3. The solution is given without any definition or justification. It is an algorithm for calculating.
4. In the first stage of the solution, the student is told to find the half of 360 and not to divide 360 by two. These two seemingly similar operations are conceptually different.
5. In the second stage, the result is given together with a unit, the yard. In principle, there should not be any units as the intention is to divide yards by yards.
6. In the third stage, the students transform a result given in yards into palms.
7. A naive interpretation of the solution could make believe that the sekt is identical with the cotangent (Smith 1958). Taking into account the earlier remarks, however, sekt and cotangent are different.
8. The sekt can be defined as the horizontal shift in palms which corresponds to a vertical shift of one yard (see figure 9.1).

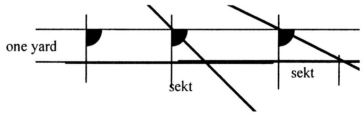

Figure 9.1: The sekt

9. The sekt can be considered as a measure of angles.

On the basis of this analysis, activities were suggested to the students, working in groups of two or three. These activities were to prompt students' reflections on trigonometrical concepts like cotangent, tangent, angle.

The goal of the first activity was to define the sekt. As we have mentioned, the spontaneous answers given by the students tended to identify the sekt with the cotangent. After they had been asked to observe and to note the position of the units in the given solution, several students succeeded in giving more appropriate definitions of the sekt.

With the objective of helping them to consider the sekt as a measure of angle, we asked them to measure the sekt of certain angles while using the metric system, a centimetre corresponding to the cubit. They were also asked to compare an angle of sekt *s* with other angles whose respective sekts were *s*/2 and 2*s*. With the same aim, we asked the students to solve other problems posed in the Rhind Papyrus in which the given and unknown properties are different, while using a reasoning analogous to that of *RP 56*.

We also asked them to guess how the Egyptians, on the basis of the sekt, might have proceeded to construct the pyramids. This question illustrates the fact that a straight line has a constant growth rate. Another and no less fascinating activity consisted in constructing an instrument to measure the sekt of angles, to provide it with a name and to compare the measurements of angles done by means of a sekt and by means of degrees. The activity of constructing real instruments was very dynamic. Indeed, the participants made great efforts to succeed. Some groups achieved classical results, while others showed more originality by providing instruments using glides (see figure 9.2). Two names were proposed for these instruments (in French, as the language of instruction): *seketeur* and *sektomètre*, the second name being maintained.

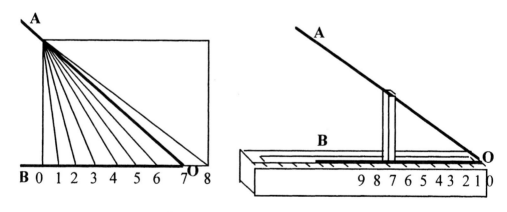

Figure 9.2: Instruments to measure sekts

Comparing degree to sekt raises the problem of the linearity of the concept of cotangent and tangent. Classes discussing the issue are led to understand the advantages of using the degree, and consequently of using to circle arcs to measure angles.   Thus, if two angles (OA,OB) and (OB,OC) are given, and S_1 and S_2 are their sekts respectively, the sekt of their sum (OA,OC) is not the sum ($S_1 + S_2$) of their sekts. Speaking trigonometrically, this signifies that the cotangent function is not linear:

ctg (*OA,OB*) + ctg (*OB,OC*) > ctg [(*OA,OB*)+(*OB,OC*)]

The same is true for the tangent function. Measuring in degrees, however, the measure of the sum of angles is equal to the sum of the measures of the angles (see figure 9.3).

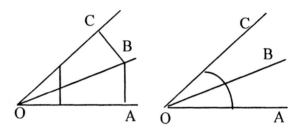

Figure 9.3: Measuring the angle, with the sekt and with degrees

These are the main activities offered to the students. We now describe how history was used and how we were able to profit from it for the education of trainee teachers.

1.   The history of mathematics is first involved in introducing the text. The extract is presented and placed into context. Some information is provided about the Egyptian culture and about research into it—the problem *RP 56* also provided an occasion for discussing the notation and concept of the unit fraction, the construction of pyramids, etc.

2.   The second part of studying the extract consists in analysing the reasoning of the answer presented in it. This analysis tries to keep as close as possible to the Egyptian way of thinking. While we cannot pretend to have identified the underlying Egyptian reasoning in all its details, an effort was made to draw the students' attention to the contextual components which are involved in analysing this reasoning. In fact, this analysis is in some respects an introduction to the reasoning of future pupils. It can be noted, for instance, that young pupils do not take great pains to justify their own reasoning altogether. They are sometimes quite content with using some ambiguous properties or operations provided these will yield correct results.

3.   History is used in the above as a pretext to work on certain practical properties from the concepts of incline, tangent, cotangent and angle. The practical interest of these properties is inspired by the ancient character of the text considered.   The historical problem RP56 enabled us to proceed to a comparison of the concepts of sekt, cotangent, and of measuring the angle in degrees.

4. The history of mathematics is in fact used in this example as a crucial motivational element for an epistemological analysis. The latter consists in analysing, from the perspective of teaching, concepts, reasonings and methods used by the ancients, and the difficulties and obstacles which have impeded the evolution of concepts or methods. Thus in this example we have complemented the historical or mathematical analysis proper by activities appropriate for the education of future teachers.

It may be concluded from the above that original texts, even in translation, may be used in a most relevant and fruitful way. To ensure the best contribution to the educational process, however, they must be carefully selected, well analysed, and presented in a dynamic and interactive way.

9.4.2   Example 2: complex numbers in geometry and algebra

A vulgar mechanick can practice what he has been taught, but if he is in error, he knows not how to find out and correct it, and if you put him out of his road, he is at a stand. Whereas he that is able to reason, is never at rest till he gets over every rub. (Newton 1694)

The course MATH 9 at Kristiansand university

This course, first put on in 1978, was intended as preparation for teaching, bearing in mind that there are different ways to integrate history into the mathematics curriculum:

(i)   Following genetically the historical development while teaching a theme;
(ii)   Using historical problems and examples as a treasure chest to illustrate a subject;
(iii) Opening the student's mind to the fact that mathematics is continually refining its theories, by seeing the historical struggle to develop solutions to problem situations, with new conceptual ideas and theories of understanding.

To read excerpts from original sources should contribute to a critical and more robust understanding of the methods of today. It enables students to work with problems from the origin of a concept, to look at historical mistakes, the etymology of words and the development of notation. The lecture notes (Bekken 1983 and 1994) were put together to help discuss

–   issues from our teaching of algebra through historical material,

–   the growth of ideas and their forms in algebra,

–   in a problem solving style,

–   with excerpts from sources, and

–   with mathematical problem studies.

   Sub-themes were developments of number concepts, like irrationals and imaginaries, symbolisation, and accepted proofs, or demonstrations.

Sources for understanding complex numbers

As Norwegians, we studied the work of a fellow countryman, Caspar Wessel. One of his concerns was how to add and multiply directed lines in the plane. Wessel's solution, first presented in 1796, provides a good introduction to the teaching of complex numbers, because in this source Wessel gave the geometric representation of complex numbers as it is taught today. It is often overlooked that this came out of his attempt to add and multiply directed line segments, vectors as we now call them. In Wessel 1797 (see Nordgaard 1959) we find:

§4. The product of two lines of length 1 in the same plane as the positive unit and with the same starting point, should be in the same plane, with an angle of direction to the unit being the sum of the direction angles of the factors.

§5. Let +1 denote the positive unit, and let a certain perpendicular unit with the same starting point be $+\varepsilon$. The direction angles of +1 $= 0°$, of $-1 = 180°$, of $+\varepsilon = 90°$ and of $-\varepsilon = 270°$. To obtain the rule of §4, we have to multiply according to:

	1	**-1**	ε	$-\varepsilon$
1	1	-1	ε	$-\varepsilon$
-1	-1	1	$-\varepsilon$	ε
ε	ε	$-\varepsilon$	-1	1
$-\varepsilon$	$-\varepsilon$	ε	1	-1

From this we see that ε *becomes* $= \sqrt{-1}$, and the product follows the usual algebraic rules.

§7. The line having direction angle v to the unit +1 is $\cos v + \varepsilon \sin v$ and when multiplied with the line $\cos u + \varepsilon \sin u$, the product becomes the line with direction angle $v+u$, denoted by $\cos(v+u) + \varepsilon \sin(v+u)$.

§9. The general representation of a line of length r and direction angle v to the positive unit +1 is $r(\cos v + \varepsilon \sin v)$.

Next Wessel demonstrates that he knows very well how this relates to imaginary numbers, and explains the fractional Euler-de Moivre formula. Thus, Wessel had found a new application of imaginary numbers: to the geometry of plane positions.

In this way he also solved another important problem of his time: to give imaginary numbers a geometric representation.   This is, in other words, to reconnect the meaning of general numbers to something geometric, but in fact this problem is nowhere mentioned by Wessel.

Glushkov (1977) points to the product of triangles (figure 9.4), introduced by Viète (1591/1983), which we can connect with Wessel's product of directed line segments. Students are asked to explore and explain this.

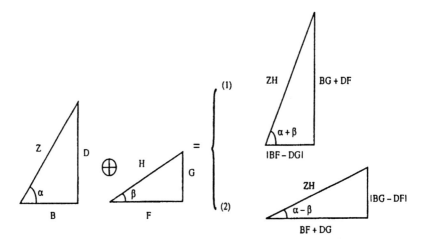

Figure 9.4: Viète's product of triangles

Impossible quantities in algebra

Earlier, imaginaries had come to be useful in algebra, first in the works of Cardano (1545/1968) and Bombelli (1572/1966), later also in Viète (1591/1983), Descartes, and Wallis. The most quoted passages in Cardano's *Ars magna* comes from his chapter 37 'On the rule for postulating a negative' (1545/1968, 219-220):

If it should be said, Divide 10 into two parts the product of which is 40, it is clear that this case is impossible.   Nevertheless, we will work thus: We divide 10 into two equal parts, making each 5.   These we square, making 25.   Subtract 40, if you will, from the 25 thus produced, as I showed you in the chapter on operations in the sixth book, leaving a remainder of -15, the square root of which added to or subtracted from 5 gives parts the product of which is 40. These will be $5+\sqrt{-15}$ and $5-\sqrt{-15}$, ... and you will have that which you seek. ... Putting aside the mental tortures involved, multiply $5+\sqrt{-15}$ by $5-\sqrt{-15}$, making $25-(-15)$. Hence this product is 40. ... This is truly sophisticated since with it one cannot carry out the operations one can in the case of a pure negative. ... So progresses arithmetic subtlety the end of which, as is said, is as refined as it is useless.

which is also worth looking at in Latin:

,ſi quis dicat,diuide 10 in duas partes,ex quarum unius in reliquam ductu,producatur 30,aut 40, manifeſtum eſt, quòd caſus ſeu quæſtio eſt impoſſibilis,ſic tamē operabimur, diuidemus 10 per æqualia,& fiet eius medietas 5,duc in ſe fit 25, auferes ex 25, ipſum producendum, utpote 40,ut docui te,in capitulo operationum, in ſexto libro,fiet reſiduum m: 15,cuius ℞ addita & detracta a 5,oſtendit partes,quæ inuicem ductæ producunt 40,erunt igitur hæ, 5 p: ℞ m: 15,& 5 m: ℞ m: 15.

5 p: ℞ m: 15
5 m: ℞ m: 15
25 m: m: 15 qd. eſt 40

This is the first known appearance of the square root of negatives, which here reads R m: 15 .

A few paragraphs later we find the following example leading to this case of working with imaginaries, or 'sophistic negatives' as Cardano called them (Cardano 1545/1968, 221):

If it be said, Divide -6 into two parts the product of which is $+24$, the problem will be one of the sophistic negative and will pertain to the second rule, and the parts will be $-3+\sqrt{-15}$ and $-3-\sqrt{-15}$.

These imaginaries are in *Ars magna* not connected to Cardano's main theme of cubic and quartic equations, but it is interesting to note his point of view on negatives (1545/1968, 154): they may be necessary for intermediate calculations toward a true, i.e. positive, answer.

The same is true for the imaginaries, but Cardano does not comment on this. Instead, we look at an example given by Clairaut in 1746, who wants to solve the cubic equation $x^3 = 63x + 162$. For this equation the Cardano-Tartaglia solution procedure leads to the formula

$$81 \pm 30 \cdot \sqrt{-3} = \left(-3 \pm 2\sqrt{-3}\right)^3$$

where the equality may be verified by direct multiplication. Then the Cardano-Tartaglia solution says that one of the solutions x is found via

$$x = \left(-3 + 2\sqrt{-3}\right) - \left(3 - 2\sqrt{-3}\right) = -6 .$$

Thus the equation has a factor $(x+6)$ and so the other solutions can be found by factoring:

$$\left(x^3 - 63x - 162\right) \div (x+6) = x^2 - 6x - 27 = (x+3)(x-9) .$$

Hence a true positive solution is $x = 9$, but Clairaut reached it only through computations involving both imaginaries and negatives.

Rafael Bombelli (1572/1966) found that in irreducible cases like the one above, there are always three real roots, but most often you are not able to do the actual reduction as simply as in the Clairaut example. Other early examples were given by Bombelli (1572) as well as Leibniz (1676). In this process we have seen Cardano computing with expressions like $\left(a + b\sqrt{-1}\right)\left(a - b\sqrt{-1}\right)$

and Clairaut with

$$\left(a + b\sqrt{-1}\right)\left(c + d\sqrt{-1}\right) = ac - bd + (bc + ad)\sqrt{-1} ,$$

just using what the English mathematician George Peacock (1842) was to call the 'principle of permanence of forms', that such new numbers behave structurally like old ones. But if so, why isn't always $\sqrt{ab} = \sqrt{a}\sqrt{b}$? Because then, we would get, as pointed out by Euler (1770/1984), that

$$-2 = \sqrt{-2}\sqrt{-2} = \sqrt{(-2)(-2)} = \sqrt{4} = +2 .$$

Resolving this apparent paradox will be a helpful discussion item for students exploring the ramifications of symbols they may have come to take for granted.

9.5   Integrating Original Sources in the Classroom

In the following we present two examples of reading original sources in school classrooms, one (§9.5.1) from Germany, the other (§9.5.2) from Italy.

9.5.1   Example 1: Greek surveying: the tunnel of Samos

The story of the tunnel

The ancient Greek historian Herodotus described a tunnel constructed on the island of Samos by the engineer Eupalinos about 530 BC. Knowledge of such a tunnel had become completely lost when it was rediscovered towards the end of the 19th century. First archaeological excavations showed that Herodotus's report was absolutely reliable. Between 1971 and 1978, the tunnel was completely excavated and examined in detail (Kienast 1986/87). The tunnel cuts through a mountain to supply the Samos fortress with water. It is 1040 metres long, 2 metres wide and 2

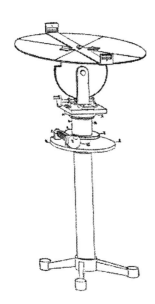

metres high, consisting of a path for inspections and a canal for the water beside it. It was mined simultaneously from both ends, and the two teams met under the mountain.

The underlying engineering feat is considerable. The standard procedure for tunnels of such length at the time was to dig several shafts to the surface in order to determine the position reached and to correct the direction of the digging. This method was not used here. Since the discovery of the tunnel, a much discussed question has been how Eupalinos surveyed the tunnel's direction with such accuracy.

A possible answer may lie in a source of some 600 years later. In a handbook describing the handling of a surveying instrument called *dioptra* (figure 9.5), Heron of Alexandria (40-120 AD) treats the problem of 'cutting through a mountain in a straight line if the entrances of the tunnel are given' (Schöne 1903, 238 ff). Heron's booklet poses a number of other interesting surveying problems which could be treated in the classroom.

Figure 9.5: Heron's dioptra

In an introduction, Heron describes the dioptra's uses, naming military applications besides land surveying and astronomy. A specially nice remark says that, frequently storm attacks on fortresses were easily repelled because the besiegers had underestimated the height of the walls, attacking with ladders which were too short. In such cases, Heron said, the dioptra had its uses, for it served to measure the heights in question "out of range" (Schöne 1903, 191).

A teaching unit about Heron's surveying text

For a long time, the experts favoured the hypothesis that Eupalinos had essentially proceeded as described by Heron (cf. Van der Waerden 1956, 168 ff), and it is also the basis of the following lesson. The above mentioned excavations, however, have led archaeologists to prefer another theory. We shall see that the students discovered both these theories on their own.

On the basis of this story about the Samos tunnel, a teaching sequence founded on Heron's text was developed and tested at various schools in the region of Bielefeld, Germany (Jahnke &Habdank-Eichelsbacher 1999).

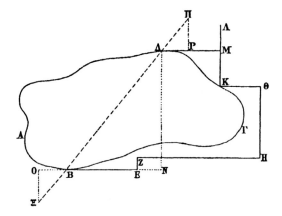

Figure 9.6: Heron's method of surveying a tunnel

While one could expect the story to be attractive to the pupils, the source might raise some difficulties. For fourteen- or fifteen-year-olds it was rather long. As is common in ancient Greek geometrical texts the essential idea is not explicitly mentioned, the argument proceeds step-by-step. Some teachers expected even problems with the Greek letters. Nevertheless, it was decided to present the source unchanged as it was printed in Schöne's Greek-German edition. The students were told that this was a section of an ancient original surveyor's handbook which had not been especially devised for them. While it might not be too easy to read, they would be able to cope with the difficulties. For a number of students, this remark proved to be quite motivating.

Mathematically, Heron's surveying method requires the notion of similarity. This had not been explicitly treated in the 9th grades where the teaching took place. The idea was to rely on the students' intuitive previous knowledge. One could expect that they had a notion of how maps work. In most classes the teaching unit on the tunnel of Samos served as an introduction to the concept of similarity.

All in all, the teaching sequence consisted of 3 + *n* lessons. An introductory lesson about the students' knowledge of history of mathematics ended with the story of the Samos tunnel. In a second lesson the problem of how the direction of the tunnel could have been determined was discussed with the students. In the third lesson the source was analysed, after a first reading had been given as homework. In further lessons other surveying problems were treated.

The classroom experience

All classes had a lot of fun establishing a map of the history of mathematics. To both their teachers' and their own surprise, the students' previous knowledge of history of mathematics was manifold. They knew a lot of facts. Above all, students have historical imagination and find questions such as why mankind started to use and write numbers, or to draw and analyse geometrical figures, quite natural and interesting.

The discussions about how Eupalinos might have determined the direction of the tunnel, under the condition that one end cannot be seen from the other, proved to be very fruitful. All classes developed essentially the same two solutions. And these are exactly those offered by the archaeologists.

The first method is that of the source. It can be understood from Heron's figure (see figure 9.6). Starting from one entrance a sequence of segments around the mountain is measured. From this one can calculate the segments BN and AN whose ratio gives the direction of the tunnel. Then, at both entrances beams are constructed showing the right direction. The second possible strategy found by the students results from the question whether it is possible to take bearings from the mountain's summit on both entrances marked by flags. If this is not directly possible, one could put up a sequence of flags connecting the entrances and then adjust the sequence until the flags lie on a straight line from one entrance to the other. Modern archaeologists found signs suggesting that Eupalinos proceeded this way, but it is possible that he used both methods.

After this preliminary and informal discussion with the students which did not end with a clear and definite result, but with a lot of ideas and a feeling for the nature of the problem, they got copies of the source which was to be read as homework. Before the next lesson, there were already discussions among the students about Heron's idea. In the lesson itself the general idea was presented by one or several students, then the source was read step by step. It was a nice experience that in one class the discussion was opened by a student with the statement "Heron has made a mistake!". In fact, if one reads Heron literally the student was right, but others argued that this is a matter of interpretation.

It was interesting to see how the students explained Heron's method without knowing the notion of similarity. In one class, they argued that his idea is the same as that underlying the determination of the slope of a straight line. As this had been treated quite a while ago, this was a compliment to them and their teacher. In the other classes, the argument was a bit vague, but intuitively correct when students argued that Heron constructed a sort of a map.

After discussion of the source, teaching was continued in various ways. It was pointed out to the students that the workers didn't meet exactly, but missed each other by about 10 metres in the middle. It was determined by drawing that the error in measuring the angle of direction had been less than 1 degree. The question how Heron coped with the difference in altitude was raised in all classes.

Written student productions

All students were assigned the task of summarising Heron's method in a small written essay. The results show that more than two thirds of the students had completely understood the text. Many students were able to free themselves from the language of the source and to express the idea in their own words, finding quite convincing descriptions which represented a mixture of everyday language and of the expert language acquired in the classroom. Such written exercises and the skills they develop and demonstrate are an important general objective of integrating historical sources into mathematics teaching.

9.5.2   Example 2: An 18th century treatise on conic sections

The teaching environment

The activity here analysed has been planned and developed in a classroom by a secondary teacher (see Testa 1996). It was carried out in an Italian Scientific Lyceum, a high school in which mathematics is an important (and difficult) subject; 16 students (11 girls, 5 boys) aged 16/17 volunteered to participate. The total time employed was 16 afternoons, after the school time.

The subject taught is conics, which in the official mathematics curriculum is suggested only as optional subject matter. In the first eight afternoons theories about conics of various classical authors such as Pappus and Eutocius were outlined. Also the means for the pointwise construction of the conics (Euclid's *Elements* book II) were discussed. The following eight afternoons were devoted to the study of De la Chapelle's *Traité des sections coniques, et autres courbes anciennes*. This text is a revision with 'didactic eyes' of classic works on optics. There is a systematic application of algebra to geometry, a unifying use of Euclid iii, 35; the links with physics are considered. The text was chosen for its clarity and elegance. The preface shows that the author was aware of the students' difficulties in learning mathematics and looked for ways of overcoming them.

The Italian teacher's choices reveal his view on the use of history in mathematics teaching: to read an ancient text is his favourite way of integrating history in classroom, and doing history of mathematics is nothing other than doing mathematics. The teacher is historically well read and experienced. Thus, to look for original sources and to work with them is not a problem to him.

The experience

In our description, we focus on the teaching of De la Chapelle's text. The main difficulty to face was the unknown language. The teacher rejected the idea of presenting a literal translation, to avoid the temptation for students to participate only passively.   Instead,   he prepared 34 worksheets containing passages of the French text, with blanks in strategic positions to be filled in by the students.   At the beginning the original text was quite fully summarised, in later worksheets the amount of original text was increased, and with the last worksheets the text was almost entirely the original.   In the worksheets De la Chapelle's symbols were kept. Since   the   original   figures   usually   contain   elements   referring   to   different propositions, the teacher drew new figures containing only the elements essential to a single proposition, during the first period; later on students were encouraged to use the original figures, and to decode the information contained in them.

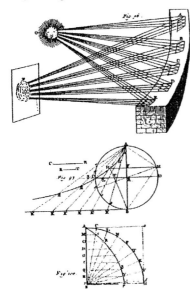

At the beginning, mediation by the teacher was important, afterwards the students' work was more and more autonomous.   Students worked in groups, and also did homework.   They devised their own strategies for handling the difficulties, using coloured pencils to decode figures, and substituting the old notations by new ones.   In order to fill the blanks, students had to understand the underlying reasoning.   This method of work roused lively discussions among students.   After they had worked at the given worksheet, the teacher showed in a transparency   the   complete   original passage and discussed the work performed by the students.

Figure 9.7: Drawings from De la Chapelle's treatise

Evaluation of the experience

After each session a questionnaire was given to the students in order to evaluate the understanding of mathematical contents and to check any difficulties.   At the end of the overall experience they answered an open questionnaire aimed at investigating how they perceived the use of history of mathematics, in particular the use of original sources.   Students were very collaborative, and gave a great deal of information; their protocols can be considered as written interviews.   The most significant points which emerged were:

– doing mathematics became more pleasant

– it was easy to see the evolution of mathematics and to become aware that there are different points of view to face problems

– the method of work led directly to seeing what there is behind a theorem

– the study of the original text was preferred since the participation in the work was more active

– it was more difficult to grasp the language than the spirit of the work

- to work directly with the text required more careful reflection on problems and better understanding of their meaning: thus apparently simple problems revealed unexpected aspects
- it was made possible to go beyond theorems and to arrive at the roots of mathematics
- the experience changed the image of school mathematics.

The words of one student point out the effectiveness of integrating original sources: "The proofs I had to complete helped me to learn working on my own. I liked working with the graded worksheets because they implied a step-by-step reasoning unlike my usual way of thinking." Working with original texts clearly produces changes in the mode of learning.

The experiment had particular features which make it difficult to draw general conclusions. Among these features were that it was an extra-curriculum activity involving only volunteer students; the source concerns a quite unknown author; it deals with a language not mastered by students; the text used was conceived as a textbook; and the teacher here possessed a remarkable competence in history of mathematics and familiarity with the use of original sources.

On the other hand for our study these elements can be seen as positive, since:

- being an optional activity allowed the teacher freedom in planning and developing the didactic procedure
- using a rarely considered author fostered originality and creativity in the experience
- the presence of an unknown language is a quite typical obstacle in using original sources and thus it is interesting to see how the teacher has faced this difficulty
- to use a text written for didactic purposes is an intermediate situation facilitating the approach to an original source
- the teacher's competence has made the experience very rich in cultural values.

The literature on the use of original sources in courses (not specifically for history of mathematics) shows that successful experiments generally refer to university level or, in the case of high school level, to optional courses (see Laubenbacher & Pengelley 1994; 1996). Other interesting examples exist, which concern limited passages in limited activities; this is the case, for example, with using of mediaeval arithmetic word problems. A wider and systematic use of original sources presents difficulties of time, souce availability, and so on. Undoubtedly the main point is the role played by the teacher. He has to really believe in the value of original sources, he has to be competent enough in order to find and to manage materials suitable to the needs of his classroom, he has to plan strategies of mediation very carefully. These strong requirements emphasise the difficulty in the transferability of good experiences from one teacher to another and in making the use of original sources a routine activity.

9.6 Didactical strategies for integrating sources

9.6.1 The triad: text - context - reader

As we explained above (section 9.3), reading a source is a hermeneutic activity and, thus, subject to the rules of hermeneutics. In every teaching where an original source is going to play a role the teacher has to consider the concrete relationship between the text, the context and the readers. Depending on the aims of teaching there should be a certain balance between the proper analysis of the source and the investigation of the context. Usually, students reading a mathematical text are not used to asking for the context. In a way, they are even educated not to consider it: mathematics should be independent of the context and understandable out of itself; the time when a text has been written, the country or the author seem to be irrelevant. Therefore, students have to be guided to asking meaningful questions about the context. Frequently, it will be necessary to do some independent investigations about the context and study the biography of the author before the source can be interpreted adequately. Also, to relate the context information to the meaning of the text under study requires some skills which presuppose some experience and have to be trained. For example, frequently it makes a difference whether a text has been written by a theoretically or a practically minded author, and it is possible to trace indications of this prevalent habit of mind in the text. This is very illuminating, but, of course, requires some experience.

It should be clear that the aim of these activities is not at all an imitation of the professional historian in regard to rigour and sophistication. Rather, the students are led to asking new questions which, in general, they had never asked before.

The use of primary sources in the classroom requires special care, to clarify the proposed objectives and the adequacy of a source to the students' needs. The concrete conditions of the students should be considered, and, of course, it makes a difference whether a source is studied with school students or with future mathematics teachers or with in-service teachers. The chosen contents need to be related to the respective student interests, the availability of texts in the mother language (or, at least, a language known to the students or the teacher) and in accordance with the objectives that the teacher intends to achieve.

9.6.2 Classroom strategies

At the moment there is no elaborated and generally accepted approach available for the reading of sources in the classroom. There are, however, some experiences, and, in the following, we want to give a generalised picture of these experiences. This may be taken as a collection of ideas and guidance (which does not pretend to be exhaustive) from which interested readers may select what is appropriate to their needs.

(i) Introducing a source

To introduce original material in the classroom, two types of strategy are imaginable: direct and indirect. Using a direct strategy, the teacher presents the text without any previous preparation. An indirect strategy is a situation where the source is consulted after some previous activities.

1. A direct strategy to a source might have as an objective to provoke a shock in the students, through perceiving the difference between their modern view of the subject and the view-point expressed in the source. This will provoke questions for study. After reading, the student is required to answer a series of questions previously established by the teacher—or it may be suggested that the student extracts questions from the text. Presenting the text in this way has the objective of challenging the student and raising a polemic around the theme.

2. An indirect strategy might result from solving problems. The teacher presents to the students a non-routine problem, to raise their curiosity and the need for a deeper study of the subject. After this the teacher might present an extract of an original text related to the questions the students had formulated.

3. Another indirect strategy could start with a historical author. The teacher begins by showing how mathematics is connected with the society of a certain time and, together with the students, he points out the mathematicians' names that stood out. The students select one or more authors and try to gather available information about them. Only after interest about the mathematician has been raised does the teacher present a source extract, and the class work culminates with its analysis.

4. Textbooks might be another point of departure. The teacher selects a theme in the textbook used in the classroom. She questions its approach. Then she presents other textbooks, or extracts from an old textbook, for analysis and comparison with the current one. It raises the students' curiosity; they feel the desire to discover who introduced that concept or theory, who formulated or solved that problem. Thus, the original text appears in a natural way and is worked on as a profound study of the text used in the classroom. Further possibilities and problems with respect to the process of interpreting and analysing ancient textbooks are discussed in Glaeser 1983 and Schubring 1987.

5. In the education of adults it might be easiest and most natural to introduce a source through a presentation from the tutor. This is a discourse within which the tutor provides information, formulates a synthesis, or introduces a new question. The tutor sketches the historical background and comments on difficulties, special features, and objectives of the text in question. Switching between different texts or different parts of the same text can also be achieved by short presentations in which the tutor provides a synthesis of the text already treated and introduces the subsequent ones. These presentations should not be over extended; a few minutes will do.

(ii) Analysis of a source and cognitive debates

The analysis of historical texts is a difficult activity in history of mathematics. Sometimes it should be supported and guided through questions from the teacher. Sometimes, it seems to be more adequate to let the students find out the right questions. An important aspect is to find suitable questions so that students become immersed in the historical context of the text under study.

To improve the conditions of analysis, some texts must be modified or translated and adapted to the general context within which they are being introduced. At the same time, they have to be modified so as to remain within the students' or trainees' grasp. Nevertheless, it is imperative that these adaptations remain as closely aligned as possible to the original author's thought (Barbin 1987).

Frequently, the analysis of a text gives rise to cognitive debates. These are discussions within which the students are called to express their own views on a concept's or method's validity and relevance, and above all to give reasons for their own choice. For this, great care is needed in selecting the texts or controversies which are to be the object of debate.

To prompt a successful debate, the educator should suggest to each group of students or trainees that they prepare to argue in favour of one or other point of view. Notwithstanding their directive character, these suggestions tend to motivate students and inspire them to find out for themselves about the advantages of a historical reasoning which at first glance might appear naive or erroneous (Desautels & Larochelle 1989, 1992; Legrand 1988; Lakatos 1976).

(iii) Construction of measuring instruments

Humankind has always been preoccupied with measuring physical or mathematical quantities and this is particularly true for mathematicians and scientists. Historical research reveals different conceptions of measurement. Although these conceptions may easily become apparent in some cases, they may not exist within a structured theory and they may not even have been used to construct instruments. Nevertheless, the ideas encountered through historical study may serve to inspire activities which can help participants to analyse their own reasoning and also encourage them to construct their own measuring instruments. For instance, mathematical machines for drawing curves may be of interest (Dennis 1997; El Idrissi 1998; Ransom 1995; also see section 10.2.2).

(iv) Verbalisation

With regard to acquainting the participants with the reasoning of mathematicians, having them verbalise this reasoning seems to be an excellent strategy. It makes students attentive to original thoughts and helps prevent them from attributing to mathematicians things they never said, and (if trainee or in-service teachers) from passing on such misunderstandings to their pupils in due course. Take care to have them distinguish in these verbalisations between things derived from the texts themselves and interpretations of the latter. This activity is also beneficial in alerting students to the difficulties which may be met when reasoning in mathematics without the support of a formal system.

(v) Translation

As with verbalisation, translations of text extracts are intended to acquaint students and trainees with the thought and conception of mathematicians in regard to mathematical reasonings and concepts. At least two types of translation can be distinguished here: translations into modern mathematical language, and translation from one language into another. While the former serves in particular to reconstruct a mathematical argument, the latter has promising educational advantages insofar as it initiates students and trainees into mastering a language and to conceptual analysis (Arcavi *et al.* 1982, 1987; Testa 1996).

(vi) Validation of reasonings

During their first studies of historical works, students and trainees sometimes disparage the mathematical value of reasonings found in historical texts, especially if they have been accustomed to continuous praise of recent mathematical progress. This attitude may prevent students from realising the educational and mathematical potential contained in ancient reasoning.

To challenge this attitude, one may ask students to validate the reasoning of the mathematicians of old. Such validations are intended to demonstrate how well-founded are the methods used in history, in the light of more elaborate mathematical knowledge. This prompts the students both to give historical reasonings the same status as present-day thought with regard to mathematical foundations and also to challenge their own conceptions of present-day methods, in particular as regards the learning of mathematics. These ancient methods often have the advantage of being within the pupils' grasp and of providing interesting hints for teaching (Arcavi *et al.* 1982, 1987).

(vii) Comparison

To compare different texts or text extracts is also a fascinating approach, in particular in history of mathematics. The comparison may include texts of the same period or of different periods, having the same or different objects. These comparisons must be accompanied by activities and questions of understanding aimed at making analysis more purposeful and more attractive.

Comparing historical texts permits students to realise how the notation and symbols of mathematics have evolved. It helps them to focus on the essential in historical mathematical writings. In addition, the comparison of mathematical textbooks is a promising approach to the history of teaching.

(viii) Synthesis

Activities of synthesis should be done by students outside of the course; for this external type of work, all the strategies mentioned above can be used. This homework should be designed either as a preparation for future courses or as a work of synthesis. It may be planned for the end of course sections.

9.7   Evaluation, research questions and issues of concern

Though the idea of integrating history of mathematics into mathematics teaching originated more than a hundred years ago, practical efforts on a larger scale beyond isolated activities of individuals have been made only in the last twenty years. Since reading sources belongs to the most demanding of possible activities, it is not surprising that up till now there is no systematic empirical research, investigating opportunities, difficulties and outcomes of sources as part of mathematics teaching.

At present, there are essentially two types of contributions to the field. On the one hand we have a number of reports reflecting personal experiences with reading original sources in various contexts, be it school teaching, or the education of future teachers (cf. Arcavi 1987; Furinghetti 1997; Jahnke 1995; Laubenbacher & Pengelley 1998; Silene & Testa 1998). On the other hand, there are quite a few papers with proposals on what could be done. However, to reach a new conceptual and practical level we do need more research. In this section we sketch some directions of work which could be followed.

First of all, we should know better whether reading a source does in fact make a difference compared to other possible activities. Given the large amount of time required for using original sources, we should be sure that the effort is really worthwhile. From theoretical reflections we are quite sure that a source will open up new dimensions of understanding. We have mentioned above experiments where the integration of historical sources has been successful. The problem is to ensure adequate conditions. It is clear that the role of the teacher/tutor is crucial for creating the right atmosphere and providing the necessary intellectual tools for students.

This leads to further questions. Reading a source demands in a specific way a feeling for the intellectual, social and cultural context in which it has been written and the ability to ask questions concerning these dimensions. This in turn presupposes that the learner has already a certain historical background and an ability which we would like to call historical imagination. Under conditions where history of science is a curricular subject neither in regular school teaching nor at universities, a historical background in science and mathematics can only result from personal reading or from the media (television, films etc.). Thus we should investigate what previous knowledge about these things our students have and how much we can rely on this as a historical foundation to build on (see Demattè 1994; Demattè & Furinghetti 1999).

Because of this context dependence, reading a source is quite different from reading a normal text of mathematics. Thus, one has to change one's reading habits, and, again, we should know more about this, theoretically and empirically.

It is very important to investigate the reading strategies and the strategies of interpretation as well as the difficulties students encounter with sources. How do students react to a text, how do they work with terms whose meaning they do not know? Are they able to identify essential elements of a text? How do they translate the meaning of a text into their own language? Only with a better understanding about this shall we be able to devise more effective teaching strategies.

One of the essential ideas connected with the reading of a source is that this will influence the students on their meta-cognitive level and contribute to their ability to reflect about mathematics. Again, we need to know more whether this is really the case, and if so, to what degree.

It would be worthwhile also to know more about processes of mathematical understanding which might not be intended, but nevertheless happen. Students, or teachers, may see in a historical document a source of insight, which may add to their understanding, regardless of the historical context, and far from the intention of the original writer. Thus the question to explore is: can original documents be the trigger for re-thinking the mathematics, even by way of erroneously attributing intentions to the text, which are not there, or misrepresenting its ideas? In other words, the source can be the motivation and inspiration for thinking differently about a mathematical topic, in a way which has nothing to do with what any historian would have seen in the source.

It is obvious that all these questions might be answered differently for young pupils or adults. Thus, the age factor is important.

There is a practical problem which will continue to remain a task of great importance for future work: the identification and editing of adequate source material.

This overview shows that we are only at the beginning of a process in which history of mathematics might become an organic part of mathematics teaching. To achieve this goal we have to solve a lot of problems. Fortunately, these problems turn out to be interesting and demanding.

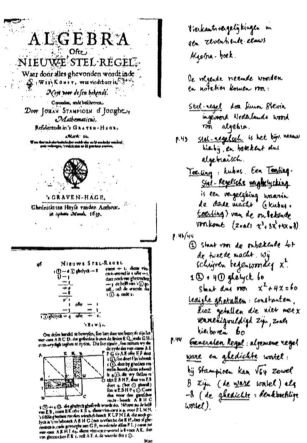

Figure 9.8: Grade 9 pupils (13 year olds) in a Dutch secondary school explored this 17th century Dutch algebra text in learning about quadratic equations. The title page, and a page with the geometrical proof of an equation-solving rule, were supplied together with the teacher's hand-written glossary to help pupils to study the text at home before the classroom discussion (from van Maanen 1997)

Bibliography, references and resources.

Abreu, G. de 1998. 'Relationships between macro and micro socio-cultural contexts: implications for the study of interactions in the mathematics classroom', in P. Abrantes, J. Porfirio, M. Baía (ed.), *Proceedings of CIEAEM 49*, Setubal, 15-26

Arcavi, A. 1987. 'Using historical materials in the mathematics classroom', *Arithmetic teacher* 35 (4), 13-16

Arcavi, A., Bruckheimer, M. 2000. 'Didactical uses of primary sources', *Themes in education*, 1 (1), 44-64

Arcavi, A., Bruckheimer, M., Ben-Zvi, R.. 1982. 'Maybe a mathematics teacher can profit from the study of the history of mathematics', *For the learning of mathematics*, 3 (1), 30-37

Arcavi, A., Bruckheimer, M., Ben-Zvi, R. 1987. 'History of mathematics for teachers: the case of irrational numbers', *For the learning of mathematics*, 7 (2), 18-23

Arcavi, A., Meira, L., Smith, J.P., Kessel, C. 1998. 'Teaching mathematical problem solving: an analysis of an emergent classroom community', *Research in collegiate mathematics education*, 7, 1-70

Barbin, E. 1987. 'Dix ans d'histoire des mathématiques dans les IREM', *Bulletin de l'APMEP*, n° 358, Avril

Barbin E. 1994. 'The meanings of mathematical proof. On relations between history and mathematical education', in: J.M. Anthony (ed.), *Eves' Circles*, Washington D.C.: Mathematical Association of America, 41-52

Barbin, E. 1995. 'Mathematical proof : history, epistemology and teaching', in: Cheung, Sui and Wong (eds.), *Retrospect and outloook on mathematics education in Hong Kong*, Hong Kong: University Press, 189-214

Barbin, E. 1996. 'On the role of problems in the history and teaching of mathematics', in: Ronald Calinger (ed.), *Vita Mathematica. Historical research and integration with teaching*, Washington D.C.: Mathematical Association of America, 17-26.

Barbin, E., Caveing, M. (eds.) 1996. *Les philosophes et les mathématiques*. Paris: Ellipses

Bekken, Otto B. 1983. *Una historia breve del algebra*, Lima: Sociedad Matematica Peruana

Bekken, Otto B. 1994. *Equáçoes de Ahmes até Abel*, Rio de Janeiro, Universidade Santa Ursula GEPEM

Bombelli, Rafael 1572/1966. *L'algebra. Opera*, Bologna, ed. E. Bortolotti, Reprint: Milano: Feltrinelli

Bruckheimer, M., Ofir, R., Arcavi, A. 1995, 'The case for and against casting out nines'. *For the learning of mathematics*. 15(2), 24-35

Bruckheimer, M., Arcavi, A. 1997. 'Mathematics and its history: an educational partnership', submitted for publication

Bühler, M. 1990. 'Reading Archimedes' Measurement of a circle', in Fauvel 1990, 43 – 58

Cardano, G. 1545/1968. *The great art, or the rules of algebra*, trans. & ed. T. Richard Witmer, Cambridge: MIT Press

Chapelle, De la 1750. *Traité des sections coniques, et autres courbes anciennes*, Paris: J. F. Quillau & fils, Paris; repr. IREM Université Paris VII, nouvelle série n.6

Dedekind, R. 1924. *Essays on the theory of numbers*, Chicago: Open Court.

Demattè, A. 1994. 'Storia, pseudostoria, concezioni', *L'insegnamento della matematica e delle scienze integrate*, 17B, 269-281.

Demattè, A., Furinghetti, F. 1999. 'An exploratory study on students' beliefs about mathematics as a socio-cultural process', in: G. Philippou (ed.), *MAVI-8 Proceedings* (Nicosia, Cyprus), 38-47

Dennis, D. 1997. 'René Descartes' curve-drawing devices: experiments in the relations between mechanical motion and symbolic language', *Mathematics magazine* 70, 163-174

Desautels, J., Larochelle, M. 1989. *Qu'est ce que le savoir scientifique? Points de vue d'adolescents*, Québec: Presses de l'Université Laval.

Desautels, J., Larochelle, M. 1992. *Autour de l'idée de science. Itinéraires cognitifs d'étudiants*, Québec: Presses de l'Université Laval.

Euler, Leonhard 1770/1984. *Elements of algebra*, trans. Rev. John Hewlett 1840, Berlin: Springer-Verlag

Farey, J.-M., Métin, F. 1993. 'Comme un fruit bien défendu'. *Repères IREM* 13, 35-45

Fauvel, J. (ed.) 1990. *History in the mathematics classroom: the IREM papers*, Leicester: The Mathematical Association

Frend, W. 1796. *The principles of algebra*, London: J. Davis, J. Robinson

Freudenthal, H. 1983. *The didactical phenomenology of mathematical structures*, Dordrecht: Reidel

Furinghetti, F. 1997. 'History of mathematics, mathematics education, school practice: case studies linking different domains', *For the learning of mathematics*, 17 (1), 55 - 61

Furinghetti, F., Somaglia, A. M. 1998. 'History of mathematics in school across disciplines', *Mathematics in school* 27 (4), 48-51.

Galilei, G. 1638/1954. *Dialogues concerning two new sciences*, New York: Dover Publ.

Gillings, R. J. 1972. *Mathematics in the time of pharaohs*, Cambridge Mass: M.I.T. Press.

Glaeser, G. 1983. 'A propos de la pédagogies de Clairaut. Vers une nouvelle orientation dans l'histoire de l'éducation', *Recherches en didactique des mathématiques*, 4, 332-334.

Glushkov, S. 1977. 'An interpretation of Viète's 'calculus of triangles' as a precursor of the algebra of complex numbers', *Historia mathematica* 4, 127-136

Hallez, M. 1990. 'Teaching Huygens in the rue Huygens: introducing the history of 17th-century mathematics in a junior secondary school.', in Fauvel 1990, 97-112

Hodgson, B. R. 1998. 'Le 'Liber quadratorum' de Léonard de Pise: morceaux choisis'. Unpublished manuscript

Horain, Y. 1990. 'Polygonal areas: a historical project', in Fauvel 1990, 113-138

El Idrissi, A. 1998. *L'histoire des mathématiques dans la formation des enseignants: étude exploratoire portant sur l'histoire de la trigonométrie*, PhD Thesis, Université du Québec à Montréal, Canada

IREM de Montpellier (ed.) 1995. *Proceedings of the First European summer university on History and epistemology in mathematics education*, Montpellier

Jahnke, H. N. 1994. 'The historical dimension of mathematical understanding: objectifying the subjective', in: *Proceedings of the eighteenth international conference for the psychology of mathematics education*, vol. I, Lisbon: University of Lisbon, 139 - 156

Jahnke, H. N. 1995. 'Historische Reflexion im Unterricht. Das erste Lehrbuch der Differentialrechnung (Bernoulli 1692) in einer elften Klasse', *Mathematica didactica* 18, 2, 30-58

Jahnke, H. N. & Habdank-Eichelsbacher, B. 1999. 'Authentische Erfahrungen mit Mathematik durch historische Quellen', in: C. Selter & G. Walther (eds.), *Mathematikdidaktik als design science. Festschrift für E. Chr. Wittmann*, Leipzig/Stuttgart/Düsseldorf: Klett, 95-104

Jozeau, M.-F. 1990. 'A historical approach to maximum and minimum problems', in Fauvel 1990, 25 – 42

Kessel, C. 1998. 'Practising mathematical communication: using heuristics with the magic square', *Research in collegiate mathematics education* 7, 42-70

Kienast, H. J. 1986/87. 'Der Tunnel des Eupalinos auf Samos', *Mannheimer Forum*, 179-241

Kleiner, I. 1989. 'Evolution of the function concept: a brief survey', *College mathematics journal* **20**, 282-300

Lakatos, I. 1976. *Proofs and refutations: the logic of mathematical discovery,* Cambridge: University Press

Lampert, M. 1990. 'When the problem is not the question and the solution is not the answer: mathematical knowing and teaching', *American educational research journal* **27** (1), 29-63

Laubenbacher, R. C., Pengelley, D. J. 1994. 'Great problems of mathematics: a summer workshop for high school students', *College mathematics journal* **25**, 112-114

Laubenbacher, R. C. & Pengelley, D. J., 1996. 'Mathematical masterpieces: teaching with original sources', in: R. Calinger (ed.), *Vita mathematica*, Washington: MAA, 257-260

Laubenbacher, R. C., Pengelley, D. J. 1998. *Mathematical expeditions: chronicles by the explorers*, New York: Springer-Verlag

Lefebvre, J. et Charbonneau, L. 1991. 'Sur quelques moyens d'accroître la diffusion et le rayonnement social de l'histoire des mathématiques', in Grant, H., Kleiner, I and Shenitzer, A. (eds), *Actes du 17 colloque de la société Canadienne d'histoire et de philosophie des mathématiques,* 211- 224

LeGoff, J.-P. 1994. 'Le troisième degré en second cycle: le fil d'Euler', *Repères IREM* **17**, 85-120

Legrand, M. 1988. 'Génèse et étude d'une situation co-didactique: le débat scientifique en situation d'enseignement', *Actes du premier colloque franco - allemand de didactique*, Paris: La Pensée Sauvage, 53-66

Logarto, M. J., Vieira, A., Veloso E. (eds) 1996. *Proceedings of the Second European summer university on History and epistemology in mathematics education*, Braga

M:ATH 1991. 'Mathématiques: approche par des textes historiques', *Repères IREM* **3**, 43 - 51

Métin, F. 1997. 'Legendre approxime π en classe de seconde', *Repères IREM* **29**, 15 - 26

Nordgaard, M. A. 1959. 'Wessel on complex numbers', trans. of Wessel 1797 p. 469-480, in: (Smith 1959, 55-66)

Nouet, M. 1992. 'Historie des mathématiques en classe de terminale', *Repères IREM* **9**, 15 - 33

Peacock, G. 1842. *A treatise on algebra*, Cambridge

Pirie, S. E. B., Schwarzenberger, R. L. E., 1988. 'Mathematical discussion and mathematical understanding', *Educational studies in mathematics* **19**, 459-470

Ransom, P. 1995. 'Navigation and surveying: teaching geometry through the use of old instruments', in: (IREM de Montpellier 1995, 227- 240)

Schöne, H. (ed. and tr.) 1903. *Heronis Alexandrini opera quae supersunt omnia*. Vol. III: *Rationes dimetiendi et commentatio dioptrica*. Griechisch und Deutsch. Leipzig: B. G. Teubner, 238 ff

Schubring, G. 1987, 'On the methodology of analysing historical textbooks: Lacroix as textbook author', *For the learning of mathematics*, **7** (3) , 41-51

Silene Thiella, C., Testa, G., 1998. 'La Geometria de figure quadre di G. A. Abate', *L'insegnamento della matematica e delle scienze integrate*, **21A-B**, 712-724

Smith, D.E. 1958. *History of mathematics*, New York: Dover, vol. II, 2nd ed

Smith, D. E. 1959. *A source book in mathematics*, New York: Dover

Testa, G. 1996. 'Conics, a teaching experience', in Logarto *et al.* 1996, ii, 449-456

Van der Waerden, B. L. 1956. *Erwachende Wissenschaft. Ägyptische, babylonische und griechische Mathematik*, (H. Habicht tr.), Basel: Birkhäuser

Van Maanen, J. 1997. 'New maths may profit from old methods', *For the learning of mathematics*, **17** (2), 39-46

Viète, Francois 1591/1983. *The analytic art*, (T. Richard Witmer tr.), Ohio: Kent State University Press

Veyne, P. 1971. *Comment on écrit l'histoire: essai d'épistémologie*, Paris: Le Seuil

Voigt, J. 1985. 'Patterns and routines in classroom interaction', *Recherches en didactique des mathématiques* 6, 69-118

Wessel, Caspar 1799. 'Om Directionens analytiske Betegning', *Nye Samling af det Kongelige Danske Videnskabernes Selskabs Skrifter*, Kiøbenhavn, 469-518

Wood, T. 1998. 'Creating classroom interactions for mathematical reasoning: beyond natural teaching', in P. Abrantes et al (eds), *Proceedings of CIEAEM 49*, Setubal, 34-43

Youschkevitch, A.P. 1976. 'The concept of function up to the middle of the 19th century', *Archive for history of exact sciences* 16, 37-85.

APPENDIX: RESOURCES

Sources of original mathematical material

The following references are to sources where original works can be found. The selection of material has to be somewhat restricted and we have chosen material that is currently, or recently, in print or material that is widely available in libraries. There are a number of excellent histories of mathematics which, while being histories, also contain a great deal of illustrative original material. These are not listed here but a selection of them will be found directly after this appendix. Nor have we included here references to Complete Works of mathematicians, assuming that the interested reader would know how to access such material.

Archimedes: Dijksterhuis, E. J., *Archimedes*, Princeton, New Jersey: Princeton University Press, 1987; Heath, T. L. *The Works of Archimedes*, New York: Dover Publications.
The Heath edition, with the 1912 supplement, presents a translation of the extant works of Archimedes, using modern notation to make the mathematics easier to follow for the modern reader, but this has the disadvantage of re-interpreting the original line of thought. Dijksterhuis uses a notation that allows the reader to come closer to the original Greek thinking. On the other hand, Dijksterhuis does not give a translation of all the propositions, preferring to guide the reader through the essential material.

Argand, R.: *Essai sur une manière de représenter les quantités imaginaires dans les constructions géométriques*. new print. Paris: Albert Blanchard, 1971.
English text books continue to use the name Argand diagram for the representation

of complex numbers on the plane and this facsimile of the 1874 second edition of Argand's 1806 essay is a clear and simple presentation of his argument. An English translation of the earlier publication by Caspar Wessel on this subject will be found in D. E. Smith.

Barrow-Green, J. 1998, 'History of mathematics: resources on the World Wide Web', *Mathematics in school* **27** (4), 16-22.
This paper annotates web addresses useful for historians of mathematics. Cf. §10.3.2.

Berggren, L., Borwein, J, Borwein, P. 1997. *Pi: a source book*, New York: Springer
Not so much a history of π – is that possible? – but a collection of articles about the number. Thus we find essays on the series formula, algorithms, computer calculations and the Gauss Arithmetic-Geometric mean. The book deserves a mention here because of the wealth of original material. Many of the original papers appear in the original Latin, French or German and without translations. The photocopies of original printed works are of variable quality and no editorial corrections of typos, etc. has been undertaken. Nonetheless, having original works in their original presentations brings its own excitement to the interested reader.

Bernoulli, Johann. *Lectiones de calculo differentialium. Mscrpt.* German edition: P. Schafheitlin (ed. & transl.), *Die Differentialrechnung von Johann Bernoulli aus dem Jahre 1691/92.* Ostwalds Klassiker der exakten Wissenschaften 211, Leipzig: Akademische Verlagsgesellschaft, 1924
This is a German translation of the first textbook on calculus ever written, though not published at its time. It can be read after some introduction into calculus.

Bibby, J. 1986. *Notes towards a history of teaching statistics*, Edinburgh: John Bibby Books
Much of interest here for projects – it includes many original sources and pictures – as well as giving the statistics teacher useful information on how the teaching of the subject has changed over that past century or two.

Cantor, G. 1915. *Contributions to the Founding of the Theory of Transfinite Numbers*, New York: Open Court Publications
An English translation of Cantor's 'Beiträge zur Begründung der transfiniten Mengenlehre'.

Cardano, G. 1545/1968. *The great art or The rules of algebra*, (T. Richard Witmer tr. 1968), Cambridge: MIT Press
First published as *Ars magna* in 1545, this a cornerstone book in the history of mathematics reveals the author's solution to cubic and biquadratic equations. Long unavailable, except in rare Latin editions, now available through a Dover reprint.

Chabert, J.-L. *et al.* (ed.), *Histoire d'algorithmes*, Paris: Belin, 1994; English tr. *A history of algorithms*, Berlin: Springer, 1999
A rich source of historical material, including many non-European works. Each chapter shows the development of a topic with extensive extracts from original writing. This would allow the teacher to introduce a topic directly from the original publications of mathematicians. Topics covered include: methods of false position, Euclid's algorithm, interpolation, approximate solutions and convergence.

Cullen, C. 1998. *Astronomy and mathematics in ancient China: the Zhou Bi Suan Jing*, Cambridge: University Press
This complete translation of an important 1st century Chinese text provides rich material for the mathematics classroom. It is also a very beautifully produced book with an easily

accessible introduction to the developing mathematical and astronomical practices of ancient Chinese astronomers and shows how the generation and validation of knowledge was closely related to statecraft and politics.

Descartes: D. E. Smith & M. L. Latham (ed., tr.), *The Geometry of René Descartes*, Open Court, 1925; New York: Dover Publications, 1954
La Géométrie, which appeared originally as an appendix to *Discours de la Méthode* (1637), presents Descartes' algebraic treatment of geometry. The English translation is in a simple and direct style, while the parallel facsimile of the first edition provides the possibility of comparison with the original French, as well the opportunity of comparing modern algebraic usage with the original French typography. The whole is enriched with numerous explanatory footnotes.

Dhombres, J. *et al.*, *Mathématiques au fil des âges*, Paris: Gauthier-Villars 1987
For readers of French this is a valuable collection of over 100 extracts, some quite extensive, grouped together to reflect ideas in the use of mathematics, arithmetic, algebra, analysis, probability and geometry. The chapters on analysis and geometry are subdivided to deal with themes, such as the origin of the infinitesimal calculus and the representation of space. The selection of material naturally reflects French interests and contributions. Here you will find Fermat's use of geometric progression to determine areas under the hyperbola and Condorcet on combining probabilities. The whole is most attractively produced, with fine illustrations, and concludes with brief biographies of more than 150 mathematicians.

Dürer, A.: *Unterwisung der Messung: Um einiges gekürzt und neuerem Sprachgebrauch angepaßt herausgegeben sowie mit einem Nachwort versehen.* Reproduction of the edition München 1908: Wiesbaden, Sändig 1970. Original edition: Nürnberg 1525, reproduced in facsimile: Nördlingen: Verlag Dr. Alfons Uhl, 1983
Contains a lot of geometrical constructions.

Eagle, R. E., *Exploring mathematics through history*, Cambridge: Univ. Press, 1995
A collection of sources, from the earliest number recordings up to Fermat and Pascal's discussion of probability, prepared to be used in the secondary mathematics classroom. Each topic contains a description of the context and a simple explanation of the mathematics for use by the teacher or by a student. The material for use in class contains brief extracts of original material. The whole is delightfully illustrated and is presented so that it can be used by a teacher who has little or no historical background knowledge.

Fauvel, J. (ed.), *History in the Mathematics Classroom: the IREM Papers*, Leicester: Mathematical Association, 1990
Nine articles by French mathematics teachers showing how they have used original material in their classrooms. Each article contains the original material in English translation, providing the teacher with lesson material. A wealth of ideas and experiences.

Fauvel, John and Gray, Jeremy, *The history of mathematics: a reader*, Basingstoke and London: Macmillan Press, 1987
This selection of over 400 extracts was originally prepared for the Open University course *Topics in the History of Mathematics* and covers mathematical writings from the earliest ideas of numbers and counting up to the mechanisation of calculation. The collection includes many comments on the nature of mathematical activity by mathematicians and other philosophers to sit alongside the original mathematical material. The contribution of Islamic mathematics is given its rightful place and of particular note is the chapter on the

mathematical sciences in Tudor and Stuart England which contains material unlikely to be encountered in other collections. The extracts have been carefully chosen to be easily accessible and include, for example, the proof by Gauss that the regular 17-*gon* is constructible.

Hay, Cynthia (ed.), *Mathematics from Manuscript to Print 1300 – 1600*, Oxford: Clarendon Press, 1988
Papers on aspects of mediaeval mathematics, containing extensive extracts from the works of Maurolico, Nicolas Chuquet and Agrippa not easily found elsewhere.

Heath, Thomas L. *Aristarchus of Samos, the ancient Copernicus: a history of Greek astronomy to Aristarchus* together with Aristarchus's *treatise On the sizes and distances of the sun and moon*; a new Greek text with translation and notes, Oxford: Clarendon Press 1966
This book contains Aristarchus' famous paper on the relative distances of the sun and the moon from the earth. The hypotheses and theorems can be discussed in a course on trigonometry.

Heronis Alexandrini opera quae supersunt omnia. Greek-German edition. Stuttgart: Teubner 1976.
Contains a lot of valuable sources on measurement, optics, geometry.

Hilbert, D. *Foundations of geometry*, Chicago: Open Court 1902.
A translation of the 1899 *Grundlagen der Geometrie*, in which Hilbert showed that it is possible to construct a geometry based on a complete system of axioms. In the first chapter, Hilbert began by stating 21 axioms involving six primitive or indefined terms. He presents five groups of axioms: incidence, order, congruence, parallels and continuity.

l'Hospital, G. M. *L'analyse des infiniment petits, pour l'intelligence des lignes courbes*. Paris: Imprimérie Royale, 1696. Reproduction Paris: ACL-éditions, 1988
The first calculus textbook ever published. See Bernoulli.

IREM: Images, Imaginaires, Imaginations, Une perspective historique pour l'introduction des numbres complexes. Paris: Ellipses 1998
Historical sources on complex numbers, and experiences with these texts in the classrooom.

IREM de Basse Normandie (ed.): Une histoire des équations par les textes. 1994
A collection of sources on the solution of equations from the Babylonians to Lagrange.

IREM de Basse Normandie (ed.): La question des parallèles: une histoire de l'émergence des géométries non-euclidiennes. 1995
Texts by Euclid, Al Khayyam, Wallis, Saccheri, Gauss, Lobatchevsky.

IREM de Basse Normandie (ed.): La création du calcul des probabilités et la loi des grands nombres de Pascal à Poisson. 1995
Texts by Pascal, Huygens, Bernoulli, de Moivre, Laplace, Poisson.

Klein, F. *et al.*, *Famous Problems and other monologues*, New York: Chelsea Publishing Company, 1955
Of the four monographs brought together in this single volume, the most useful from our point of view is the translation of Klein's *Famous Problems of Elementary Geometry*. Not only do we have the presentation of the three classical problems – the duplication of the cube, the trisection of an angle and the quadrature of the circle – as well as a detailed explanation

for the construction of a 17–*gon*, but also, in part II, a discussion of the transcendence of π and a very nice presentation of the countability of algebraic numbers. This last is at a level that could be used as a rich source, accessible to school mathematicians.

Lietzmann, W. Bd. 1: *Aus der Mathematik der Alten: Quellen zur Arithmetik, . . .* 1928, Bd. 2: *Aus der neueren Mathematik: Quellen zum Zahlbegriff und zur Gleichungslehre, zum Funktionsbegriff und zur Analysis* 1929, Leipzig: Teubner
A useful collection, unfortunately no longer in print.

Midonick, H. (ed.), *A treasury of mathematics*, New York: Philosophical Library, Inc., 1965.
An attractively produced volume of fifty four original sources selected to illustrate contributions which changed or altered the course of the development of mathematics. More extracts from non-European sources than in other comparable collections. Each selections is preceded by a short introductory essay.

Newman, J. R., *The world of mathematics*, London: George Allen & Unwin, 4 vols. 1960
Described as a 'small library of mathematics', this four volume collection of articles contains many examples of original mathematical writing. Here will be found, for example, Newton's letters of 1676 in which he explains the extension of the binomial theorem to fractional and negative exponents (as well as his use of $a^{1/2}$ and a^{-1}), *The Sand Reckoner* by Archimedes, Euler's original article on the seven bridges of Königsberg and Alan Turing's article 'Can a Machine Think?'

Newton, I., *The mathematical papers of Isaac Newton*, ed. D. T. Whiteside: Vol. V: *Lectures on Algebra*. Cambridge: University Press, 1972
Newton's lectures on algebra, from 1683 to 1684. A bilingual edition, Latin and English, containing a full commentary by Whiteside and facsimiles of Newton. It starts with 'First book of universal arithmetic', where it is possible to detect the author's conception of algebra. Particularly interesting is Newton's didactic approach to show the use of algebra in a mathematical problem, translating a word problem from natural everyday discourse to mathematical symbolism. Clearly expressed, the text can easily be read by mathematical beginners.

Open University, *Topics in the history of mathematics,* (General ed. John Fauvel), Milton Keynes: Open University Press, 1987
The Open University course material for the degree level unit of this title contains 17 booklets, each of which can be obtained separately. While being a teaching course, each booklet contains extracts of original material. Video materials are also available.

Pappas, T., *Mathematics appreciation*, John Bibby Books, 1988.
A source book containing ten lessons, each with photocopiable assignment pages. Historical material at the level of elementary mathematics.

Rhind mathematical papyrus: Chace, A. B. *et al*: Oberlin, Ohio: Mathematical Association of America, 1927, 1929; reprt. National Council of Teachers of Mathematics, 1978; Robins, G & Shute, C., London: British Museum Publications, 1987, 1998.
The Chace edition includes almost all of the problems from the Rhind Papyrus, with attractively presented text in hieroglyphic and hieratic writing alongside the English

translation. The British Museum publication only has some sample problems but contains attractive full colour plates of the papyrus.

Riese, Adam, *Rechenbuch*, facsimile of 1574 edition, Hanover: Th. Schäfer, 1992
This book is perhaps the most famous of the early printed arithmetics and the name Adam Riese has come into the German language to signify accurate calculation. The fact that it is in German, and in Gothic script as well, makes it difficult for the non German reader to use, but the beautiful woodcut illustrations alone recommend the book as an important stage in the change from abacus calculation to written methods.

Schneider, I. *Die Entwicklung der Wahrscheinlichkeitstheorie von den Anfängen bis 1933 : Einführungen und Texte.* Darmstadt: Wissenschaftliche Buchgesellschaft 1988
A comprehensive collection of sources from the history of probability theory, translated into German.

Smith, David E., *A source book in mathematics*, New York: Dover Publications, 1929, 1959.
A collection of 125 extracts, mostly not available in English elsewhere. The book is divided into five sections (number, algebra, geometry, probability and calculus/functions). The extracts have been chosen to illustrate significant incidents or 'discoveries'. Some of the extracts, such as Cardan on imaginary roots or the correspondence between Pascal and Fermat on the notion of probability, are capable of being used in upper secondary school mathematics.

Struik, Dirk J., *A source book in mathematics, 1200–1800*, Princeton, New Jersey: Princeton University Press, 1969, 1986
A selection of mathematical writings of authors from the Latin world who lived between the thirteenth and the end of eighteenth century. By Latin, Struik means that there are no Arabic or Oriental sources, except where much used Latin translations are available, for example in the case of Al-Khwarizmi. Struik intersperses helpful explanatory commentary on the selected texts but substantial blocks of original writing remain intact. There is a great deal of rich material here, ranging from Stevin's description of decimal notation to Euler's theory of zeros of different values.

Swetz, F. (ed.) *Learn from the masters!*, Washington, DC: Mathematical Association of America, 1995
A collection of twenty-three articles by contributors who are actively engaged in using history in the teaching of mathematics. The intention is to show how one can use history in mathematics teaching and many of the articles contain direct extracts from original material which could be used by the teacher. An excellent starting point for the interested mathematics teacher.

Swetz, F., *Capitalism & arithmetic: the new math of the fifteenth century*. La Salle, Illinois: Open Court Publishing Co., 1987
The Treviso Arithmetic of 1478 is the earliest known dated printed arithmetic book and this English translation from the Venetian dialect comes with a useful commentary. Many of the problems, for example on the rule of three or problems of inheritance, could be used directly in the mathematics classroom. Students will also benefit from seeing so many ways of setting out 'long' multiplication.

Thomas [=Bulmer-Thomas], Ivor, *Greek mathematical works*, Cambridge, Mass & London: Harvard University Press, vol. 1, 1939, 1980, vol. 2, 1941, 1993.
This valuable collection of writings is arranged roughly chronologically with the first volume dealing with the mathematics up to Euclid (*fl.* 300 BC) and the second volume taking the story on as far as Pappus of Alexandria (*fl.* 300 AD). Thomas arranges his material around themes so some later writings appear in the first volume, when giving examples of Greek writing on arithmetic, for example. The whole work is set with the Greek original alongside the English translation and helpful footnotes are used to explain the text. Among the gems for use in the mathematics classroom are: Nicomachus on figurate numbers and his description of the sieve of Eratosthenes, selections from Archimedes and early ideas of trigonometry, including Ptolemy's table of chords and Diophantus on types of equations.

Wieleitner, H. Mathematische Quellenbücher. Bd. 1: Rechnen und Algebra, 1927, Bd. 2: Geometrie und Trigonometrie, 1927, Bd. 3: Analytische und synthetische Geometrie, 1928, Bd. 4: Analysis, 1928. Berlin: Salle
A useful collection, unfortunately no longer in print.

Viète, François, *Introduction to the Analytical Art*, in: J. Klein, *Greek Mathematical Thought and the Origin of Algebra*, Cambridge: MIT Press, 1968, 313-353
Klein's study of the revival of Greek mathematics, via Arabic science, in the 13th to 16th centuries, contains an English translation (by J. Winfree Smith) of Viète's important work which marks the beginning of the use of symbolism in mathematics.

Chapter 10

Non-standard media and other resources

Ryosuke Nagaoka

with June Barrow-Green, Maria G. Bartolini Bussi, Masami Isoda, Jan van Maanen, Karen Dee Michalowicz, María Victoria Ponza, Glen Van Brummelen

Abstract: *The integration of history is not confined to traditional teaching delivery methods, but can often be better achieved through a variety of media which add to the resources available for learner and teacher.*

10.1 Introduction

Jan van Maanen

10.1.1 Why other media?

Can we still speak about 'traditional' ways of teaching mathematics? If so, would this be the chalk-and-blackboard manner which will be familiar to most readers of this book from their own school days? In his address to the 7th International Congress on Mathematical Education (Howson 1994) Geoffrey Howson showed slides of classes learning mathematics. The range of conditions in which classes worked was enormous, from open-air teaching in Africa to spacious western classrooms, from barefoot kids to strictly disciplined Asian classes. Despite the wide variety of teaching and learning conditions, in many cases the blackboard was the centre of activities and chalk was the medium. In many countries of the world, the majority of schools have no electricity supply let alone telephone cables, and are often too poor even to provide pencils and notebooks for schoolchildren. And although it is clear that in many parts of the world the situation is now changing, and that other media, notably computers, are coming into classrooms both as

John Fauvel, Jan van Maanen (eds.), *History in mathematics education: the ICMI study*, Dordrecht: Kluwer 2000, pp. 329-370

presentation tools for the teachers and as working tools for the students, the usual method of teaching mathematics is still with blackboard and chalk (or sometimes whiteboard and marker). At any rate this is the starting point for this chapter. If in the life-time of this book the centrality of focal board—whether blackboard, whiteboard or indeed overhead projector—and board-writing implement (chalk or marker pen) no longer holds, perhaps the reader will consider this chapter as a historical document, foreshadowing things to come.

During the past century or more, blackboard and chalk were identical with mathematical activity. Thus in 1925 the Dutch teacher and historian of mathematics E. J. Dijksterhuis characterised and defended deductive mathematical thinking by presenting the prototype mathematician as one who begins with nothing but chalk and then starts to create new mathematics (translated from Van Berkel 1996, 132):

The man comes and stands in front of you; he has a blackboard and a piece of chalk; he has seen nothing nor experienced anything that he comes to report about; he does not need apparatus in order to give life to phenomena that lead to questions, but he builds an immaterial world for you, apparently from nothing.

The traditional way of teaching is notable for the way it focuses the class's attentions on a vertical surface (board or screen) at the front, whose content is controlled by the teacher. Present teaching practices are strongly governed by technology and media; they are just so familiar that we may not think of them in that way. This technique is highly effective in the sense that it conveys messages, from the teacher to the student, quickly and with little cost of material and personnel. Some students (often those who go on to be teachers in their turn) learn very well through this process. On the other hand, it appears that this way of teaching does not invite all learners. It may be, as researchers are beginning to realise, that the difficulties experienced by many students are as much to do with the traditional teaching mechanisms as with any innate lack of competence or application. Making an appeal to many learners seems to require other means than just passing accurate information. It is here that the non-standard media discussed in this chapter may have an important role to play in educating young people broadly across the whole range of students and institutions.

The underlying issues here are investigated in some more depth in the remainder of this introductory section. Then in section 10.2 some specific cases of using non-standard media in connection with the history of mathematics, to improve educational experiences and opportunities, will be presented and investigated. Section 10.3 explores the educational value of one of the most rapidly developing uses of new technology, the World Wide Web, in the context of support for mathematical learning from historical resources.

10.1.2 And which media?

Non-standard media for teaching mathematics have been listed already in Ch. 7, within the broader framework of possible ways of implementing history in the mathematics classroom. Some of them fit well within traditional methods, such as having 'historical snippets' in textbooks (§7.4.1); using 'worksheets' (§7.4.4); and working on errors, alternative conceptions, and other instances where history presents a contrast to the usual perspective (§7.4.6). Several of the items surveyed within §7.4 will not be considered here. This chapter restrict itself to explorations in more depth of domains, such as working with mechanical instruments or doing dramathematics, that the authors have experience with themselves.

The media chosen for discussion here form but a small selection of those which are possible and have been used by imaginative teachers. An important earlier survey of the range of possibilities was the eighteenth yearbook of the US National Council of Teachers of Mathematics (NCTM), published in 1945 under the title *Multi-sensory aids in the teaching of mathematics*. Its opening paragraph provides an interesting historical record (NCTM 1945, vii):

> Teaching aids in mathematics are not new. The last hundred years have brought us the telephone, the phonograph, the radio, television, the silent and sound motion picture, the stereoscope, the three-dimensional coloured pictures on lenticulated film and the Polaroid three-dimensional pictures, and motion pictures in colour. These inventions and developments are being used in many forms in our schools at the present time. It is only natural that mathematics teachers, too, consider the possible adaptation of these materials to the improvement of instruction in their field.

It is interesting to compare this list with a list one might draw up now under a similar rubric. Perhaps discussion of precisely this point may provide a useful exercise for trainee teachers who are exploring the uses of history and media in their future mathematics classroom.

A powerful example of how insights from new media can be integrated into learning is the innovative book by Eduardo Veloso (1998). Veloso discusses, among a wealth of other things, how historical problems can be taken as a source of inspiration for investigations with computer programs like Sketchpad and Cabri. The publisher of this book, which aims to present materials to teachers, is the governmental institute for educational innovation. Apparently media like computer programs are still seen as an 'educational innovation', at least by the government. This is but one instance of how the range of ideas developed in this chapter can be integrated into the development of progressive mathematics education.

10.1.3 Affect and effect

The work done in preparation for this ICMI study has produced very positive reports about the greater affectiveness that non-standard methods can bring about. That is, the way students warm to mathematical learning through the range of methods such as doing projects, watching films, constructing models, researching history in libraries, devising dramatic presentations, surfing the World Wide Web. In part, of course, the perception of the learning benefits arise from a 'new technology' effect: something seems good just because it is new, a perhaps welcome change from traditional lessons. But even here the concerned educator can learn a lesson: pupils and students today have a far wider range of outside influences than once they did, and a pedagogy that does not take that into account will fail to inspire and carry with it an ever-growing proportion of the young people concerned. On this argument, the development of using non-standard media, as a delivery mechanism for teaching, is essential part for mathematics teachers in the years ahead.

The general experience is that at first the media discussed in this chapter take more time to produce learning effects than the traditional teaching methods, but that in the long run teacher and learner will earn their earlier investment back. This is true, of course, of the use of history as a whole in mathematics education. Some teachers are fearful of using historical resources because they expect them take up more precious class time. This fear may be justifiable in the short run. It is the long-term growth in understanding, however, which is at stake. The advantages argued for history in general, and non-standard media support in particular, are to do with the overall educational experience and the development of the learner over time.

As an example, consider the experience of Argentinian fourteen-year-olds described by Vicky Ponza in §10.2.1. There is a narrow sense in which they might have been taught more mathematics more quickly by conventional means than they learned by researching and constructing a play on the life of Galois. This view, however, fails to understand that the students were somewhat disengaged from mathematics beforehand and showed no promise of mathematical learning achievements in any event; the subject had seemed too remote from their concerns and too alienating as an emotional experience. Involving the students affectively did take time, but had a possibly lifelong benefit for them in securing their engagement with the idea of developing mathematical strengths.

This has long been the experience of teachers experimenting with non-standard media activities. One medium that has been explored, for example, especially in primary and middle schools, is that of curve-stitching. This was a technique popularised in English education circles at the end of the nineteenth century, notably by Mary Boole (the widow of the mathematician George Boole), for constructing curves by stitching their tangents on cards. When these practices were tried in the US in the 1940s it was found that hitherto-alienated students were attracted to mathematics as a result of the novelty and interest of this practical activity. One of the teachers involved recorded the following observations (McCamman 1945, 85):

An intricate and lacy design was made by a boy noisy in voice and manner, who was so unfamiliar with sewing that he thought the needle had to be tied to the end of the thread. A particularly striking design was made by a Chinese boy who at one time had been considered incapable of taking the regular mathematics courses. His chief difficulty was his inability to express himself in a strange language. [. . .] Some students who have not been doing well in geometry find in this work a new opportunity to be successful and to earn the praise of their classmates. In many such cases, the increased interest in geometry seemed to carry over to subsequent work.

Not that non-standard media benefit only previously under-achieving students. One of the major arguments of this book, for the media discussed in this chapter no less than elsewhere, is that there are benefits across the full range of student abilities, ages and institutions. A second message from the accounts here is that multi-media teaching ideas work best when the teacher is personally committed to and enthusiastic about the technology in question. That is one reason why what we present is a range of possible resources, and ideas for using them, and in no sense a recommendation that teachers 'should' use this, that or the other. A teacher who is personally excited about sundials, say, or dynamic geometry software, or the patterns and rhythms of folk dances, can do wonders in the classroom with that material which another teacher could not. We intend that this chapter will also provide useful ideas and stimulus for a teacher whose particular enthusiasms are something else again from the particular examples discussed here: for film-making, perhaps, or basket-weaving, or architecture, or devising mathematical trails, or encouraging poster production.

The question of the effectiveness of the various teaching modes, technologies and experiments described here is not easy to evaluate scientifically by the norms of modern experimental science. Elsewhere (Ch. 3) there is a discussion of research criteria and techniques, and there it is argued that anthropological or sociological research methods are better paradigms for evaluation in situations with as many uncontrollable variables as the learning situations in question here. Hence observational reports from teachers, and the sharing of subjective experiences by both teachers and students, are an important way of the researcher's gaining confidence in the effectiveness of the procedures.

10.1.4 Media and cognitive aspects of learning

One of the main benefits of having a range of media resources available is that this enables the cognitive needs of a greater number of students to be met. Through recognising more explicitly that students are very different, their individuality is allowed for and addressed more than traditional teaching methods are supposed to do. There have been serious concerns from many teachers in recent years, in two different directions. One is that traditional mathematics education preferentially benefits students with particular cognitive skills. The other is that all students are affected by the range of stimuli in their lives today, notably on their attention spans, in a way which has consequences for learning. Of course this is a complex and contentious area on which much work has been and continues to be done. But the implication for the present context is that students whose needs are not well met in the present system, for whatever reason, may find renewed learning possibilities in a range of other approaches.

This viewpoint again has deep historical roots: the French writer Jean-Jacques Rousseau was among the foremost advocates of learning aids, in the eighteenth century. Yet they have always remained at the periphery of pedagogic strategies. Pre-echoing the discussion of traditional educational techniques at the beginning of this chapter, a New York teacher at the end of World War II described the classrooms she saw as follows (Carroll 1945, 16):

In the high school, classical subjects [. . .] are all too commonly taught by the medieval methods of lecture, question, and answer. Except, perhaps, for the differences in dress and attitude of the students, a casual visitor might be unable to tell the difference between many 1944 classrooms in mathematics and their prototypes of the Middle Ages.

From the historian's perspective, these concerns (like the perennial concerns about the value of history in mathematics teaching) are always with us. Each generation needs to confront afresh the ways in which contemporary technology and media can support the role of the mathematics teacher.

Various other aspects are likely to play a role here, but need further research before their effect on mathematical learning can be established. In an arbitrary order I list some of these. Reading skills, particularly skills to understand ideas that are not familiar to the student, are probably well stimulated through the study of historical problems and methods. Given the effects of socialisation upon gender roles, girls and boys may respond differently to mathematics taught with history than to mathematics taught without. And students vary dramatically in what grips their imagination: understanding and describing what Fibonacci asked about the rabbits may be a goal that is better realisable for some students than proving general statements about Fibonacci numbers. This does not mean that teachers should forget about general proofs, but rather that we could think about setting a richer variety of goals among mathematics learners. Mathematics is a difficult area to reach the level of producing independent results, or otherwise feeling 'ownership' of the subject. History of mathematics may be helpful here. In this area students may sooner have the feeling that they have done valuable independent work, and be proud of it.

10.1.5 Media and assessment

An increasingly important aspect of mathematics education, in many countries, is the assessment procedure. Examinations as a critical component of the school experience began to be developed in the early nineteenth century, with the educational reforms following the French Revolution, and have come over the past two centuries to attain great significance. Sometimes teachers' pay has depended on the examination results of their pupils, and sometimes the status or remuneration of the school, besides the familiar fact that students' future progress, through education or the outside world, depends upon the results of their mathematics and other examinations. Not all teachers, still less students, view these developments with equal enthusiasm, but nevertheless this dimension of school experience is with us for the foreseeable future and can be made to yield benefits for the participants as well as for the wider political-economic system.

In this context, the form which examinations take is of great importance. Traditionally, these consist of students writing down answers to mathematical problems, generally within a fixed time limit. In recent decades other forms of assessment have been explored, such as relaxing the time limit and assessing work done over the whole of the study period (see Niss 1992a, 1992b, especially Izard 1992; Swan 1992). Here it can be useful to explore the way in which the combination of history and non-standard media can provide a much richer assessment environment, in which the skills and talents of a wider range of students can be represented and given credit.

Many of the resources discussed in this chapter give rise to assessment opportunities which can satisfy both the expectations of the wider system, for a ranking of students for various public purposes, and the expectations of students for a fulfilling and relatively stress-free mode of assessment. The Argentinian experiences recounted by Vicky Ponza (§10.2.1), for instance, concern the construction of a drama in such a way that every member of the class is concerned in some aspect of the production. The instruments whose classroom use is described by Maria Bartolini Bussi (§10.2.2) again offer opportunities for non-standard assessment in terms of mathematics classrooms, but which are familiar in creative arts. Students can be asked to assemble their own productions in a portfolio, or to present it to fellow-students or to an outside audience. This will be a way of assessing progress in mathematics of increasing importance.

References for §10.1

Carroll, L Grace 1945. 'A mathematics classroom becomes a laboratory', in: *NCTM 1945*, 16-29

Howson, Geoffrey 1994. 'Teachers of mathematics', in C. Gaulin et al., *Proceedings of the 7th international congress on mathematical education*, Sainte-Foy: Les Presses de l'Université Laval, 9-26

Izard, John 1992. 'Challenges to the improvement of assessment practice', in: Niss 1992b, 185-194

McCamman, Carol V 1945. 'Curve-stitching in geometry', in: *NCTM 1945*, 82-85

NCTM 1945. *Multi-sensory aids in the teaching of mathematics*, NCTM 18th yearbook, New York: Columbia University

Niss, Mogens 1992a. *Cases of assessment in mathematics education: an ICMI Study*, Dordrecht: Kluwer

Niss, Mogens 1992b. *Investigations into assessment in mathematics education: an ICMI Study*, Dordrecht: Kluwer

Swan, Malcolm 1992. 'Improving the design and balance of mathematical assessment', in: Niss 1992b, 195-216

Van Berkel, Klaas 1996. *Dijksterhuis Een biografie*, Amsterdam: Bert Bakker

Veloso, Eduardo 1998. *Geometria: temas actuais: materiais para professores*, Lisboa: Instituto de Inovação Educacional

10.2 Learning through history and non-standard media

10.2.1 Mathematical Dramatisation

Vicky Ponza

There is a clear tendency in the world today to use sophisticated electronic media in education. In many countries however, such as Argentina, such facilities are not widely available to students in educational institutions. Even where computers are

John Fauvel, Jan van Maanen (eds.), *History in mathematics education: the ICMI study*, Dordrecht: Kluwer 2000, pp. 335-342

available, in some cases teachers and students cannot profit wholly by them owing to the lack of resources such as funding for telephone bills. In the face of these needs, in my country teachers usually assume one of the following attitudes: either they come to terms with technological limitations, making use of wholly traditional methods, or they try to explore new ways to make up for these deficiencies. What new ways are these? We might, for instance, mention mathematical dramatisation. By this I mean the search for elements which may touch the students' sensibility and turn mathematics into a warmer, friendlier subject. Thus mathematical dramatisation means working on mathematics by involving intuition, creativity and the human body.

The need for developing such dimensions of mathematics has long been known by more insightful commentators. In a lecture delivered at the beginning of the twentieth century to the Psychological Society of Paris, Henri Poincaré commented on the psychological dimension of mathematical activity in these words (Poincaré 1914, 49-50):

A mathematical demonstration is not a simple juxtaposition of syllogisms; it consists of syllogisms *placed in a certain order* [. . .]. If I have the feeling, so to speak the intuition, of this order, so that I can perceive the whole of the argument at a glance, I need no longer be afraid of forgetting one of the elements; each of them will take place itself naturally in the position prepared for it, without my having to make any effort of memory. [. . .] It is time to penetrate further, and to see what happens in the very soul of the mathematician.

Searching deep into the mathematician's soul is an approach in which teachers can help encourage their students. In Argentina I have worked with students in a number of ways in order to help them develop their feelings in harmony with their mathematical skills and interests.

1. Intuition and creativity as related to the body allow us to go from dance (the choreography of the Argentinian dance music the 'zamba') to the drawing of described curves, and from these to the discovery of graphs of continuous, discontinuous and quadratic functions, as well as their characteristics.

2. They may also lead us to what I call 'corporised geometry'. Geometry is perhaps the part of mathematics most closely related to the natural and artificial surrounding created by the human being. Bodies, planes, straight lines, angles are to be found in the human body and we can discover them or make them up among several individuals.

3. Some students have a natural talent for performing as mimics. Bodily expression may be orientated from and towards geometry and a theorem may be proved, or at least made plausible, without resorting to either verbal or written language at all.

4. And by fusing these talents with the act of performance we will be able to produce 'mathematical theatre', that is to say, write and act mathematics.

These four activities are among those made with students between 13 and 17 and evaluated at Mariano Moreno School, Rio Ceballos, Cordoba, Argentina. For example, two students whose interest in human relations is more vivid than their command of mathematical language may nevertheless be encouraged to relate the two: Juliet may ask Romeo to leave, appealing to mathematical terms: "You have to leave. My mother interposes as a transversal between the parallels of our love. We are alternative exterior. Although we may be congruent we will never be together." Mathematical dramatisation becomes an open door towards interdisciplinary work and consequently towards a wide range of possibilities for all subjects.

Interdisciplinary trials were made all through one year (1994) at an Argentine school, which involved seven subjects including mathematics. The work revealed the need to start from history and take history as its leading theme. Mathematics was involved through working on a play with the students, whose scene was the Alexandrian Library. Students carried out research into the history of mathematics, which opened the way for introducing a dialogue between Euclid and Eratosthenes, which included references to several mathematical topics which formed part of the curriculum. The biographical allusions served the purpose of humanizing concepts. (Similar dramatic activity, also relating to ancient Alexandria, is seen in the 'Museum Strategy' of Pennington and Faux 1999, described below, §10.2.1 Annex.)

Once the potential of these teaching and learning techniques are explored, it will be clear that they are of value whatever the original motivation for their introduction, whether or not there are economic or other difficulties about the use of other media. Dramatisation is an important tool in the repertoire of every teacher, in whatever circumstances they work. Other teachers have worked on mathematical dramatisation, notably Gavin Hitchcock from Zimbabwe who in a series of contributions (Hitchcock 1992; 1996a; 1996b; 1997) has devised dramatic pieces for humanising and contextualising the development of mathematical concepts. Hitchcock 1997, for example, dramatizes the development of negative numbers between 1870 and 1970. In his introduction to that Hitchcock notes that "it might be good to expose a form of children or grown-up students to a variety of different approaches to this topic or others and encourage them to become active critics instead of passive receivers".

The effectiveness of this working method became evident in the course of a trial made in 1997 with students between 12 and 13 years of age (2nd school year) of the above mentioned Argentine school and its follow-through in 1998, which I now outline here (for fuller details see Ponza 1998).

Dramatisation of the life of Galois

1. Divided in task-groups from the beginning of the class year, the pupils looked for information about the history of mathematics. They were instructed to search in their own library (which in general had rather poor resources), in the school library and in the town during a period of two weeks. They did this work outside school times. They had to bring the material and note carefully the name of the work and of the author.
2. Once all the material had been gathered, it was sorted out in the school during mathematics class-time according to the people and themes they had researched. Despite the fact that in the whole village there was no specific bibliography on the history of mathematics, some of the pupils surpassed my expectations on the quality and quantity of the information found. I made use of my personal library, contributing as a member of the research group.

3. I made a second selection of the material contributed by us all, relating it to the contents to be developed in the second year, and I planned the classes from the starting point of history wherever possible.

4. When starting a unit, I would distribute amongst the groups the relevant material. The pupils would look up the historical elements appropriate to the specific themes and they would then read out the researched information to the whole class. Each one would do their own introductory summary, emphasizing distinct aspects, according to their personality and inclinations. This introduction served to motivate and would be enriched by new additions discovered in the course of the unit's development.

5. In some units pupils were captivated by the life of particular mathematicians. Evariste Galois was a case in point, whose life and work arose when the pupils were researching into the history of equations. They decided to write a small dramatic piece and show it first within their own course and then before all interested pupils in the school. In the writing and performance of the play they took the following steps:

a) The groups looked for details related to the life of Galois.

b) Each group informed all others about the details they had compiled.

c) Each group wrote a play and read it to fellow classmates.

d) A new task-team made up of representatives from each of the original task-groups synthesised the different plays into the final work to be shown.

e) They shared out responsibilities of every aspect of the production: between those acting, those in charge of stage design, wardrobe and other helpers, always bearing in mind and with due regard for personal idiosyncrasies and leanings. This enabled all the pupils on the course to participate in one form or another.

Here are extracts from the play. We give here the first and second scenes, and the concluding statement (the full text is in Ponza 1998; the translation is by Antonio Luque).

FIRST SCENE

(Classroom in the Ecole Louis Le Grand. Various students, Galois and François, a friend, conversing. A tutor is close by his writing desk, talking in private with a student).

GALOIS: Dear François, I can't stand this place any longer. It is all so strict! They don't let us think for ourselves, nor have our own opinions, one cannot sleep or eat.

FRANÇOIS: Yes, the food is meagre, all is dull, but you need patience! Here we learn Greek and Latin and all else to prepare us for the future.

GALOIS: I am not interested in Greek or Latin. What is more the tutors... (*the tutor talking to the other student raises his voice, grabs him by the scruff of his neck and starts slapping him*)

TUTOR: So you can't repeat the phrase which I gave you in Latin, nor in Greek, huh?

GALOIS: *(Jumping towards the tutor)* Enough! Stop hitting him, you have no right!

TUTOR: Don't I? Of course I do. Just as I'm allowed to tell you to shut up. Particularly you, who are repeating the course, and still can't pass Greek or Latin.

GALOIS: I am not interested! And you can't punish us for that!

TUTOR: *(Calmer now, as he has come across someone who stands for himself)* I find you strange, Galois. Your parents were very studious in those matters. You will not go far. You will remain mediocre all your life. Let's see, is there *anything* which interests you?

GALOIS: I enjoy geometry: there I can see, feel and think. Legendre is a master. And also algebra. So it is that I chose the subject of mathematics, and now I know what I want to be in life: I shall be a mathematician. I am studying analysis and the algebra of Abel and will enrol in the Ecole Polytechnique.

TUTOR: Bah! ... Just a simple optional subject! You are a day-dreamer and you will be nothing but a failure all your life.

SECOND SCENE

(Other students leave the stage. Galois moves forward and sits down on a writing desk. While he is talking the scene changes to the Ecole Polytechnique.)

GALOIS: I have got to prepare really well. It is difficult to obtain a place in the Ecole Polytechnique but from it have proceeded the best mathematicians. I will also write down my discoveries on equations and will ask those geniuses to present them to the Academy of Sciences.

(Continues working. Meanwhile in the Ecole Polytechnique, Dinet and Cauchy are now set up. Galois takes up his note book and walks towards them)

GALOIS: Morning, Profs! Are you the great sage mathematician Cauchy?

CAUCHY: Yes, I am. And this is the great professor Dinet, who marked your exam paper.

DINET: *(He is very old and a bit deaf)* Tell me young man, what did you mean when you wrote here that it is possible to define which equations can be solved using roots?

GALOIS: Ah, yes. It appears that looking for the conditions necessary for the coefficients to have a formula which would give the solutions, I discovered what you have just read, because if you analyze an algebraic equation...

DINET: *(Interrupting)* No, no, no. Don't come to me with weird things, inventions, discoveries. What can a young man like you ever discover! Why don't you show all your rough work, all the steps you've taken? Why did you not move each term one at a time as you have been taught?

GALOIS: But professor, this is something obvious, allow me to pursue my explanation of the solutions that I discovered.

DINET *(Interrupting)* And besides, what is this about logarithms? These are but hieroglyphs *(looks at Cauchy)* Plainly you lack a systematic training.

GALOIS: (Looking at Cauchy as if asking for help) Gentlemen please.

DINET and CAUCHY: Dismiiiiiiiiiiiiiiissed.

GALOIS: *(despairingly)* Wait gentlemen, please wait *(he stops and shows them his notes)*. Listen to my theories and you will see that I am right. I have discovered the solution to equations of degree beyond two and three: in order to solve them by roots, it is necessary ...

CAUCHY: *(takes the notebook and leafs through it)* Rubbish, rubbish. Why have I got to read the work of a 17 year old youngster like you? Besides, I don't understand any of it at all *(throws it into a paper bin)*

GALOIS: You are, you are ... I can't stick ... you two idiots. And you look it! *(he throws all loose sheets on the writing desk over their heads)*.

DINET and CAUCHY: Dismissed, dismiiiissed for good and e-ver.

[...]

Concluding statement

NARRATOR: We identify ourselves with many aspects of the life and experience of Galois, and so ask: for how much longer will the superficial prevail over the deeper? Will we forsake the opportunity of developing ourselves in school? Will we lose our humanity? He was only twenty years old when he died. He suffered the greatest injustices of the so called geniuses of the epoch. Notwithstanding the theft, envy, indifference they could not stop him becoming, years later, one of the most important men in mathematics. We believe *[Creemos]* that justice was finally done, but we request *[queremos]* that justice be realised in real time. It is not good enough years later.

Evaluation

This play is only one of the many issues of a school year devoted to working at mathematics with history as a starting-point, supplying evidence for the possibilities of mathematical dramatisation. For further discussion of the issues, see § 6.2.2.

Can mathematics be learned by means of dramatisation? If so, is it just intuitive learning? The trials mentioned in this section have proved that mathematics can, in fact, be learned in this way. By starting from the intuitive-emotional as a means of approaching pure and abstract matter, we overcome the resistance often put up against mathematics. This method is endorsed by biological research which, as concerns mathematics, holds the existence of an almost total lack of formation of emotive memories (right cerebral hemisphere). In his essay 'El saber sí ocupa un lugar' (Knowledge *does* occupy a space), published in 1994, Osvaldo Panza Doliani bases his judgment on multi-comprehension, that is to say, on the development of a scientific discipline supported by all sciences, the epistemological basis of which is natural evolutionary rigour. He holds that

all the events that take place, be they orderly or disorderly, turn out to teach the human being a lesson. Therefore, when teaching, the incentives are not just pedagogical, but various. The three issues related to biological findings which should lead the way to a new pedagogy are: a) The fact that, so far, teaching has scarcely considered the biological tempo needed for the organisation of memories, b) Disregard of the fact that sensory perception is the life of the brain and, consequently, the latter depends on it, c) Disregard of the fact that sensory perception depends on teaching.

This method of teaching is not just intuitive. When students write or dance or perform mathematics they work out, they analyze, organize and solve.

As well as inquiring into the potentialities, it is well to consider the risks and limits of mathematical dramatisation. The main risk lies in remaining at the initial phase of the process, that is to say, at its intuitive, motivating stage, at the stage of mere fun. This means overlooking the educational factor and, accordingly, failing to transmit to the students the need of arriving at the pure matter, of attaining mathematical rigour and an appropriate language. This would reveal that the teacher has not understood the real motivating meaning of the history of mathematics. We have to work out strategies which may help us to approach practical as well as pure and abstract mathematics, and that may have students realize that they are also building up their knowledge when they spend long hours studying by themselves.

Regarding its limits, I consider the main problem to be time, worsened by bureaucratic hindrances and disorganisation of the educational system. This makes it very difficult to develop a coordinated task among parallel courses, which might permit us to go deeper into this method in lower courses of high school, so as to be free to work on pure, abstract mathematics in the higher courses. The economic factor is a corollary of the time issue since, among other things, teachers' training depends on it. The qualification of *a few* teachers and the effort invested by them is not enough to guarantee the fulfillment of these aims.

Annex: The Museum Strategy

The work of the UK mathematics educators Eileen Pennington and Geoff Faux is another example of exploring the potential of mathematical dramatisation. In their ten lesson project *No royal road to geometry* (named after Euclid's reported response when King Ptolemy asked for a shorter route than by studying the *Elements*) Pennington and Faux encourage children to act the roles of museum designers who are working to bring ancient mathematics and its context to life in a modern museum setting. By engaging the children in a second-level dramatisation, as it were, playing the part of museum workers who may decide to play the part of Greek mathematicians, the situation is both more realistic and better able to meet other pedagogic aims such as explicitly encouraging research activities, design and other cross-curricular work and an even wider range of roles for different children to adopt. In addition, the focus on needing to carry out a range of research activities in order to prepare the museum exhibit usefully helps pupils attend both to their sources of information and to transmuting it for the benefit of others. The teacher participates explicitly in an unforced way as the museum curator. This strategy of dramatisation provides much opportunity for children in middle schools (the work of Pennington and Faux is aimed particularly at 10-11 year olds) to develop their understandings across a range of mathematical, historical and design subjects. For example, there might be lively and valuable discussion among the museum designers as to whether ancient Alexandria would be likely to have had a public statue in the market-place labelled "Aristarchus of Samos, 320 – 250 BC"!

References for §10.2.1

Boero P., Pedemonte B. and Robotti E. 1997. 'Approaching theoretical knowledge through voices and echoes: a Vygotskian perspective', *Proceedings of the 21st International Conference on the Psychology of Mathematics Education*, Lahti, Finland, ii, 81-88

Durán, Antonio José 1996. *Historia, con personajes, de los conceptos del cálculo*, Madrid: Ed. Alianza Universal, 17-22.

Hitchcock, Gavin 1992. 'The "grand entertainment": dramatising the birth and development of mathematical concepts', *For the learning of mathematics* 12 (1), 21-27

Hitchcock, Gavin 1996a. 'Dramatizing the birth and adventures of mathematical concepts: two dialogues', in R Calinger (ed), *Vita mathematica: historical research and integration with teaching*, Washington: MAA 1996, 27-41

Hitchcock, Gavin 1996b. 'A window on the world of mathematics, 1871: reminiscences of De Morgan—a dramatic presentation', in E Veloso (ed), *Proc. HEM, Braga* ii, 35-42

Hitchcock, Gavin 1997. 'Teaching the negatives, 1870-1970: a medley of models', *For the learning of mathematics* 17 (1), 17-25, 42

Muñoz Santoja, José, Carmen Castro, María Victoria Ponza 1996. 'Pueden las matemáticas rimar?', *Suma* 22, Zaragoza: Federación Española de Sociedades de Profesores de Matemáticas, junio, 97-102.

Panza Doliani, O, Ponzano, P., 1994. *El saber, sí ocupa lugar*, Córdoba: Ciencia Nueva, 13-24

Pennington, Eileen & Geoff Faux 1999. *"No royal road to geometry": a ten lesson project in mathematics, history and drama for year 5 or 6*, Dalston: Education Initiatives

Poincaré, Henri 1914. 'Mathematical discovery', in *Science and method*, (tr. Francis Maitland), London: Nelson; repr. New York: Dover 1952, 46-63

Ponza, María Victoria 1996. 'La experiencia interdisciplinaria en la realidad educativa de hoy', *Suma* 21, Zaragoza: Federación Española de Sociedades de Profesores de Matemáticas, febrero, 97-101

Ponza, María Victoria 1998. 'A role for the history of mathematics in the teaching and learning of mathematics', *Mathematics in school* 27 (4), 10-13

Ruiz Ruano, Paula & Pérez, Pilar 1996. 'Hipatia en el país de las empatías', *Revista Centro de profesores de Linares*, Jaen: Consejería de Educación y Ciencia, 9-18.

Savater, Fernando 1997. *El valor de educar*, Barcelona: Ariel SA

10.2.2 Ancient instruments in the modern classroom

Maria G. Bartolini Bussi

The history of mathematics can enter classroom activity, besides the other ways discussed in this chapter, by investigating also copies of ancient instruments and other artefacts, reconstructed on the basis of historical sources. In museums of the history of science throughout the world there are to be found beautiful collections of original instruments. We may refer to the Museo di Storia della Scienza, in Florence (Italy), to the Hilbert Raum of the Mathematics Institute in Göttingen (Germany) and to the Emperor Collection, stored in the Palace Museum of the Forbidden City in Beijing (China), to mention just a few in different parts of the world. Because of the delicacy of those precious artefacts, visitors are not usually allowed to touch them. Hence an important part of the experience, namely the visual tactile feedback while handling the instrument, is not accessible to teachers and students. It would be really more useful to have rough yet working copies of them in the classroom (preferably as physical objects, although computer simulations increasingly have a role to play in this area). To provide even rough working models is not an easy matter, especially for complex ones, but some specimens representative of important class of instruments could be built by teachers or students themselves (see CIEAEM 1958; Cundy & Rollet 1952). Here we present briefly some ideas for instruments in the modern classroom, categorized by the mathematical subject matter involved.

Arithmetic

Modern copies of arithmetic historical artefacts comprise for instance Mesopotamian tablets, different kinds of abaci and Napier's rods. Mesopotamian tablets were made of clay, but copies may be built now by plasticine, so allowing the same material to be used several times. The system of signs for numerals in their sexagesimal counting system, an early positional system, may be introduced in the classroom by means of realistic re-creation of copies of the original tablets (Robson 1996; 1998) and the tablets can be used for various problem-solving and investigation activities (Burns 1998; MacKinnon 1992).

Figure 10.1: 987 654 321 on the soroban

John Fauvel, Jan van Maanen (eds.), *History in mathematics education: the ICMI study*, Dordrecht: Kluwer 2000, pp. 343-350

The positional system of notation up to the construction of algorithms for arithmetic operation is embodied by several kinds of abaci: instances are given by the dust abacus of the Babylonians, the line abacus of the Greeks, the grooved abacus of the Romans, the bead abaci in the Chinese (*suan pan*), Japanese (*soroban*) or Russian (*s'choty*) versions (Metallo 1990; Yoshinko 1963; Boyer 1968 ch. 12; Smith 1958, 156-196).

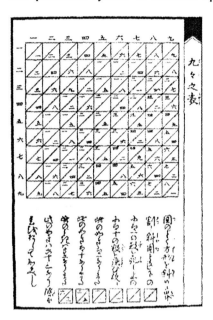

Figure 10.2: Napier's rods in Japan

Multiplication in the decimal positional system can be carried out by using Napier's rods (early 17th century), a way of utilising, in wood, bone or cardboard, the principles of the ancient 'gelosia' method of calculating on paper (Smith 1958, 101-128; Swetz 1994, 179-192). This may be thought of as an idea about carrying out arithmetic calculations with carefully designed physical objects. Later devices to the same end, true calculating engines in the modern sense, are less easily reproduced in the classroom. Even the 17th century engines (eg Schickard's, Pascal's or Leibniz's calculating machines, cf. figure 10.3) involve quite complicated machinery, let alone later developments such as Babbage's difference and analytical engines (Swade 1991).

Algebra

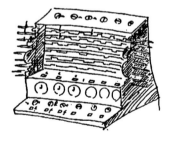

Figure 10.3: The design of a calculating machine which Schickard sent to Kepler in 1624

Solving equations by mechanical artefacts dates back to the classical age. The problem of finding two or more mean proportionals between two given segments, that is equivalent to solving an equation of degree at least three, lead to the production of meanfinders or, according to the ancient name, of examples of mesolabon. Instances are offered by the mesolabon of Eratosthenes (Fauvel & Gray 1987, 83-85), of Plato, of Dürer (Dürer 1525/1995), of Descartes (Fauvel & Gray 1987: 344-5). A mechanical method of solution of equations is given by D'Alembert's machine (Diderot 1751):

two working copies at least have been built, in Pisa by Franco Conti and in Grenoble by Jean Marie Laborde.

Geometry

A very rich collection of more than one hundred and seventy geometrical models and instruments have been constructed in Modena (Italy) by a group of secondary school teachers, under the scientific direction of M. Bartolini Bussi. A virtual visit might be done by surfing one of the following sites:

      http://www.museo.unimo.it/labmat/ or http://www.museo.unimo.it/theatrum/.

The former is a trilingual (Italian, English, French) site updated to February 1997. The latter is a monolingual site (Italian) updated to December 1998 with dozens of photos, animations and interactive simulations: a copy on CD-rom may be delivered free to interested people who send a message to the author of this section (bartolini@unimo.it). The artefacts are made with wood, plexiglass, brass, lead and threads, on the basis of historical sources from the classical ages to the nineteenth century. For each model, historical sheets and activity sheets for secondary and university students are available.

    A visit to the collection could start from the visual tactile exploration of the most ancient instruments, namely Platonic and Archimedean solids and big size static models which illustrate the three-dimensional theory of conics, from the ancient static models, to the compasses of the XVII century. The classical problems of doubling the cube or trisecting the angle are considered. For the former, meanfinders can be used (see above). For the latter, several trisectors are available (see also Yates 1945b).

    There are also several instruments related to the two trends that characterise modern geometry from the seventeenth century onwards and came together in the

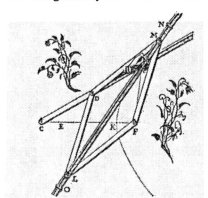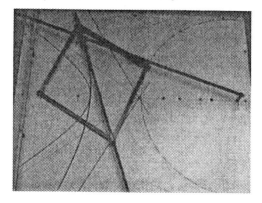

Figure 10.4: Instrument designed by van Schooten (1646) to draw an hyperbola. On the left the design from the Exercitationum mathematicarum libri quinque, 1656; *on the right the instrument as it is reconstructed by the Laboratorio di Matematica of the museum of the university of Modena and Reggio Emilia. A Java-simulation is available on CD and on the internet.*

late nineteenth century: the mechanical-analytical style initiated by Descartes, and the projective-synthetical style initiated by Desargues. In the former area, there are on show several curve-drawing devices to draw conics, cubics, quartics and curves of higher degree and pantographs to realise linear or more generally birational transformations. In particular, the problem of drawing straight lines (i. e. of transforming a circular motion into a linear motion) is illustrated by means of the linkages which realise the most relevant solutions proposed in the nineteenth century. The history of the representation of curves (Bos 1981) can be explored through the instruments from the classical age (eg Nicomedes compass), to the works of Descartes and van Schooten, to the multifaceted study of organic generation of curves (with Newton's contributions), up to the theoretical proof of the possibility of drawing any general algebraic curve, offered by Kempe in 1876. In this case the story reaches forwards to today's development in pure and applied mathematics (Bartolini Bussi, 1998).

In the case of the projective-synthetical branch of geometry, working models of practical perspectographs can be used by students and teachers. These instruments (dating back to Dürer, Niceron, Scheiner, Lambert and others) allow the exploration of various geometrical themes. In art, for instance, the production of real life paintings that give the illusion of reality, or anamorphoses that can return an understandable image of something only when they are looked at from a very particular and unexpected point of view. Desargues' and De la Hire's projective approach to conics is illustrated by dynamical models that explain the genesis of plane definitions, and Newton's study of cubics by shadows is presented.

A complete interactive catalogue, in Italian, is in the CD Rom realised by Bartolini Bussi & al. (1999) (see also the webpage reference). A historical excursion through the models conveys the idea of the progressive expansion of geometry that goes along with the introduction of more and more theoretical

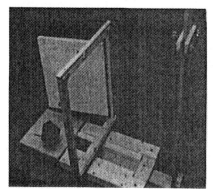

Figure 10.5: Dürer's 'perspectograph' with three strings, in the version from the Underweysung *(1525) and as a model in the Modena university museum*

instruments over the centuries. An example is offered by the shift from Euclid's geometry, based on the straightedge and compasses—in which context problems like the trisection of angle and the duplication of cube were (it eventually turned out) theoretically unsolvable—to Descartes' geometry. Here the same problems became solvable by means of other kinds of mechanical devices to be conceived as theoretical instruments. This shift paves the way to a critical approach to the more sophisticated electronic instruments that are today available: the computer is much more flexible than ancient instruments, yet the understanding of the underlying theoretical assumptions that make it possible to solve problems (approximately or rigorously?) is more difficult and hidden inside the black box.

Applied mathematics

Nearly all the instruments from ancient and modern technology embody a lot of mathematics, hidden in the instrument itself and accessible only through a careful and intentional analysis. Just to quote some examples, we refer to astronomical instruments such as sundials and astrolabes (Ransom 1993), instruments for navigation (Albuquerque 1988a, 1988b; Ransom 1993), for surveying (Kiely 1947; Eagle 1995, 65-74) and mechanical instruments (Gille 1978). A very rich catalogue is in the 1972 reprint of Bion's classical treatise (1758).

Examples of classroom activity

Classroom activity with instruments of the kind described in this section can take place in several ways, of which there are two main categories:
1. visiting the instruments, either in reality or a virtual visit (eg. by means of videos, computer simulations, CD roms, or websites). This can address in an agreeable way the cultural dimension of mathematics, such as the link between visual, tactile and intellectual activity in mathematics, and the dialectic interaction between pure and applied mathematics over the centuries.
2. specific classroom activities for helping students experience the tasks of working mathematicians (eg constructing proofs) or develop the understanding of some specific piece of knowledge.

Pupils can handle instruments with mathematical goals from very early. An interesting example is given by a Portuguese project. In the last decade a number of teachers of mathematics in Portugal have been exploring in their lessons the theme of 15th and 16th century Portuguese voyages. In the past, this theme was used mainly by general history teachers, and the relevance of mathematics as a major way to understand the processes used in high sea navigation by the Portuguese navigators was almost neglected. But during the school year 1991-1992, a national project was set up which involved around three hundred pupils of 8th and 9th grades. They interpreted marine rules, studied maritime principles, built and graduated models of nautical instruments, and learned how to use them to measure the altitude of the stars. They used and developed their knowledge of mathematics to understand the basic principles of astronomical navigation (Veloso 1992, 1994).

This is but one example of the way an informed teacher can make use of historical ideas involving instruments, devices and artefacts from the past to enrich their mathematics lessons and attain various education goals. I give references here for some other examples, at a range of educational levels:

– perspectographs (see figure 10.5) in primary school (Bartolini Bussi 1996)
– gears in grades 3-6 (Bartolini Bussi & al. 1999)

- Napier's rods in grades 5-6 (Navarra 1994)
- the kaleidoscope for middle-school geometry classes (Graf & Hodgson 1990)
- linkages in grades 6-8 (Damiani et al. 1998; Yates 1945a)
- cross-staff and sundials in years 10-12 (Ransom 1993)
- pantographs in grade 11 (Bartolini Bussi & Pergola 1996)
- curve drawing devices in grades 12 (van Maanen 1992; Dennis 1995; Dennis 1997)
- pulleys in calculus teaching (van Maanen 1991)
- 3-dimensional theory of conics in grade 12 (Bartolini Bussi & Pergola 1994)
- abaci for prospective teachers (Metallo 1990)

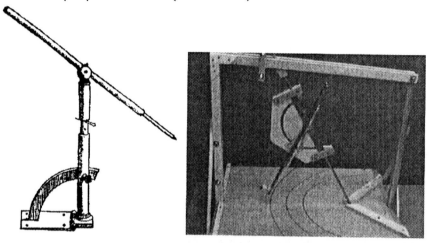

*Figure 10.6: A 'compasso perfetto'. The first leg can be placed at a fixed angle towards the plane, the length of the second leg is variable through a pin which can shift into a tube. Here the design by Barozzi (*Admirandum illud geometricum problema, Venice 1586*) and the model built in Modena.*

It is quite difficult to compare different methodologies in such a short space. What seems to be shared is the focus on manipulative activity: it means that instruments are not only looked at but really handled by the students. This tactile dimension add something specific to the historical dimension of all the activities that are described throughout this whole book. We can wonder why tactile activity turns out to be so important not only with young pupils but also with high school students and adults.

Surely a part of the answer is in motivation: not least, people who do not like mathematics (we could say, *especially* people who do not like mathematics) enjoy recourse to physical objects, closer to their everyday experience than blackboards full of symbolic equations (hence the success of 'hands on' scientific exhibitions, all over the world). But this is only a part of the story and, maybe, the less important. In tactile experience there is an important part of the cognitive foundations of mathematical activity.

An interesting epistemological analysis of the abacus (and of other important media) is offered by Brian Rotman. The abacus is a machine which keeps track of the process of counting, and Rotman (1993, 33) points out that

to move from abacus to paper is to shift from a gestural medium (in which physical movements are given ostensively and transiently in relation to an external apparatus) to a graphic medium (in which permanent signs, having their origin in these movements, are subject to a syntax given independently of any physical interpretation).

A similar analysis could be done for most of the instruments described above. This cognitive aspect is analysed by Bartolini Bussi & al (1999) for the genesis of the sign 'arrow' to denote orientation. In a similar way, Dennis (1995) studies the genesis of the idea of variable and of singular points of a curve in the manipulative activity with curve drawing devices of the geometers of the 17th century. In this way many important experiences and conceptual transitions from the past may be replicated in today's mathematics classroom.

References for §10.2.2

Albuquerque de L. 1988a. *Astronomical navigation*, Lisboa: National Board for the Celebration of the Portuguese Discoveries

Albuquerque de L. 1988b. *Instruments of navigation*, Lisboa: National Board for the Celebration of the Portuguese Discoveries

Bartolini Bussi M. & Pergola M. 1994. 'Mathematical machines in the classroom: the history of conic sections', in Malara & Rico (eds.), *Proc. of the 1st Italian-Spanish symposium in mathematics education*, Modena: Dipartimento di Matematica, 233-240

Bartolini Bussi M. & Pergola M. 1996. 'History in the mathematics classroom: linkages and kinematic geometry', in Jahnke H. N., Knoche N. & Otte M. (eds.), *Geschichte der Mathematik in der Lehre*, Goettingen: Vandenhoeck & Ruprecht.

Bartolini Bussi M. 1996. 'Mathematical discussion and perspective drawing in primary school', *Educational studies in mathematics* 31, 11-41.

Bartolini Bussi M. 1998. 'Drawing instruments: theories and practices from history to didactics', *Documenta mathematica - extra volume ICM 1998* iii, 735-746.

Bartolini Bussi M., Boni M., Ferri F. and Garuti R. 1999a. 'Early approach to theoretical thinking: gears in primary school', *Educational studies in mathematics* 39, 67-87

Bartolini Bussi M., Nasi D., Martinez A., Pergola M. Zanoli C. & al. 1999b. *Laboratorio di matematica: theatrum machinarum*, I CD rom del Museo (1), Modena: Museo Universitario di Storia Naturale e della Strumenntazione Scientifica

Bion M. 1758. *The construction and principal uses of mathematical instruments*, (repr. 1972), London: The Holland Press

Bos H. J. M. 1981. 'On the representation of curves in Descartes' *Géométrie*', *Archive for history of exact sciences* 24, 295-338.

Boyer C. B. 1968. *A history of mathematics*, John Wiley & Sons.

Burns, Stuart 1997. 'The Babylonian clay tablet', *Mathematics teaching* 158, 44-45

CIEAEM 1958. *Le matériel pour l'enseignement des mathématiques*, Neuchatel: Delachaux

Cundy H. Martyn & Rollet A. P. 1952. *Mathematical models*, Oxford: Clarendon Press.

Damiani A. M *et al.* 1998. 'De l'étude d'un "modèle dynamique" aux définition: un parcours interactif', Proc. CIEAEM 49 (Setùbal), 377-384.

Dennis, David 1995. *Historical perspectives for the reform of mathematics curriculum: geometric curve drawing devices and their role in the transition to an algebraic description of functions*, doctoral thesis, Ithaca: Cornell University

Dennis, David & Jere Confrey 1997. 'Drawing logarithmic curves with Geometer's
    sketchpad: a method inspired by historical sources', in: J.R. King and D. Schattschneider
    (eds) *Geometry turned on! (...)*, Washington: MAA, 147-156
Diderot, Denis & d'Alembert, 1751. 'Constructeur universel d'equations', *Encyclopedie*
Dürer, Albrecht 1525/1995. *Géométrie* (1525), présentation et traduction de J. Pfeiffer, Paris:
    Seuil (translation of *Underweysung der Messung*)
Eagle, M Ruth 1995. *Exploring mathematics through history*, Cambridge: University Press
Fauvel John & Gray Jeremy 1987. *The history of mathematics: a reader*, London: Macmillan
Gille B. 1978. *Histoire des techniques*, Paris: Gallimard
Graf, Klaus-Dieter and Hodgson, Bernard R. 1990. 'Popularizing geometrical concepts: the
    case of the kaleidoscope', *For the learning of mathematics* **10** (3), 42-50
Kiely, Edmond R 1947. *Surveying instruments: their history and classroom use*, NCTM 19th
    yearbook, New York: Columbia University
Maanen, Jan van 1991. 'L'Hopital's weight problem', *For the learning of mathematics* **11**
    (2), 44-47
Maanen, Jan van 1992. 'Seventeenth century instruments for drawing conic sections',
    Mathematical gazette **76** (476), 222-230
MacKinnon, Nick 1992. 'Homage to Babylonia', *Mathematical gazette* **76** (475), 158-178.
Metallo F. R. 1990. *The abacus: its history and applications*, Himap: Module 17
Navarra G. 1994. 'Dalla moltiplicazione a "gelosia" ai bastoncini di Genaille', *Atti I
    Internuclei Scuola dell'Obbligo (Salsomaggiore Terme)*, 23-28
Ransom, Peter 1993. 'Navigation and surveying: teaching geometry through the use of old
    instruments', in , IREM de Montpellier (ed.) Actes de la 1re Univ. d'été Europ., 227-239
Robson, Eleanor 1996. 'From Uruk to Babylon: 4500 years of Mesopotamian mathematics',
    in *Proc. HEM (Braga)* **i**, 35-44
Robson, Eleanor 1998. 'Counting in cuneiform', *Mathematics in school* **27** (4), 2-9
Rotman, Brian 1987. *Signifying nothing: the semiotics of zero*, Stanford University Press
Smith, David Eugene 1958. *History of mathematics* **ii**: *special topics of elementary
    mathematics*, New York: Dover
Swade, D. 1991. *Charles Babbage and his calculating engines*, London: Science Museum
Swetz, Frank 1994. *Learning activities from the history of mathematics*, Portland: Walch
Veloso, Eduardo 1992. 'Portuguese discoveries: a source of interesting activities in the
    mathematics classroom', paper presented to Toronto meeting of HPM Study Group.
Veloso, Eduardo 1994. 'Practical uses of mathematics in the past: a historical approach to the
    learning of mathematics', *Proc. XVIII PME Conference (Lisboa)* **i**, 133-136
Yates, Robert 1945a. 'Linkages', in *Multi-sensory aids in the teaching of mathematics*,
    NCTM 18th Yearbook, New York: Columbia University, 117-129
Yates, Robert 1945b. 'Trisection', in *Multi-sensory aids in the teaching of mathematics*,
    NCTM 18th Yearbook, New York: Columbia University, 146-153
Yoshinko, Y. 1963. *The Japanese abacus explained*, New York: Dover

Web references

Conti
    http://www.sns.it/html/OltreIlCompasso/Mostra-
        Matematica/mostra/macchina.htm
Bartolini Bussi
    http://www.museo.unimo.it/labmat/
    http://www.museo.unimo.it/theatrum/

10.2.3 Inquiring mathematics with history and software

Masami Isoda

Many mathematical instruments are discussed in the previous section. Each instrument can also be represented by some computer software. This section discusses software from the viewpoint of integrating traditional instruments and computers for mathematical inquiry in the classroom. As an example of conceptual integration, in what follows the word 'tool' may stand both for instruments and for mathematical software.

Mathematics software enables us to represent mathematics on a computer and change this representation depending on mathematical rules. Figure 10.7 shows the recent history of software innovation for general users of mathematics: Graphing Software (Algebraic Expresser, Function Probe, Calculus Unlimited, etc.); Dynamic Geometry Software (DGS) (Cabri, Geometer's Sketchpad, etc.); Spreadsheets (Excel, Lotus, etc.); Computer Algebra Systems (CAS) (Derive, Mathematica, Maple, etc). Using functions or macros, some of these packages can be extended to design special instruments. Some were developed for research, but the evolution of the interface has made such software more accessible to general users. These days, many mathematical software packages incorporate multiple representation features (Yerushalmy & Schwartz 1993, 47) and enable us to use it on the world wide web with Java (Cabri Applet, Sketchpad Applet etc).

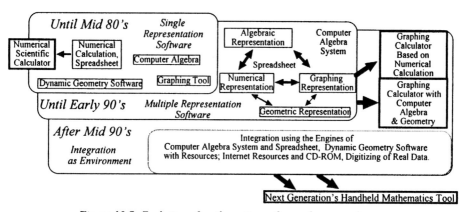

Figure 10.7: Evolution of mathematics software for general users

Several studies have already indicated the power of multiple representation tools for knowledge construction. Through the use of these tools, we can assist students to translate and interpret concepts through various representations (Lesh, Landau &

John Fauvel, Jan van Maanen (eds.), *History in mathematics education: the ICMI study*, Dordrecht: Kluwer 2000, pp. 351-358

Hamilton 1983, 271; Kaput 1989, 171; Isoda 1998a, 270), and help students' inquiry into mathematical ideas.

10.2.3.1 Inquiry using multiple representation tools: a historical view

To use software and instruments for mathematical inquiry, we should identify some features of tools used in such inquiries. In the following, the roles of tools and the context for using them are discussed from a historical viewpoint. In addition, an example of students' inquiry is presented.

A lot of historical examples indicate the following roles of tools in inquiry:

a) determining the subject of the mathematical inquiry;
b) giving a method for the mathematical inquiry;
c) revealing the epistemological obstacles inherent in using such tools in the specific context.

The classical tools of ruler and compasses are well known examples for all three of these roles.

As David Dennis and Jere Confrey (1997) discussed, the problem of using tools is closely related to the problem of representation because any tool can be used to represent an idea. For example, in the 17th century Descartes (1628) lamented the loss of geometric intuition possessed by the ancients. So, in 1637 he applied an algebraic representation to the ancients' geometry and tried to develop a new analysis of how to carry out mathematical research. He felt the restriction of ruler and compass was strange as he could use many tools outside the context of Euclid. We should recognise the following points about the context for using tools:

1. we can change the role of tools depending on the context;
2. we can support students' understanding through the changing of tools and representations.

History tells us that the tools used for mathematical inquiry are themselves reformulated using mathematical (especially algebraic) representations, as mathematicians in the scientific revolution tried to select, find or construct convenient representations and instruments for their research. Even though Pascal (1640) tried to retain the ancients' geometry for the discussion of truth, we find that 200 years later his projective geometry was reconstructed using algebraic representation. Such computational contexts have enabled mathematicians to develop innovative software for mathematics, so that everyone can use multiple representation software on computers. In the area of education, the increasing dominance of algebraic formulation of mathematical ideas strongly influenced school mathematics until the age of modernisation. The positives are balanced by some negative aspects. Today, in many countries, students, and even teachers, have no opportunity to learn about the higher concept of geometric representation of curves because they have only learned about curves through algebraic representation. To guarantee student inquiry, we should avoid anti-didactic inversion (Freudenthal 1973, 122). Thus we should add the following additional point about the context for using tools:

3. although the generality or viability of a mathematical ideas depends on the representation, we should give students the opportunity to select, find or create new tools or representations for constructing knowledge.

In the last decade, the multiple representation environment of tools has encouraged new laboratory approaches and has changed learning contexts (Zimmermann *et al* 1990; Leinbach *et al* 1991). This environment helps students in their mathematical inquiry through multiple representations (see figure 10.8, Isoda 1998a, 269). In this environment, no undue emphasis should be given to a particular representation or tool so that students will better appreciate the power and beauty of various representations or tools.

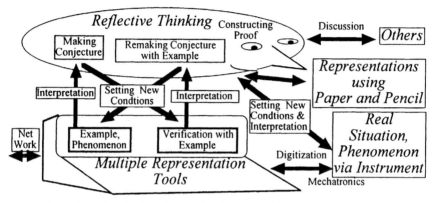

Figure 10.8: Inquiry based on tools which allow Multiple Representation

For example, Jan van Maanen (1991) discussed his classroom teaching activity based on L'Hôpital's weight problem (L'Hôpital 1696, 62) with physical instruments (figure 10.9).

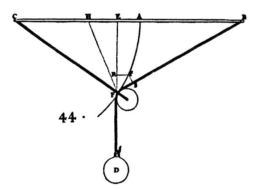

Figure 10.9: The problem from the Analyse des infiniment petits *(1696) that L'Hôpital used to show the power of the differential calculus*

Let *F* be a pulley, hanging freely at the end of a rope *CF* which is fastened at *C*, and let *D* be a weight. *D* is hanging at the end of the rope *DFB*, which passes behind the pulley *F* and is suspended at *B* such that the points *C* and *B* are on the same horizontal line. One supposes that the pulley and the ropes do not have mass; & one asks at what place the weight *D* or the pulley *F* will be.

Using this physical problem, L'Hôpital demonstrates the significance of the method of calculus by showing that the result is same as that obtained by the method of geometry. The problem can be investigated in today's classroom using a concrete model, through the means of computer algebra (CAS) or dynamic software (DGS). Masami Isoda observed undergraduate students' mathematical inquiry: the roles a, b, c and the contexts 1, 2, 3 were confirmed. Using such tools, students experienced the visual correspondence between geometric representations of motion and graphical representations of motion, the emergence of the same equations by differentiation and by geometrical reasoning, the correspondence between data from measurement and the results of mathematics and so on. These correspondences are not the same as in L'Hôpital's discussion but students are able to experience the methodological correspondence between geometry and calculus that L'Hôpital wished to highlight. Thus students appreciate the power and beauty from these correspondences.

10.2.3.2 Technology and history
Technology can help students to understand history better, and thus mathematics more deeply. When students read Descartes' *Geometry* based on their knowledge of school algebraic geometry, they cannot understand it very well because they are not starting from where Descartes was starting. Descartes was trying to make a new universe of mathematics beyond classical geometry, including moving beyond the limit of three geometrical dimensions. If students think that, say, multiplication is represented only numerically and graphs are sets of ordered pairs, then it is difficult for them to understand Descartes' geometrical reasoning. But if they try to draw each figure in his *Geometry* using Dynamic Geometry Software, they easily find out why Descartes had to discuss the geometric representation of multiplication from the beginning. So technology help us to understand the history more appropriately. But there remains a distinct cognitive difference between Descartes and the students. By using DGS, students' understanding may well come closer to that of Descartes. But Descartes' lament that the ancient intuitions had been lost could not be understood by students, not least because DGS gives them alternative intuitions for inquiry. Descartes had to reconstruct mathematics based on algebra as a new way of knowing. So the use of technology is not putting students back into Descartes' frame of mind, but is broadening their awareness of the richness of mathematics and its roots. One of the major pedagogical concerns for many years has been that students have lost the opportunity to experience classical geometrical intuitions, which are not replaced by a haze of algebraic symbols; DGS begins to offer a chance to re-experience some age-old intuitions.

In a similar way, modern programmable calculators enable today's students to redo calculations of former times, often to greater accuracy and far further into the calculation. In capturing in a few seconds a calculation which may have taken a sixteenth century astronomer days or months, students are arguably not recapturing the experience of old but generating a fresh one. In some cases today's students may be able to find things in the figures which their predecessors could not.

Another case in point is the Japanese mathematics *wasan*, or traditional mathematics (Smith and Mikami 1914). Some of the high level numerical methods developed in Japan before the influx of western mathematics in the nineteenth century have been lost; all that is known is that they were different from western proof-based mathematics, and that the results are correct. The comparison of results tells us the correctness of Japanese lost methods, but we could not know the method by this comparison. To explore what their methods might have been, computer

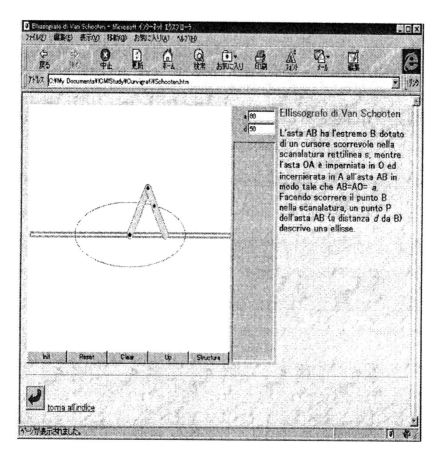

Figure 10.10: Java applet of ellipse compass by van Schooten, on a Japanese web-page that visits Bartolini Bussi's 'museum' (http://www.museo.unimo.it/labmat/)

software is available to track through the calculations under different hypotheses. It is thus a powerful tool for checking conjectures about historical methods, as well as doing mathematics, but also it introduces the possibility of misunderstanding, just as much as if we make conjectures about ancient methods of sand-board calculation by using paper and pencil.

10.2.3.3   Integrating approach with tools
There are many research projects which have been designed to examine the integration of mathematics with tools. Some ongoing projects are aimed at curriculum development of mathematics with tools and others at the development of a curriculum which integrates mathematics and history, but each of them adopts history in the classroom.

   As examples of projects that focus on curriculum development with tools, Jere Confrey and David Dennis (1995, 1997) in the US, and later Masami Isoda in Japan (1997, 1998b), have designed projects for the integration of geometry, algebra and calculus using drawing instruments, and multiple representation software including

DGS. Their physical instruments are made from a changeable parts set, so new instruments are easy to construct. To allow students to construct instruments, Jere Confrey and Masami Isoda (see Web reference) began to use LEGO. Their projects used tools for integrating multiple representations which are supported by history.

Here are two examples of projects that focus on the development of a curriculum which integrates mathematics and history. Recently in Italy, the project of Maria G. Bartolini Bussi developed Java tools (figure 10.10) for her virtual mathematical laboratory and now in her project, many kinds of representation tools, instruments and software are available for the teaching of mathematics and history (see §10.2.1). Arzarello's project in Italy originally named EuCart (Euclid & Descartes) is focused on the teaching of proof. The project uses the multiple representation tools of DGS and CAS. In the project, there is a focus on three historical periods, Euclid, Descartes and Hilbert, with an introduction of original sources in the classroom, framed by the teacher's introduction. DGS is oriented to developing the semantics of proof whilst CAS is oriented to develop the syntax of proof.

10.2.3.4 Beyond each tool's disadvantages

An instrument can be made from many kinds of representations. Each representation of it has advantages and disadvantages. One aim of integrating various tools or representations is to develop the student's competence for selecting and creating appropriate tools or representations. For example, in the *Algebra, geometry and calculus for all* project by Isoda (1999), students were asked to explore ellipses, with original pictures by Van Schooten (Maanen 1992; see also figure 10.11), using various representations. When students used physical pieces of LEGO, they commented on the changing of physical resistance when they tried to draw an ellipse. In the case of DGS, they did not. With physical tools, students discussed the difficulty of using them for drawing. In the case of DGS, students could draw some parts of an ellipse quite easily, but to draw other parts they needed additional constructions and

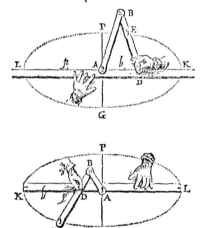

Figure 10.11: Van Schooten's ellipse-drawers (1646); a screen showing a Java simulation is shown in figure 10.10.

this advantage led to misunderstandings by some students. A student reported that we must first solve equations if we are to represent an ellipse using BASIC. Students began to change parts without the teacher's intervention because they had experience of changing LEGO parts when they were young. But students did not try to change equation parameters until the teacher suggested it. By using LEGO and DGS, students could find the general equation of an ellipse. When teacher asked the students to make a drawing tool from LEGO with different parameters, some students changed the parameters of the figure on DGS first; which led to success. In short, various representations support students' multiple reasoning abilities and the development of their relational understanding.

References

Arzarello F., Micheletti C., Olivero F., Paola D. & Robutti O., to appear. 'The transition to formal proofs in geometry', in: Paolo Boero (ed.), *Theorems in school: from history and epistemology to cognitive and educational issues*, Dordrecht: Kluwer

Arzarello, F., Olivero, F., Paola, D. & Robutti, O., in press. I problemi di costruzione geometrica con l'aiuto di Cabri: L'insegnamento della matematica e delle scienze integrate

Bartolini Bussi M., Nasi D., Martinez A., Pergola M. Zanoli C. *et al.* 1999. Laboratorio di Matematica. Theatrum Machinarum, I CD rom del Museo (1), Modena : Museo Universitario di Storia Naturale e della Strumenntazione Scientifica

Dennis, David & Jere Confrey, 1995. 'Functions of a curve: Leibniz's original notion of functions and its meaning for the parabola', *College mathematics journal* **26**, 124-131

Dennis, David, and Jere Confrey, 1997. 'Drawing logarithmic curves with *Geometer's sketchpad*: a method inspired by historical sources', in James R. King and Doris Schattschneider (eds,) *Geometry turned on! Dynamic software in learning, teaching and research, MAA*, 147-156

Descartes René 1628. 'Rules for the direction of the mind', in E. Haldane, G Ross (tr), *The philosophical works of Descartes*, Cambridge, 1-77

Descartes, R. 1637/1954., *The Geometry of René Descartes*, (D. E. Smith & M. L. Latham ed., tr.) Open Court, 1925; reprint New York: Dover Publ. 1954

Freudenthal, H. 1973. *Mathematics as an educational task*, Dordrecht: Reidel.

Isoda M. (1997), 'Connecting mathematics with machine engineering and art: perspectives for calculus and geometry for all via technology', in W. Yang & Y. Hasan (eds), *Proceedings of the 2nd Asian Technology Conference in Mathematics*, University of Malaysia, 60-70

Isoda M. (1998a), 'Mathematical inquiry enhanced by harmonized approach via technology', in H. Park *et al*, (eds), *Proceedings of ICME-EARCOME* **3**, 267-278

Isoda M., 1998b. 'Developing the curriculum for curves using history and technology', in W. Yang *et al* (eds), *Proceedings of the Third Asian Technology Conference in Mathematics*, Springer, 82-89

Isoda M., 1999. Annual report of 'Algebra, geometry and calculus for all' project, 6 vols (written in Japanese)

Kaput J., 1989. 'Linking representations in the symbol system of algebra', S. Wanger and C. Kieran (eds), *Research issues in the learning and teaching of algebra*, Lawrence Erlbaum Associates, 167-194

Lesh R., Landau M. & Hamilton E., 1983. 'Conceptual models and applied mathematical problem-solving research', in R. Lesh and M. Landau (eds), *Acquisition of mathematical concepts and processes*, Academic Press, 263-343

Leinbach L, et al, 1991. *The laboratory approach to teaching calculus*, Washington, DC: Mathematical Association of America

L'Hôpital, Guillaume Francois Antoine de, 1696. *Analyse des infiniment petits, pour l'intelligence des lignes courbes*

Maanen, Jan van, 1991. 'L'Hôpital's weight problem', *For the learning of mathematics* **11.2**, 44-47

Maanen, Jan van, 1992. 'Seventeenth century instruments for drawing conic sections',
    Mathematical gazette **76**, 222-230
Smith, David Eugene & Yoshio Mikami 1914. *A history of Japanese mathematics*, Chicago:
    Open Court
Yerushalmy M. & Schwartz J, 1993. 'Seizing the opportunity to make algebra
    mathematically and pedagogically interesting', in T. Romberg et al (eds), *Integrating
    research on the graphical representation of functions*, Lawrence Erlbaum Assoc., 41-68
Zimmermann W. and Cunningham S. (eds), 1991. *Visualization in teaching and learning
    mathematics*, Washington: Mathematical Association of America

Web reference

Confrey, Jere, and Masami Isoda
http://130.158.186.11/mathedu/mathedu/forAll/index.html

10.3   Resources for history of mathematics on the World Wide Web

10.3.1   Teachers, learners and the World Wide Web

Glen Van Brummelen

Massive, epoch-making revolutions in communications technology are so rare that only a couple (the printing press, and perhaps the telephone) occurred before our lifetimes. But in the last thirty years we have witnessed a cascade of revolutions, all products of computer technology. New developments often seem more dramatic from within than from without; nevertheless, the computer revolution seems to be changing, genuinely and substantially, who we are, and how and what we think. The World Wide Web, for example, has evolved from non-existence only five years ago to become the central storehouse of human knowledge today. The Web is more than the next step in a larger process; it is the first major realisation of the alteration of ourselves by our own machines. The Web in its current incarnation may not be recognisable ten years from now, but our children will identify it as the birth of a new era.

As is typical for technological advances, the Web thrusts its new tools upon educators much more quickly than we can adjust to them. The mathematics education community has only recently come to terms with the use of pocket calculators, and debate still rages over more sophisticated tools like graphing calculators and computer algebra systems. The Web's revolution is more fundamental: it affects not just mathematics but all disciplines, jarring into motion the previously static media of paper, chalk and books, and expanding globally the communities within our reach. We shall examine what is now available for the use of educators in the history of mathematics on the Web, but perhaps more important

John Fauvel, Jan van Maanen (eds.), *History in mathematics education: the ICMI study*,
Dordrecht: Kluwer 2000, pp. 358-362

to our vision of the future is an exploration of the medium itself. The best way to learn what the Web represents is not to read about it, but to use it. Hence, readers are encouraged to set this chapter aside for now and visit the site that accompanies this article:

http://nmc.bennington.edu/faculty/gvanbrum/luminy/title.htm

(As a self-referential hint to the possibilities of global communication of this new technology, the author presented this work from Vancouver, Canada, to the ICMI Study Meeting in Luminy, France, during the main drafting of this section.)

The new medium

The Web's innovation, deceptively simple and over thirty years old, is *hypertext*. A hypertext document is a book freed of its binding: one may link the pages any way one likes, unlike the sequential ordering of a conventional book. Rather than pursuing topics according to the author's design, readers follow their own path through the content, guided by their own purposes and interests. Hypertext has been used to some effect in computerised reference sources such as Windows Help and CD-ROM encyclopaedias. However, hypertext alone does not revolutionise the planet. The power of the Web is not primarily in its navigational flexibility, but in its universality. It is a system with few barriers to communication. Anyone with modest financial resources may access, contribute, and alter it. The Web thus spreads ideas globally by removing the physical constraints of publishing, marketing and dissemination. The result is a virtually endless supply of information with the absence of imposed structure. Unbounded access to knowledge frees the user to construct her own intellectual environment, uninhibited by other perspectives: an Enlightenment nightmare, and a postmodern paradise.

The recent advent of Web applets (programs that can be executed through the Internet) promises to extend the medium substantially. CGI and Java applets now run seamlessly, often unnoticed by the user, to produce the distracting graphics and banner advertisements at commercial sites. More positive educational uses include instructional aids that permit the user to interact with simulated environments, tapping learning channels that recall the tactile explorations of childhood.

Developments on the horizon include increasingly seamless interaction not just with machines, but with expert systems and with other people. Videoconferencing and virtual whiteboards will transform the Web into a means whereby one could hold a seminar where every participant is in a different country, yet the communication will be as smooth as if all were in the same room. Through our own natural choices in such a fluid environment, the existing trend for our communities to be shaped by our interests, rather than geographical barriers, may accelerate.

Innovations seldom live up to the unbounded expectations of instantaneous reform, or to the apocalyptic warnings of detractors. Consider, for example, the real place occupied by pocket calculators in today's classroom compared to the over-enthusiastic predictions of thirty years ago. Some problems with the Web are already surfacing. For instance, the very democracy treasured by Internet enthusiasts produces a number of undesirable side effects. Relevant to education is the spread of misinformation. With printed material, the financial overhead involved in its production ensured that publishers had some commitment to the quality of their works. With the Web, one can be no surer of what one reads than of what one overhears in a conversation at a dinner table. The Web's fluidity implies that teachers can never be sure that a resource which exists during their preparation period will exist at class time. Advertising is becoming so prevalent that many people find it annoying, although the Web is hardly the only medium to suffer from this particular vice. Finally, class distinction issues raise worrying problems: often, only wealthier students with computers—in wealthier countries—have access to the new technology. This is partially resolved, in wealthier countries, through free Internet services sponsored by advertising and access at public libraries.

Some good uses of Web technology in the history of mathematics

The rest of this section highlights some of the most creative uses of Web technology for classroom use with respect to the history of mathematics. In some cases these sites are chosen because of their effective use of the new medium; in many other cases they are chosen merely because they make globally available a powerful resource. Most of the sites are aimed at high school to undergraduate students.

Two large sites are good comprehensive sources of information in the history of mathematics. The St. Andrews MacTutor History of Mathematics Archive contains biographies of hundreds of historical mathematicians, and a number of survey articles on a variety of historical topics. David Joyce's history of mathematics site contains a great deal of useful bibliographic information. An index of Web sites related to the history of mathematics is maintained at the Canadian Society for History and Philosophy of Mathematics site. Due to the fluidity and expansion of the Web, indices like this one are becoming more difficult to maintain. Search engines such as Altavista and Yahoo can be useful for finding new sites, and britannica.com regularly reviews sites of interest in addition to providing content from the *Encyclopedia Britannica*.

David Joyce's *Euclid's Elements Online* is an excellent example of the interactivity made possible by the Web. Joyce provides the entire text of the *Elements*, but what makes the site special are its geometric figures. The points in the diagrams may be moved while the geometric relations between the objects in the diagram are preserved, much like Geometer's Sketchpad or Cabri Géométrie but without the need for software. The meaning of a theorem can be made clear with a few simple experiments with the handsomely rendered figures. The Famous Curves Index at the St. Andrews site contains explorations of a similar nature, suitable for mathematical experimentation with historically important curves.

Although the technological advances displayed at the sites listed above are intriguing and entertaining, equally important is the increased availability of less technologically-oriented material. PBS's Nova Online, for example, includes a site devoted to the recent television programme on Andrew Wiles' solution to Fermat's Last Theorem. In addition to a complete transcript of the program, users will find a biography of Sophie Germain, teacher's guides to using the program in class, and exercises in Pythagorean mathematics appropriate for use by students before and after viewing. The Museum of History of Science (Oxford, U.K.) has several 'virtual exhibitions', providing an experience that comes close to an in-person visit. Gary Stoudt maintains a page designed for his history of mathematics course that allows student access to a variety of primary sources. The Galileo Project (Rice University) contains an impressive amount of detail on Galileo's life, work and times, stored in an unconventional and innovative manner. Much of this material would have been very difficult for a teacher to locate and distribute to his or her class prior to the advent of the Web.

A number of brave attempts have been made to place entire courses on the Web. While this has not yet occurred in the history of mathematics (to my knowledge), several sites contain material of at least tangential interest. Brian Martin, at The King's University College, Edmonton, has converted his astronomy course, taught to a great extent from an historical perspective, to the Internet. The site is used as a supplement, not as a replacement for student-teacher interaction. From his and others' experiences, readers tempted to follow suit should be warned that the effort required for such an enterprise is enormous, well beyond any of the authors' initial expectations. I have heard estimates that the commitment is up to ten times that required for a standard course.

A number of instructors have initiated projects whereby students' efforts in the course are placed on the Web. Where this occurs gradually over successive years of the offering of a course, the results can be quite impressive. Larry Riddle's site honouring and exploring the work of women mathematicians at Agnes Scott College, written by students in his classes, has won a number of Internet awards. A more general effort is underway with Len Berggren's history of mathematics course at Simon Fraser University, Vancouver, Canada. A number of sites contain student papers done at a variety of schools and universities, for instance Tufts University's course *Inventing Science*. It would be as well to be aware of these sites not only for their potential for benefit, but also for the potential they provide for students' academic dishonesty.

Where are we going?

Clearly, extensive efforts have already generated creative and helpful Internet resources in the history of mathematics, but much more could be done. The terrain is covered haphazardly and the quality varies considerably from site to site. As communication speed increases, students will come to expect much greater interactivity, and educators may have difficulty coming to grips with how it might be used effectively. Finally, the potential for classes and seminars that transcend geographical boundaries has not yet been realised. Within several years students in France, for example, may be able to share a virtual classroom with students in South Africa and North America, with instructors or leaders from Australia and Germany. For a small, geographically scattered community like those of us dedicated to the history of mathematics, this will be ground-breaking. In the end, however, computers do not organise such co-operation on their own. It is incumbent on us to begin thinking how this might be accomplished and to increase our collaboration, so that when the tools are placed in our hands, we will be ready.

References for §10.3.1

The following list provides the addresses of sites referred to in this section. For a categorised listing, see the next section (§10.3.2.3).

Canadian Society for History and Philosophy of Mathematics
`http://www.kingsu.ab.ca/~glen/cshpm/home.htm`

Euclid's Elements Online (D. Joyce)
`http://aleph0.clarku.edu/~djoyce/java/elements/elements.html`

Famous Curves Index (St. Andrews)
`http://www-groups.dcs.st-and.ac.uk/~history/Java/index.html`

The Galileo Project
`http://es.rice.edu/ES/humsoc/Galileo/index.html`

History of Mathematics (J. L. Berggren, Simon Fraser University)
`http:/www.math.sfu.ca/histmath`

Altavista
`http://www.altavista.com`

Encyclopedia Britannica
`http://www.brittanica.com`

Inventing Science (Tufts University course)
`http://www.perseus.tufts.edu/GreekScience/`

David Joyce's History of Mathematics Site
`http://aleph0.clarku.edu/~djoyce/mathhist/mathhist.html`

Brian Martin's Introduction to Astronomy Course
`http://www.kingsu.ab.ca/~brian/astro/a200home.htm`

Museum of History of Science (Oxford)
`http://www.mhs.ox.ac.uk`

Nova Online: The Proof
`http://www.pbs.org/wgbh/nova/proof/`

Primary Sources for the History of Mathematics (G. Stoudt)
`http://www.nsm.iup.edu/ma/gsstoudt/history/ma350/sources_home.html`

St. Andrews MacTutor History of Mathematics Site
`http://www-groups.dcs.st-and.ac.uk/~history/`

Women Mathematicians (L. Riddle)
`http://www.agnesscott.edu/lriddle/women/women.html`

10.3.2 Web historical resources for the mathematics teacher

June Barrow-Green

The general argument of this book is that mathematics teachers and learners can be greatly empowered by having historical resources available to them. It is argued in chapter 4 that mathematics teachers will benefit from some historical training, either

John Fauvel, Jan van Maanen (eds.), *History in mathematics education: the ICMI study*, Dordrecht: Kluwer 2000, pp. 362-370

before or during their teaching career. Of course, one must be realistic about what is possible within the time constraints of life, and the time commitment that this suggests will not fit the current needs or options of all teachers or future teachers. One can treat history of mathematics as an area to become acquainted with gradually, as and when time permits, but one can begin to use resources sooner and develop the skills on the hoof, so to speak. In this context the internet is a tremendous boon. But there is a problem, as Glen Van Brummelen pointed out in the previous section (§10.3.1): information on the Web is of such variable quality that a responsible teacher must both be aware of that and seek to train students in techniques of critical judgement (see also Fauvel 1995). This section consists of some ideas to help both teachers and students in this process.

As with all resources, gaining a sense of what kind of thing they comprise and gaining some practice in using them is strongly desirable as a preparation for exploiting them later on in an educational context. What this section offers is firstly (§10.3.2.1) some guidance on web searching and determining the likely reliability of information; secondly (§10.3.2.2) a listing of sample questions which could help you to use the internet with your students; and thirdly (§10.3.2.3) a listing of some sites which could provide answers or generate material for helping the teacher who is seeking to use history as a resource.

10.3.2.1   Searching and reliability

There is now a wide variety of material on history of mathematics available on the Web, ranging from interactive texts and museum catalogues, through to high school student projects, and including a wide range of reference sources. However, wonderful though it is to have access to huge quantities of information, it is not much use if you cannot find what you want. Moreover, the quality of Web pages varies widely and so far there is no common standard. Anyone, anywhere, can place material on the Web and so it is important to develop a strategy for assessing the quality of the information. This is especially important for students, given the importance (for general education, not only in mathematics) of web users learning to assess the reliability of web information. To make the best use of the Web you need to be able both to search the Web efficiently and to recognise a good site when you find one. Furthermore, it is recommended that you check the status of a site before you recommend it someone else. Just because a site was located at a particular address last time you looked, that is no guarantee that the next time you look it will still be at the same address, or indeed that it will still exist! What follows is a summary of some general points for using the Web with a critical awareness of its possible flaws.

Searching the Web

1. Let someone else do the work! Start from an annotated list of Web resources, such as provided in § 10.3.2.3 below.
2. Be as precise as possible. Remember that computers are very literal-minded!
3. Try different search engines.

4. Look for indicators of downloading time, eg graphics, video, software applications, etc. The site may have been developed on a more powerful computer or network than you have at home or school, and what seemed a rapid image to the site's constructor can clog up your computer or communication links for minutes on end, or indeed cause it to freeze.
5. Be aware of your browser's limitations.
6. Make bookmarks for future reference.

Evaluating a Web Site

1. Establish the authority (or not!) of the author: check any links to the author's home page; look for evidence of other publications; look for reason/motivation for the site's creation.
2. Accuracy of information: look for references to established sources; try to cross-check some information (but beware of the proliferation of errors through web pages copying each other!).
3. Currency of information: look for the dates when the site was created and when it was last revised.
4. Links to other sites: external links need separate evaluation.
5. General characteristics: consider the standard of the prose, the ease of navigation around the site, the completeness of the information, and any evidence of commercial interest.

10.3.2.2   Some sample questions in exploring Web resources for history of mathematics

Here are seven examples of questions which teachers could use either to find out information themselves or to use as a project (or component of a project) with students. For speed, some 'answers': that is, the addresses of sites which the search might lead to, are given here. Note that addresses which are too long are split over two lines. It would be useful to gain practice by starting with the question (without yet knowing a suitable site) and recording the stops along the way while getting closer (hopefully!) to the goal. The aim of this exercise is three-fold, therefore: to answer particular questions or follow particular leads, to indicate the kind of inquiries that could be put or followed, and also to gain experience in searching and evaluating the Web.

1. Find three different types of numeral systems (apart from the Hindu-Arabic 1, 2, 3 ...). Compose a sum which includes one or more of the basic operations (+, -, x, ÷) in each of the different systems.
   Chinese:
   ```
   http://www.mandarintools.com/numbers.html
   http://Aleph0.clarku.edu/~djoyce/mathhist/china.html
   ```
   Egyptian:
   ```
   http://eyelid.ukonline.co.uk/ancient/numbers.htm
   ```
   Mayan:
   ```
   http://www.vpds.wsu.edu/fair_95/gym/um001.html
   ```

Old Babylonian:
   `http://it.stlawu.edu/~dmelvill/mesomath/index/html`
Roman:
   `http://www.deadline.demon.co.uk/roman/front.htm`

2. When were the words quadratic and cubic first used?
    `http://members.aol.com/jeff570/mathword.html`
    Who was the earliest person to solve a quadratic and/or cubic equation? When?
    What method did they use? What did the equations look like? When were the
    formulae for solving quadratic and/or cubic equations first used? By whom?
    `http://www-history.mcs.st-and.ac.uk/history`

3. Find an illustration of a page from *Euclid's Elements*. What does it show? How
    old is it? How does it differ from a modern geometry text?
    `http://metalab.unc.edu/expo/vatican.exhibit/exhibit/Main`
       `_Hall.html`
    `http://rsl.ox.ac.uk/imacat/ino5.html`
    `http://cccw.adh.bton.ac.uk/schoolofdesign/MA.COURSE/17/L`
       `OB.html`

4. Who was the first female mathematician who we know about?
    `http://www.agnesscott.edu/lriddle/women/kova.htm`

5. Alan Turing was responsible for cracking the Enigma cipher in World War II.
    Who was Alan Turing? What were the essential features of the Enigma machine
    and where can you see one?
    `http://www.turing.org.uk/turing`
    `http://www.cranfield.ac.uk/ccc/bpark/`

6. In October 1998 an Archimedes manuscript was sold at Christie's in New York.
    What was on the manuscript? How old was it? How much did it fetch?
    `http://www.thewalters.org/archimedes.html`
    `http://www.mcs.drexel.edu/~crorres/Archimedes/contents.h`
       `tml`

7. What is 'Fermat's Last Theorem'? Why did it take so long to prove?
    `http://www.pbs.org/wgbh/nova/proof/`
    `http://www-groups.dcs.st-and.ac.uk/%7Ehistory/HistTopics`
       `/Fermat's_last_theorem.html`

10.3.2.3   Examples of internet sites as a resource for history of mathematics
Below is a selection of internet sites which may be found useful as a resource. For
ease of reference, and to emphasise the wide range of available sites, they have been
put into twelve categories: B1   General History of Mathematics Sites; B2   Web
Resources; B3   Biography; B4   Regional Mathematics; B5   Web Exhibits; B6
Books on-line; B7   Student Presentations; B8   Bibliography; B9   Societies; B10
History of Computing; B11   Education; B12   Miscellaneous.
   The listing here is fairly compact, with the main purpose of illustrating the kind
of historical resource available on the Web. Any given site is only mentioned no
more than once, although the larger sites could be cited under several of the above

headings, and those flagged in Glen Van Brummelen's discussion (§ 10.3.1) have not always been included again.   Slow-loading sites or those requiring special browsers are indicated with an asterisk (*). Fuller descriptions and further sites may be found in Barrow-Green 1998, and on the BSHM Resources website given in B2.

B1 General History of Mathematics Sites

Large sites have a gateway page which give an indication of the type of resources that are available on other pages of the site. The following are the addresses of gateways to some of the best known of the general sites on history of mathematics. (Some of the pages on these sites are also included in other sections.)

The Math Forum Internet Resource Collection

`http://forum.swarthmore.edu/~steve/steve/mathhistory.html`

This site is part of The Math Forum, an on-line mathematics education community centre, hosted by Swarthmore College, and provides an extensive list of annotated links to other sites. The sites are ordered alphabetically and the collection can be viewed in outline or annotated form. There is a well designed search engine which allows for a variety of searches, i.e. keywords, categories and dates.

Trinity College, Dublin, History of Mathematics archive

`http://www.maths.tcd.ie/pub/HistMath/HistMath.html`

This site, created and maintained by David Wilkins, includes biographies of some seventeenth and eighteenth century mathematicians, material on Berkeley, Newton, Hamilton, Boole, Riemann and Cantor, and an extensive (but unannotated) directory of history of mathematics websites.

B2 Web Resources

Many sites contain pages that are devoted to links to other related sites. Provided they are kept up to date, these can be extremely useful. However, there is a tendency to provide lists of links with no annotation, which means that there is no way of telling whether a site is, for example, hypertext, interactive, image intensive, or indeed whether it has been prepared by scholars or students. You can therefore end up wasting a lot of time waiting for a site to be downloaded which turns out not to have the information you require. Until you have visited a lot of sites and know your way around enough to recognise sites only by name, it is generally better to use an annotated list of resources.

BSHM Resources

`http://www.dcs.warwick.ac.uk/bshm/resources.html`

An annotated guide prepared on behalf of the British Society for the History of Mathematics which includes a fuller range of sites under the same categorisation as in this section. Useful to bookmark!

The Mathematical Museum - History Wing

`http://elib.zib-berlin.de:88/Math-Net/Links/mathe-museum.hist.html`

The 'History' wing of The Mathematical Museum is part of the Math-Net Links to the Mathematical World and contains links to exhibitions, hyperbooks, information systems, museums and pages of interest for the history of mathematics and associated fields. It consists of sections on history of mathematics, history of computing and communication, and related history information. It is well annotated and includes some sample illustrations.

WWW Virtual Library. History of Science, Technology & Medicine
`http://www.asa.unimelb.edu.au/hstm/hstm_ove.htm`
A gateway into a wide range of resources covering the history of many scientific fields including mathematics. A particular feature of the site is its rating system that, with given criteria, evaluates the depth, content and design of each site listed.

B3 Biography

There is plenty of material on the Web concerned with lives of mathematicians. It comes in a wide variety of guises, much of it excellent, and, on the whole, the Web is a very good place to start looking for biographical material. These 'personal' sites generally contain a broad spectrum of material about the individual and have good links to other relevant sites.

Archimedes
`http://www.mcs.drexel.edu/~crorres/Archimedes/contents.html`
An extremely rich collection of Archimedean miscellanea produced by Chris Rorres of Drexel University, Philadelphia, including a pages on different aspects of Archimedes' mathematics, books on Archimedes, information on Syracuse, and links to other related sites, eg a bibliography of Archimedean literature.

Hypatia of Alexandria
`http://www.polyamory.org/~howard/Hypatia`
An extensive and partially annotated list of web resources connected with Hypatia. The internal links include a long list of published books and articles which contain information on Hypatia, as well as transcriptions from 18th and 20th century texts.

The Alan Turing Home Page
`http://www.turing.org.uk/turing`
A large and well structured site on Turing's life and work maintained by Andrew Hodges, the author of *Alan Turing: The Enigma*. It includes material on the history of the computer as well as links to other related sites.

B4 Regional Mathematics

There are many sites on regional, particularly ancient, mathematics. Many of the general sites include good regional pages. The following is a short list of some smaller (in terms of their history of mathematics content) sites which illustrate a variety of presentations.

Mathematicians of the African Diaspora
`http://www.math.buffalo.edu/mad/mad0.html`
An excellent site created and maintained by Scott Williams of the State University of New York at Buffalo to exhibit the accomplishments of the people of Africa and Africa Diaspora within the mathematical sciences. The history pages include the

mathematics of Ancient Egypt, Pre-Colonial Nigeria, and Swaziland (the Lemombo Bone). There are good links to other related sites.

Egyptian Mathematics Problems

http://eyelid.ukonline.co.uk/ancient/maths1.htm

Also the addresses that end on /maths2.htm and /maths3.htm They present some basic mathematical problems for high school pupils, produced by artist Mark Millimore as part of his extensively illustrated Ancient Egypt site.

Mesopotamian Mathematics

http://it.stlawu.edu/~dmelvill/mesomath/index.html

Plenty of interesting and historically accurate material, collected and written by Duncan Melville for his undergraduate history course, and an extensive annotated bibliography by Eleanor Robson.

B5 Web Exhibits

These are sites which make use of a variety of devices available to those with sufficiently enhanced browsers, and also tend to be very image intensive. However, it is not always necessary to enable all the tools in order to get a good sense of the potential of the sites and they are well worth visiting, if only to get an idea of what sort of things are possible.

*The Archimedes Palimpsest

http://www.thewalters.org/archimedes/frame.html

An excellent site created by the Walters Art Gallery, Baltimore to complement their exhibition of The Archimedes Palimpsest (20 June-5 September 1999). It contains richly illustrated pages on the life of Archimedes, the history (past, present and future) of the Palimpsest and a commentary by Reviel Netz.

B6 Books on-line

Texts on-line come in two forms: straightforward copies of original texts-- particularly useful if the text in question is otherwise difficult to obtain--or copies which have been annotated or translated in order to increase accessiblility.

*Euclid's Elements

http://aleph0.clarku.edu/~djoyce/java/elements/toc.html

A full text interactive version of Euclid's Elements with historical and mathematical comments produced by David Joyce. With a Java enabled browser (Netscape or Internet Explorer Version 3 or higher) it is possible to dynamically change the diagrams. A remarkable site which makes the *Elements* accessible in a completely new way (see § 10.3.1 above; Van Brummelen 1998). Highly recommended.

B7 Student Presentations

Student projects are now well established across the curriculum, and the Web can be an extremely useful resource in this context. It can be used not only as a means of supplying information for the project, but also as the medium by which the students present their work. In the latter case students not only have the chance to share the fruits of their labours with others, but also have the opportunity to receive feedback too.

*Galileo Project at Rice University
http://es.rice.edu/ES/humsoc/Galileo/Student_Work/
An excellent collection of student projects which derived from a course on Galileo in Context.

B8 Bibliography

These sites contain lists of published books and/or articles which are relevant to using history of mathematics in an educational context.

BSHM Abstracts
http://www.dcs.warwick.ac.uk/bshm/abs.html
Brief abstracts, sorted alphabetically by author, of papers published in journals and books. There is a separate education section covering abstracts of papers on the uses of history of mathematics in education, history of mathematics courses, and the history of mathematics education.

B9 Societies

Most academic disciplines have societies which exist to help the promotion of their subject at a variety of levels. The Web now provides a very convenient way to find out what such Societies offer their membership, how much it costs to join, as well as supplying a means of obtaining an application form. Both of the following societies actively promote the use of history of mathematics in education at all levels.

The British Society for the History of Mathematics
http://www.dcs.warwick.ac.uk/bshm/
The site includes membership details, BSHM abstracts (see Bibliography), an archive containing a list of talks given to the Society, and a page of links to other sites.

The Canadian Society for the History and Philosophy of Mathematics
http://www.kingsu.ab.ca/~glen/cshpm/home.htm
The site includes membership details, free access to the *History and Pedagogy of Mathematics Newsletter*, and a page of links to other sites.

B10 History of Computing

The Virtual Museum of Computing
http://www.comlab.ox.ac.uk/archive/other/museums/
          computing.html
A site, developed and maintained by Jonathan Bowen of Reading University, made up of an extensive collection of links to sites connected with the history of computing and computer-based exhibits. The site is divided into galleries covering a variety of topics such as general historical information, on-line exhibits etc.

Charles Babbage's Analytical Engine
http://www.fourmilab.ch/babbage/contents.html
An excellent site containing texts of historical documents, including Menebrea's description of the Engine translated by Ada Lovelace, and a detailed description of an Analytical Engine emulator which runs as a Java applet.

B11 Education

Some of the most interesting sites are emerging from teachers with an interest both in history and in using computers. See also Sharp 1998.

Teaching with Original Historical Sources in Mathematics
`http://math.nmsu.edu/~history/`
The experiences of Reinhard Laubenbacher and David Pengelley of New Mexico State University in using original historical sources in teaching mathematics. The site is well referenced and there are links to several of the articles mentioned (mostly in the form of .dvi or .ps files) as well as links to other resources.

B12 Miscellaneous

Other sites which are helpful or interesting to visit but which do not fall naturally into any of the categories above.

Earliest Uses of Various Mathematical Symbols
`http://members.aol.com/jeff570/mathsym.html`

Earliest Known Uses of Some of the Words of Mathematics
`http://members.aol.com/jeff570/mathword.html`
The above two sites, which are the product of multiple contributors, are very high quality and provide an excellent resource. They are maintained by Jeff Miller of Gulf High School, Florida and contributions are welcomed.

References for 10.3.2

Barrow-Green, June 1998. 'History of mathematics: resources on the world wide web',
    Mathematics in school **27** (4), 16-22
Brummelen, Glen Van 1998. 'Books, the next generation', *British Society for the History of Mathematics Newsletter* **36**, 48 – 50
Fauvel, John 1995. 'History of mathematics on the web', *British Society for the History of Mathematics Newsletter* **30**, 59 – 62
Sharp, John 1998. 'History observed as it happens: computers and the revival of geometry', *British Society for the History of Mathematics Newsletter* **37**, 51 – 53

Chapter 11

Bibliography for further work in the area

John Fauvel

with Éliane Cousquer, Fulvia Furinghetti, Torkil Heiede, Chi Kai Lit, Harm
Jan Smid, Yannis Thomaidis, Constantinos Tzanakis

Abstract: *A considerable amount of research has been done in recent decades on the subject
of this study, which is here summarised, in the form of an annotated bibliography, for works
appearing in eight languages of publication.*

11.1 Introduction

John Fauvel

The final chapter of the Study illustrates something of the range and scope of work
in recent decades on the relations between history and pedagogy of mathematics,
across a number of countries. Through annotated bibliographies of work published
in eight languages some impression is given of how much and how varied the
activity has been, mostly in the last two decades, of which the present work is in part
a consolidation. One of the aspects of international activity which the chapter
highlights is how many different countries have been engaged in internal discussions
about supporting mathematics teachers through the integration of history. Note that
for the purposes of this book, the annotations are given in English, with the
exception of the Chinese bibliography (§11.2) which is written in Chinese.

This chapter does not form this ICMI Study's bibliography, in the usual sense,
since individual chapters and sometimes sections of the book have their own
bibliographical list of references. Nor is it for the most part a list of primary source
or other printed resources for classroom use, such as works of the great
mathematicians or secondary histories of mathematics. Help on these issues is given
elsewhere. The resource appendix to Chapter 9 contains an annotated list of original
sources which may be helpful to teachers, other useful works are mentioned in

John Fauvel, Jan van Maanen (eds.), *History in mathematics education: the ICMI study,*
Dordrecht: Kluwer 2000, pp. 371-418

various places throughout the book, and other resources are discussed in chapter 10, for example the guide to internet resources in §10.3. In making the selection of papers to include in this chapter, we had to put to one side most of the very considerable number of excellent pieces of work which mediate *between* history and the mathematics classroom, providing resources for classroom teachers to use and learn from. (While recognising that there is a continuum of work here, and that it is difficult to draw a rigid line between classroom resources, on the one hand, and discussions about the value and use of those resources, on the other.)

Of course, for future research in this area to build upon the work already done it is not sufficient merely to list that work bibliographically: it has to be categorised and evaluated for its relevance to particular research questions. In the preparation of this Study quite some discussions were held on various possible categorisations. Notable work was done by Harm Jan Smid, who divided the themes addressed in the Dutch and German reports he had surveyed as falling into five categories (a refined version of the 'continuum' referred to above).

(i)   Discussions and/or advocacy in general of the possibilities and advantages of the use of history of mathematics in teaching and learning mathematics.

(ii)   Examples, didactical guidelines and hints for teachers on how to use historical material in their lessons.

(iii)   The provision of more or less ready made historical resources to use in the classroom.

(iv)   Empirical descriptions of the use of historical material in the classroom;

(v)   Research focussing on the results and effects of the use of historical materials, both affective and/or cognitive.

Any particular article might well incorporate several of these themes. In the event it seemed best to present the bibliographical data in this chapter, covering a yet wider range of approaches from different countries and linguistic traditions, in a more unmediated form (that is to say, without an elaborate classification), while urging that some such categorisation would be a good starting point for future work in the area.

The point of this chapter is, then, to provide in capsule form an impression of the kinds of work published in eight languages on the topic of the study, namely discussions of the relations between history of mathematics and the teaching and learning of mathematics. While no complete coverage is achievable, even in the languages here let alone across the rest of the world, it is hoped and intended that a fairly representative selection of work has been noticed here. Reading the annotations provides a further reinforcement of various messages and arguments put forward in the rest of the book, about the many different ways of integrating history, reasons for doing so, and the different benefits to mathematics curricula and learning experiences across the world.

11.2 Chinese

Chi Kai Lit

「數學史與數學教育」中文文獻目錄*

列志佳編

本目錄盡量搜集在一九九八年中或以前已發表的「數學史與數學教育」中文文獻．純數學史文獻則不包括在內．由於有關的材料散見於中、港、台的學術期刊、科普雜誌等，加上曾未有一全面而有系統的檢索系統，故本目錄相信未能網羅所有的素材．本人謹對並未編入的文獻作者致歉．盼望有識之士不吝賜正，使本目錄日漸完善，方便有興趣的人士查閱、研究．

* 本目錄蒙香港大學蕭文強教授提供寶貴的意見，謹此致以衷心謝意．

John Fauvel, Jan van Maanen (eds.), *History in mathematics education: the ICMI study*, Dordrecht: Kluwer 2000, pp. 373-383

「 數學史與數學教育 」 中文文獻目錄

1. 列志佳 (1995) 。 推廣及深化數學史在本港數
學教育中的運用 。 發表於 5 月 13 日「香港
數學教育：轉變的時機」研討會上 。 後
列於 莫雅慈．馮振業 (1998) 。《香港數
學教育會議論文集 95 – 97 》．頁 35 – 40．
香港 香港數學教育學會 。

2. 列志佳 (1996) 。 運用數學史於數學教育的初
步調查研究 。《數學教育 》．第三期．
頁 21 – 22 。

3. 列志佳 (1996) 。 本港中學教師對運用數學史
於數學教育之 調查研究 。 發表於「 香港
數學教育會議 – 96 」

4. 列志佳 邵慧爽 (1996.) 。「 數學史與數學教
育」資料分類目錄的編製與構想 。《數
學教育》 第三期．頁 55 – 58 。

5. 朱學志. (1984) . 關於在高等師範院校開設 "
   數學史、數學方法論 ' 課的幾點看法.
   《數學通報》. 第三期, 頁 20-23, 29.

6. 吳文俊 (1993) . 數學教育現代化問題. 列於
   21 世紀中國數學教育展望——大眾數學
   的理論與實踐課題組《21 世紀中國數學
   教育展望 (第一輯)》, 頁 16-27. 北京:
   北京師範大學出版社.

7. 何世君 (1987) . 在數學教學中要重視傳授數
   學史.《寧夏教育》, 1, 頁 27-29.

8. 李迪 (1997) . 數學史教育在中國. 外於《中
   國近現代數學教育史》第一卷. 八ニカ
   イ出版印刷株式會社.

9. 洪萬生 (1984) 。數學史與數學教育 。《科學月列》，15 (5) ，頁371 – 376 。

10. 洪萬生 (1991) 。數學史與數學教育 —— 數學教育研究的一個新面向。《孔子與數學》，頁47 – 53 。台北：明文書局 。

11 洪萬生 (1992) ，數學史教學與師範院校 。《科學月列》，23 (12) ，頁 945 – 946 。

12. 洪萬生 (1996)。數學史與代數學習 。《科學月列》，27 (7) ，頁 560 – 567 。

13. 施確探 (1997)。古法求開方 。《數學教育》，第四期。頁 86 – 89 。

14. 袁小明 (1992) 。論數學教育中歷史材料的應用。《數學教育學報》，第一期。頁 119 – 123 。

15. 孫名符．呂世虎．傅敏、王仲春（1997）．數學方法與數學史的教育功能．《數學．邏輯與教育》，頁153－162．台北：建宏出版社．

16. 莫雅慈（1993）．「站在巨人的肩之上」——立體的體積．《課程論壇》，3（3）頁43－49．

17. 梁鑑添．蕭文強（1980）．一門與數學發展史有關的課程．《抖擻》，41，頁38－44.

18. 黃志華（1997）．閒談古董與另類計算工具．《數學教育》，第五期．頁41－42．

19. 張孝達（1993）．大眾數學與中國古代數學思想．引於21世紀中國數學教育展望大眾數學的理論與實踐課題組．《21世紀中國數學教育展望（第一輯）》．頁1－15．北京：北京師範大學出版社．

20 曹亮吉 (1986) . 數學教學與歷史 . 《科學月刊》. 17 (1) , 頁 39 − 41 .

21 張祖貴 (1992) . 數學史與近代中國數學教育 . 《數學傳播》. 16 (3) , 頁 3 − 15 .

22 馮振業 (1998) . 來自古埃及的教學靈感 . 《數學教育》. 第六期 , 頁 31 − 33 .

23. 黃毅英 (1998) . 從課程角度探討數學史在課堂中之運用 . 《數學教育》. 第六期 , 頁 8 − 9 .

24 傅海倫 (1998) . 《九章算術》與小學數學教學 . 《數學教育》. 第二期 , 頁 24 − 30 .

25 傅敏 . 張維忠 . 楊勇 . 王仲春 (1997) . 在中學數學中強化數學史的意義與作用 . 《高觀點下的中學數學教育研究》. 頁 181 −188 , 台北 : 建宏出版社 .

26. 楊淑芳 (1992) . 數學史在數學教育中的重要性 .《數學傳播》. 16 (3), 頁 16 - 22 .

27. 楊淑芳 (1992) .《從皮亞傑的認識論端談數學史與數學教育的關聯》. 碩士論文 . 臺灣師範大學數學研究所 .

28. 駱祖英 (1990) . 談數學教師的數學史素養 .《高師函授》, 第六期 . 後列於駱祖英 (1996).《數學史教學導論》, 頁 391 - 398 .

中國：浙江教育出版社 .

29. 駱祖英 (1994) . 數學史教學的若干問題 .《成人高等教育》, 第五期 . 後列於駱祖英 (1996) .《數學史教學導論》. 頁 348 - 354 . 中國：浙江教育出版社 .

30. 劉隆華 (1993) 試論中學數學教學中的數學史教育價值 .《數學史研究文集 (第四輯)》. 頁 148 - 153 . 中國： 內蒙古大學出版

社，台北：九章出版社．

31. 歐陽絳 (1998)．數學 - 數學史 - 數學教育 - 素質教育．《數學教育》，第六期．頁2-7．

32. 蕭文強 (1976)．數學發展史給我們的啟發．《抖擻》，17，頁 46 - 53

33. 蕭文強 (1977)．從幾何發展史看幾何在中學教育的作用．《教與學》，第九期，頁27-33．

34. 蕭文強 (1978)．數學發展史和數學教學．《教與學》，第十四期，頁31 - 34

35. 蕭文強，林建 (1978)．概率論是源於賭博嗎？《抖擻》，25，頁16 - 21．

36 蕭文強 (1979). 從數學發展史重整習數學的方法和意義.《數學教學途徑的探討》, 頁 18 - 21.

37 蕭文強 (1981). 活用數學史.《數學教學季刊》, 第二期. 頁 6 - 9.

38 蕭文強 (1981). 數學教學上如何 " 古為今用 '《抖擻》, 44, 頁 70 - 73.

39. 蕭文強 (1982). 微積分的故事 —— 數學發展史的一個範例.《抖擻》, 49, 頁 68 - 76.

40. 蕭文強 (1983). 數學, 數學史, 數學教師.《抖擻》, 53, 頁 67 - 72.

41. 蕭文強 (1983). 從方程到零的故事.《抖擻》54, 頁 58 - 68.

42. 蕭文強 (1984) . 「縱橫圖」縱橫談 —— 從幻方到拉丁方 . 《數學教學季刊》, 第八期 , 頁 6 – 13 .

43. 蕭文強 (1984). 我們的祖先怎樣求圓球的體積 ? . 《數學教學季刊》. 第八期 , 頁 4 – 5 .

44. 蕭文強 (1985). 數學 —— 科學的語言 ? 《數學傳播》. 第九期 , 頁 43 – 47 .

45. 蕭文強 (1987), 誰需要數學史 . 《數學通訊》, 第四期 , 頁 42 – 44 .

46. 蕭文強 (1992). 數學史和數學教育：個人的經驗和看法 . 《數學傳播》, 16 (3), 頁 23 – 29 .

47. 蕭文強 (1992) . 數學＝證明 ? : 《數學傳播》. 16 (4) . 頁 50 – 58 .

48. 顧泠沅 (1990) . 在中學數學教學中貫穿愛國主義思想 . 《上海中學數學》, 5 . 頁

Figure 11.1:delegates at the ICMI Study meeting take an evening walk: Chi Kai Lit, Fung Kit Siu, Chun Ip Fung, Man Kung Siu, Masami Isoda, Wann Sheng Horng, Ryosuke Nagaoka

11.3 Danish

Torkil Heiede

This section is an annotated list of some of the works written in, or translated into, Danish which are suitable for use in teacher education or in the classroom: not comprehensive but indicative of the kind of material available.

Aaboe, Asger, *Episoder fra matematikkens historie* ['Episodes from the history of mathematics'], Copenhagen: Munksgaard 1966, repr. Borgen, 1986
Both in its original American edition and in this Danish translation by the author, this book has been an inspiration for many teachers of the history of mathematics. Chapters on Babylonian mathematics, Greek mathematics and Euclid's *Elements*, Archimedean mathematics, and Ptolemy's trigonometry.

Andersen, Kirsti, et al., *Nogle kapitler af matematikkens historie* ['Some chapters of the history of mathematics'], Aarhus: Matematisk Institut 1979, 2 vols.
Nineteen essays on different historical topics by Danish historians of mathematics, with emphasis on the history of analysis.

Andersen, Kirsti (ed.), *Kilder og kommentarer til ligningernes historie* ['Sources and commentaries to the history of equations'], Vejle: Forlaget Trip 1986

John Fauvel, Jan van Maanen (eds.), *History in mathematics education: the ICMI study*, Dordrecht: Kluwer 2000, pp. 383-386

Nine chapters (by six different authors) on the history of algebra from Babylonian antiquity to Descartes and Newton, with a short epilogue on the fundamental theorem of algebra, and on equations of degree higher than four. With long excerpts from original sources (in Danish translation), exercises, notes and references.

Andersen, Kirsti, Henk Bos & Jesper Lützen, *Træk af den matematiske analyses historie: En antologi af kilder og sekundær litteratur* ['Features of the history of mathematical analysis: an anthology of sources and secondary literature'], Aarhus: Institut for de Eksakte Videnskabers Historie, Aarhus Universitet 1987
The first part contains 18 substantial excerpts from original sources, some of them in the original language (Latin, French, German, English etc.), all with translations into Danish, and with introductions, notes and exercises in Danish. The second part contains 11 excerpts from secondary literature, in English or Danish. With an annotated list of references.

Beck, Hans Jørgen, et al., *Matematik i læreruddannelsen: Kultur, kundskab og kompetence* ['Mathematics in teacher education: culture, knowledge, and competence'], Copenhagen: Gyldendal 1998-, vols. 1-
A textbook (planned in two volumes) for the education of primary and lower secondary mathematics teachers, with emphasis on the historical and cultural dimension. The first volume treats numbers, geometry and probability. Two volumes of work cards are also planned, and a volume on the didactics of mathematics.

Bomann, Gunnar, *Talsystemerne og deres udviklingshistorie* ['Number systems and the history of their evolution'], Copenhagen: Danmarks Lærerhøjskole 1992
The history of the number concept and the 19th century construction of the natural, rational, real, and complex numbers. Written mainly for the further education of teachers in the *folkeskole* (grades 1-10 in the Danish school system); many portraits and references.

Clausen, Flemming, Poul Printz & Gert Schomacher, *Ind i matematikken* ['Into mathematics'], Copenhagen: Munksgaard 1989-1994, 6 vols.; 2.ed. under publ.
The mathematics curriculum of the Danish *gymnasium* (grades 10-12) is here presented in a full-size historical and cultural frame, permeating the whole exposition. Written as school texts, but eminently useful also in teacher education. The six volumes are: *Numbers and geometry; Analytic geometry and functions; Differential calculus; Probability theory and statistics; Vectors and solid geometry; Integral calculus and differential equations.* Beautifully and richly illustrated, each volume contains hundreds of exercises.

Euklids Elementer I-XIII. Copenhagen: Gyldendal 1897-1912, 6 vols.: repr. of 1-IV, Vejle: Forlaget Trip 1985
Heiberg's Greek text of Euclid's *Elements*, translated by one of his students, Thyra Eibe, herself a gymnasium mathematics teacher: the standard text for generations of mathematics students and their teachers at Danish universities.

Heiede, Torkil, *Matematisk analyse: hvad er det for noget?* ['Mathematical analysis: what is it?'], Copenhagen: Matematisk Institut, Danmarks Lærerhøjskole 1991
Heiede, Torkil, *Differentialregning: hvad er det for noget?* ['Differential calculus: what is it?'], Copenhagen: Matematisk Institut, Danmarks Lærerhøjskole 1991
Heiede, Torkil, *Integralregning: hvad er det for noget?* ['Integral calculus: what is it?'], Copenhagen: Matematisk Institut, Danmarks Lærerhøjskole 1992

Heiede, Torkil, *Logaritme- og exponentialfunktioner: hvad er det for noget?* ['Logarithmic and exponential functions: what is it?'], Copenhagen: Matematisk Institut, Danmarks Lærerhøjskole 1993
Aimed at in-service education of teachers and at evening classes at introductory university level, these four volumes treat differential and integral calculus with emphasis on the history of the subject. With exercises and references.

Huygens, Christiaan, *Om regning på lykkespil* ['On the calculus of fortune games']. Tr. Kirsti Andersen, Aarhus: Videnskabshistorisk Museums Venner 1986
A translation of Huygens's classic treatise, with commentaries, exercises, and references.

Høyrup, Jens, *Algebra på lertavler* ['Algebra on clay tablets'], Copenhagen: Matematiklærerforeningen 1998
Detailed readings—with substantial explanations and commentaries—of 18 original Babylonian sources in Danish translation, with 10 others left to the reader as exercises. More demanding and more rewarding than many other treatments, this book argues that the Babylonians' starting point was geometric rather than algebraic.

la Cour, Poul, *Historisk Matematik* ['Historical mathematics'], Copenhagen: P.G.Philipsen 1888, later eds. 1899,1909, 1942, 1962
A classic among Danish historical presentations of elementary mathematics. The book was originally written for the authors' own teaching at one of Denmark's folkehøjskoler (folk high schools), a special sort of historically-minded schools for young adults (in those days mostly from rural surroundings).

Lobatjevskij, N.I., *Geometriske undersøgelser over teorien for parallelle linier* ['Geometrical investigations on the theory of parallel lines']. Transl from German by Lars C. Mejlbo, Aarhus: Matematisk Institut, Aarhus Universitet 1988
An annotated translation of the first publication in a modern western language on non-Euclidean geometry. The first Danish translation of this important source.

Lund, Jens, *Regn med en skriver: Matematik i det gamle Ægypten* ['Calculate with a scribe: mathematics in ancient Egypt'], Copenhagen: Munksgaard 1997
A detailed presentation of Egyptian mathematics, mainly in the form of exercises involving problems from original sources (the Rhind and Moscow and other papyri, the leather roll, inscriptions on stones), with very full commentaries and references.

Lützen, Jesper, *Cirklens kvadratur, vinklens tredeling, terningens fordobling: Fra oldtidens geometri til moderne algebra* ['The squaring of the circle, the trisection of the angle, the doubling of the cube: from the geometry of antiquity to modern algebra'], Herning: Forlaget Systime 1985
A history of the whole of mathematics, seen as inspired by the three great classical problems and all their ramifications—in geometry, algebra, and analysis—from antiquity to the transcendence of π. With many exercises and a full list of references.

Lützen, Jesper & Kurt Ramskov, *Kilder til matematikkens historie* ['Sources for the history of mathematics'], Copenhagen: Universitet 1998, 2nd ed.1999
A collection of 36 sources, in Danish or English translation, with commentaries and exercises. They are chosen so as to be of central importance for the history of mathematics and cover the whole range from Babylonian calculations to Dedekind on irrational numbers. One of them is from the secondary literature and consists of extracts from the Unguru - van der Waerden discussion of the 'geometrical algebra' in Euclid's *Elements* Books ii and vi.

Mejlbo, Lars C.., *Uendelige rækker: en historisk fremstilling* ['Infinite series: a historical exposition'], Aarhus: Matematisk Institut, Aarhus Universitet 1983
Infinite series from antiquity till modern times, including the contributions of the Indians, and those of mediaeval and Renaissance Italians.

Mejlbo, Lars C., *Om den elementære geometris historie* ['On the history of elementary geometry'], Aarhus: Matematisk Institut, Aarhus Universitet 1989
A very detailed and full presentation of the whole history of geometry, from antiquity to Hilbert and Poincaré; with annotated sources and extracts from secondary literature, portraits, biographical sketches, exercises and copious references.

Niss, Mogens, *Matematikkens udvikling - op til renæssancen: Skitse med pointer* ['The evolution of mathematics to the Renaissance: a sketch with highlights'], Roskilde: IMFUFA 1985
A very short history of mathematics (in large format) from antiquity up to but not including the European Renaissance, emphasising the relations between mathematics and society.

Nordisk Matematisk Tidskrift (from **27**, 1979, also called *Normat*)
Since 1953 this journal has been published jointly by the mathematical societies and associations of teachers of mathematics in the five Nordic countries (Denmark, Finland, Iceland, Norway, Sweden). Since its beginning, a main interest has always been the history of mathematics, and over the years many papers of historical or biographical content have been published, in Danish, Norwegian, or Swedish. This journal has always been an important source of material relevant to the education of teachers.

Zeuthen, Hieronymus Georg, *Mathematkens Historie: Oldtid og Middelalder* ['The history of mathematics: antiquity and the middle ages'], Copenhagen: Høst & Søn 1893, new ed. rev. by Otto Neugebauer, 1949

Zeuthen, Hieronymus Georg, *Mathematikens Historie: 16de og 17de Aarhundrede* Copenhagen: Høst & Søn 1903
These two volumes constitute the classical Danish exposition of the history of mathematics up to Newton and Leibniz, in their time translated into both German and French. They are now dated, but Neugebauer's revision of the first volume is still useful.

11.4 Dutch

Harm Jan Smid

The papers reviewed and annotated in this section are chosen from those on the relation between history of mathematics and the teaching and learning of mathematics which have appeared in Dutch over the past three-quarters of a century.

Amerom, Barbara van, *Geschiedenis van de wiskunde in de klas*, Masters Thesis University of Groningen, 1994
Based on original sources, two booklets on differentiation and integration were composed for use in the classroom. The booklet on differentiation was used in five classes, and took some

John Fauvel, Jan van Maanen (eds.), *History in mathematics education: the ICMI study*, Dordrecht: Kluwer 2000, pp. 386-389

five lessons to work through. Pupils' reactions were determined by means of classroom observations, interviews with the teacher and a questionnaire. The results were that the pupils found the texts too difficult and they were not interested in historical materials.

Auwera, N. van der., 'Diophantus in de klas', *Wiskunde en onderwijs* **21** (1995), 207-211.
Two simple problems of Diophantus in the original Greek were given to 17-year-old pupils (who were studying Greek). Beforehand they studied a worksheet with an explanation of Diophantus' notation and symbolism, then answered questions.

Barbin, Evelyne, 'Het belang van de geschiedenis van de wiskunde voor de wiskundige vorming', *Uitwiskeling* 10 (1994), 1-7
Educational interest in history of mathematics originated in part from resistance to the 'new math'. Learning deductively only makes sense if mathematics has a meaning for the pupil. A historical perspective, for instance by reading historical texts, gives insight into the development of mathematics, and provides a teacher with more understanding of the pupil's problems.

Beckers, D.J., 'Historia magistra vitae: de geschiedenis als inspiratiebron voor een rekenles', *Euclides* **72** (1997), 259-262
History of mathematics should be a source of inspiration. There is no simple analogy between the history and the learning processes of children. History of mathematics should not become a part of the mathematics curriculum in school, but ideas and examples from the history of mathematics could enrich lessons.

Breugel, K. van, 'Van kleitablet tot overhead', *Euclides* **63** (1987), 117-118
There are three main reasons for using the history of mathematics in mathematics teaching. When teaching mathematical concepts, it can be helpful to know something about the historical development of the concept. History can explain why some definitions or notations originated, like the division of a circle in 360 degrees. And the history of mathematics has interesting stories that can arouse the interest of the students.

Bunt, L.N.H., *De geschiedenis van de wiskunde als onderwerp voor het gymnasium A*, Groningen 1954
During the years 1951-1953 experiments were held in five classical gymnasia on the teaching of the history of Egyptian, Babylonian and Greek mathematics. These experiments took place in the classes 5 and 6 (the two highest classes), where usually solid geometry was taught. These experiments can be regarded as highly successful and satisfactory. One result was the publication of a textbook on the history of mathematics for this type of school. Due to these experiments the teaching of the history of mathematics was made a optional subject (which was widely chosen) in classical gymnasia. (This possibility disappeared from the curriculum in the 1970s when the Dutch educational system underwent a major change.)

Grootendorst, A.W., 'De geschiedenis van de wiskunde en het onderwijs in de wiskunde', *Wiskunde en onderwijs* **8** (1982), 287-306
One task of a mathematics teacher is to pass a cultural inheritance to future generations. History is very appropriate for that purpose. Also, it is nowadays difficult for a teacher to remain an active mathematical scientist. Studying history of mathematics is a good way to remain active in mathematics, apart from its help in teaching.

Gullikers, Iris, *Geschiedenis van de wiskunde in het onderwijs: literatuurlijst*, Report University of Groningen, 1996

A list describing 23 articles for the use of the history of mathematics in teaching, mainly from *Euclides*, *The Mathematical Gazette*, *Nieuwe wiskrant* and *Wiskunde en onderwijs*.

Hairs, E de, 'Het cultuur-historisch element in het wiskundeonderwijs', *Euclides* **4** (1927), 106-117
The cultural-historical element in mathematics teaching should be more than just an illustration to regular teaching. There is an international movement going on to reform mathematics teaching. The teaching of the history of mathematics fits in that movement; the genetic-historical method of teaching, especially, can profit very much from the use of the history of mathematics.

Huisjes, J. and Langeland, J, 'Wat deed een Egyptenaar 4000 jaar geleden met een differentiaalvergelijking?', *Nieuwe wiskrant* **11** (1992), 32-35
In 1992, a questionnaire was send to 600 mathematics teachers of all levels about their knowledge of history of mathematics and their interest in using it in the classroom. Most teachers did not know very much about the history of mathematics; 90% sometimes mentioned history in class, varying from just a casual remark to extensive treatment of a historical topic. Many teachers would like to do more on history of mathematics, but are impeded by lack of knowledge, time and appropriate materials.

Kool, M., 'Waarom kort als het ook lang kan?, Wiskundige notaties in zestiende-eeuwse rekenboeken', *Nieuwe wiskrant* 18 (1998), 5-9
16th century arithmetic books hardly used modern symbolic mathematical notations. Trying to understand 16th century solutions and abbreviations can be a challenging and interesting learning experience for today's students.

Looij, H. van, 'Het nut van de geschiedenis van de wiskunde', *Wiskunde en onderwijs* **6** (1980), 429-444
History of mathematics can help pupils to discard the idea that mathematics is a completed and faultless edifice, instead of a human project with many new developments and unsolved problems. With the historic-genetic method the teacher can help the pupil to gain a better understanding and to experience mathematics as a living entity. History of mathematics also teaches the pupil to see mathematics as a part of human culture.

Maanen, Jan van, 'Over het verdelen van aangeslibd land: een brugklaspoject', *Euclides* **60** (1984), 161-168
English version ('Teaching geometry to 11 year old "mediaeval lawyers"') cf. §11.5.1.

Maanen, Jan van, 'Een gewichtig probleem van L'Hôpital', *Nieuwe wiskrant*, 10 (1990), 6-9
English versiom ('L'Hôpital's weight problem') referred to in §11.5.

Meskens, Ad, 'Zestiende-eeuwse wiskunde doorheen het middelbaar onderwijs', *Wiskunde en onderwijs* **18** (1992), 232-248
A number of examples from 16th-century arithmetic books for schools are presented, mainly from trading applications. Such problems could be used in the classroom.

Mooij, H., 'De geschiedenis van de wiskunde en de didactiek', in: *Over de didactiek van de meetkunde benevens benaderingsconstructies ter verdeling van een hoek in gelijke delen*, Amsterdam 1948, chapter 2
Incorporating the history of mathematics in teaching, especially in plane geometry, is useful because pupils gain a better understanding of the necessity of doing mathematics, the

importance of intuitive and inductive reasoning is illuminated and the mutual influencing of mathematics and society can become more clear.

Schrek, D.J.E. 'Het cultuurhistorisch element in het wiskunde-onderwijs', *Euclides* 1 (1924) 29-46

Paying attention to history narrows the gap between the exact sciences and the liberal arts. History of mathematics has become a full grown branch of science; in several countries, e.g. Germany, history of mathematics has been recommended as a school subject. Historical examples, problems or the study of theorems in their original Greek formulation can broaden the cultural horizon of children. Nowadays there is enough material available for teachers to interweave elements from the history of mathematics into their mathematics lessons. It should not be taught as a separate subject.

11.5 English

John Fauvel

Arcavi, Abraham, 'Two benefits of using history', *For the learning of mathematics* 11.2 (1991), 11

One benefit lies in using history to unpack the automatic quality of mathematics, to re-examine known and taken-for-granted mathematical ideas. Another is to sensitise the teacher to possible difficulties of student understanding, and help in listening to students' arguments.

Arcavi, Abraham, Maxim Bruckheimer and Ruth Ben-Zvi, 'History of mathematics for teachers: the case of irrational numbers', *For the learning of mathematics* 7.2 (1987) 18-23

The development and implementation of a course on irrational numbers, taught through worksheets with further materials and answer sheets. The course objectives were to strengthen the teachers' knowledge, pursue other pedagogic issues, develop work around primary sources, and foster an image of mathematics as creative human endeavour.

Arcavi, Abraham, Maxim Bruckheimer and Ruth Ben-Zvi, 'Maybe a mathematics teacher can profit from the study of the history of mathematics', *For the learning of mathematics* 3.1 (1982) 30-37

A two-day teacher workshop was designed to create a picture of the development of a topic (negative numbers), with details of worksheets and of the reception of the event.

Barbin, Evelyne, 'The reading of original texts: how and why to introduce a historical perspective', *For the learning of mathematics* 11.2 (1991) 12-13

Reading original texts allows the teacher or learner to study mathematical activity, and gain access to the concepts permeating mathematical texts. This process changes the image of mathematics and enables learners to see it as an activity, illustrated by comparing the way Euclid and Clairaut approach angles of a triangle.

Barbin, Evelyne, 'The role of problems in the history and teaching of mathematics', in R. Calinger (ed), *Vita mathematica: historical research and integration with teaching*, Washington: MAA 1996, 17-25

John Fauvel, Jan van Maanen (eds.), *History in mathematics education: the ICMI study*, Dordrecht: Kluwer 2000, pp. 389-404

Introducing history of mathematics to future teachers transforms the practice of teaching mathematics, through changing the epistemological concepts of mathematics; in particular by emphasising the construction of knowledge out of the activity of problem solving. This is seen in the examples of the concept of angle and the concept of curve.

Bartolini Bussi, Maria, and Maria Alessandra Mariotti, 'Semiotic mediation: from history to the mathematics classroom', *For the learning of mathematics* 19.2 (1999), 27-35
Whether a section of a cone is the same as a section of a cylinder, and whether either is egg-shaped, has long been debated. Students trying to find the flaw in historic arguments such as those by Witelo (c.1200) and Dürer (1525), need help in harmonising the figural and conceptual aspects of the problem. The teacher has a key role in helping them to master the conflict and achieve a new conceptual control.

Bos, H. J. M., 'Mathematics and its social context: a dialogue in the staff room, with historical episodes', *For the learning of mathematics* 4.3 (1984) 2-9
The history of mathematics can inform both pupils and teachers about the social context of mathematics, and help them to decide what position they hold in debates about it.

Brummelen, Glen Van, 'Jamshid al-Kashi: calculating genius', *Mathematics in school* 27.4 (1998), 40-44
The remarkable and beautiful insights of the C15 Iranian astronomer al-Kashi, working in Samarkand in the 1420s, led to unprecedentedly accurate values of π and the sine of 1° (the equivalent of 16 and 17 decimal places, respectively). His method for sin 1° is essentially that of fixed-point iteration which can be done on a calculator in class.

Brummelen, Glen Van, 'Using ancient astronomy to teach trigonometry: a case study', *Histoire et épistémologie dans l'éducation mathématique,* IREM de Montpellier (1995), 275-281
Students who perform well on technical examinations at the end of a course may still not grasp why the subject exists, what the mathematics means, or how to ask mathematical questions. Carefully planned use of history can help address these problems. A guided case study of ancient astronomy has proved fruitful in evoking greater trigonometric confidence and understanding.

Burn, Bob, 'What are the fundamental concepts of group theory?', *Educational studies in mathematics* 31 (1966), 371-377
The conventional way of teaching the notion of group, as a set with a binary operation satisfying four axioms, is more logically than psychologically satisfying. Starting from the historical origins of permutation and symmetry may have pedagogic benefits.

Burn, R.P., 'Individual development and historical development: a study of calculus', *Int. J. Math. Educ. Sci. Technol.* 24 (1993), 429-433
The rigour of undergraduate analysis was introduced by Cauchy and Weierstrass during the 19th century, and the conventions of pre-19th century calculus are close to the conventions of pre-university calculus in England today. The analogy between personal development and historical development in calculus is richly suggestive—but may not be pressed too far.

Burns, Stuart, 'The Babylonian clay tablet', *Mathematics Teaching* 158 (1997), 44-45
Investigations of a Babylonian tablet by middle school pupils revealed some remarkable differences, from those who discovered what it was about without realizing what they had

achieved, to those who used a book to learn what all the numbers were—to give them the 'right' answer—without gaining any idea what the tablet was about.

Carvalho e Silva, Jaime, 'History of mathematics in the classroom: hopes, uncertainties and dangers', Sergio Nobre (ed), *Proceedings of HPM Meeting, Blumenau, Brazil* 1994, 129-135
Portugal provides an example, in the 1950s-60s textbooks of José Sebastião e Silva, of historically-informed school textbooks whose successors, when fashion changed, contained no history. Although the Portuguese syllabus now pays vague lip-service to history of mathematics, proponents must safeguard against changes of fashion and political whim.

Cooper, Amira, 'Integration of the historical development of mathematics in mathematics teaching in the high school using self reading', Eduardo Veloso (ed), *Proceedings of HEM Meeting, Braga, Portugal* 1996, vol II, 3-10
Providing historical material for students, to read on their own at home, contributes to a significant change in students' attitudes towards mathematics, as well as increasing the number who saw individual reading as an important part of the learning process.

Crawford, Elspeth, 'Michael Faraday on the learning of science and attitudes of mind', *Science and education* 7 (1998): 203-211
Faraday's ideas about learning are relevant to scientific learning in general. It is central to learning in science to acknowledge that an inner struggle is involved in facing unknowns. Following Faraday, for teachers to understand their own feelings while teaching is essential to enable empathy with the fears and expectations of learners.

D'Ambrosio, Ubiratan, 'Ethnomathematics and its place in the history and pedagogy of mathematics', *For the learning of mathematics* **5.1** (1985) 44-48
A suggestion for looking at the history of mathematics in a broader context, to incorporate practices which are mathematical in their nature without constituting mathematisation in the traditional sense. Such an approach has implications for curriculum development, particularly in third world countries.

D'Ambrosio, Ubiratan, 'Where does ethnomathematics stand nowadays?', *For the learning of mathematics* **17.2** (1997), 13-17
History is critical to ethnomathematical studies; conversely ethnomathematics calls for a broader concept of sources and a new historiography for the history of mathematics, which in turn affects mathematics.

Deakin, Michael A. B., 'Women in mathematics: fact versus fabulation', *Australian mathematical society gazette* **19** (1992), 105-114
Many historical accounts of women in mathematics, some recent, overlook Theon's instruction to his daughter Hypatia "*To teach superstitions as truths is a most terrible thing.*" To pursue the truth about mathematical women in the past leads to recognition of the diversity of role models they provide, in their very disparate talents and interests.

Deakin, Michael, 'Boole's mathematical blindness' *Mathematical gazette* **80**, no. 489 (1996) 511-518
George Boole never solved a particular problem in operational calculus, despite working on it throughout his life and holding the key to its solution. Analysing the factors which prevented him—a technical deficiency, coupled with failing to conceive of the solution as being a solution—helps us understand difficulties encountered by today's students.

Dennis, David, and Jere Confrey, 'Drawing logarithmic curves with *Geometer's sketchpad*: a method inspired by historical sources', in James R. King and Doris Schattschneider (eds,) *Geometry turned on! Dynamic software in learning, teaching and research, MAA* 1997, 147-156
A mechanical linkage device from Descartes' *Geometry*, which can be used for finding any number of points on a logarithmic or exponential curve, can be simulated on computer. Such a tool helps populate the dialogue between grounded activity and systematic inquiry, between physical investigations and symbolic language, in mathematics learning.

Dennis, David, and Jere Confrey, 'The creation of continuous exponents: a study of the methods and epistemology of John Wallis', *CBMS Issues in mathematics education* 6(1996) 33-60
History provides rich sources of alternative conceptualization and diverse routes to the development of an idea. This deepens the close listening of teachers and researchers to student mathematics, and leads to reconceptualizing the epistemology of mathematics. History is seen as the coordination of multiple forms of representation. How geometry and ratio supported Wallis's development of exponents is explored in depth.

Dorier, Jean-Luc, 'On the teaching of the theory of vector spaces in the first year of French science universities', *Edumath* 6 (1998), 38-48
Historical analysis enables us to explain the specific meaning which formalism has in the theory, and thus the teaching, of vector spaces. Other pedagogical issues, including students' mistakes, can be understood and acted upon better through the study of history.

Downes, Steven, 'Hypatia versus the National Curriculum', *Mathematics teacher* 153 (1995), 8-9
Reflections on the tensions between the demands of a national curriculum and attempts to help pupils enjoy mathematics through historical activities.

Downes, Steven, 'Women mathematicians, male mathematics: a history of contradiction?', *Mathematics in school* 26.3 (1997), 26-27
It is not sufficient to show pupils that some women (the familiar few names from history) can do mathematics; rather, it is necessary to educate girls into seeing that they as women are not 'other' to mathematics, through a historical analysis of how women's participation in mathematics has been constructed.

Eagle, Ruth, 'A typical slice', *Mathematics in school* 27.4 (1998), 37-39
Exploring Archimedes' *Method* with trainee teachers reveals a method for determining volumes which kindles interest and is well within the grasp of secondary school pupils.

Ernest, Paul, 'The history of mathematics in the classroom', *Mathematics in school* 27.4 (1998), 25-31
Examples of classroom worksheets devised by student teachers drawing upon the history of mathematics.

Fauvel, John, 'Algorithms in the pre-calculus classroom: who was Newton-Raphson?', *Mathematics in school* 27.4 (1998), 45-47
The so-called Newton-Raphson method (due in its present form to Thomas Simpson) provides insights into algorithms and iterative processes which can be useful for pupils before as well as after they learn calculus.

Fauvel, John, 'Empowerment through modelling: the abolition of the slave trade', in R. Calinger (ed), *Vita mathematica: historical research and integration with teaching*, Washington: MAA 1996, 125-130
An example of the use of a historical artefact—a diagram from Thomas Clarkson's *History of the abolition of the Africa slave-trade* (1808)—in order to help students to think and learn about graphical modelling techniques.

Fauvel, John, 'Platonic rhetoric in distance learning: how Robert Record taught the home learner', *For the learning of mathematics* 9.1 (1989) 2-6
The textbooks of Robert Record (c.1510-1558), the first writer of mathematics textbooks in English, show astonishing freshness and pedagogic insight, not least how to empathise with the reader at a distance.

Fauvel, John, 'Using history in mathematics education', *For the learning of mathematics* 11.2 (1991) 3-6
A survey of ways history can be used in the mathematics classroom, the reasons advanced for doing so, and political and other issues surrounding the introduction of a historical dimension to mathematics education.

Fernandez, Eileen, 'A kinder, gentler Socrates: conveying new images of mathematics dialogue', *For the learning of mathematics* 14.3 (1994) 43-47
Revisiting the celebrated encounter between Socrates and the slave-boy, in Plato's *Meno*, with a view to drawing out its implications for mathematics teacher training: in particular, how it might be used to promote an image of teachers and students empowering one another.

FitzSimons, Gail, 'Is there a place for the history and pedagogy of mathematics in adult education under economic rationalism?', Eduardo Veloso (ed), *Proceedings of HEM Meeting, Braga, Portugal* 1996, vol II, 128-135
Before the growth of economic rationalism and the adoption of industrial values to the exclusion of others, further education classes in Australia enabled adults returning to study to learn about the history of mathematics and recreate parts for themselves.

Fowler, David H., 'A final-year university course on the history of mathematics: actively confronting the past', *The mathematical gazette* 76 (1992), 46-48
History is the active confrontation of the past and the present. Students on this course are encouraged in this by (a) reading a selection of texts and writing a short description and a short essay on their reactions; (b) giving a 15-minute talk to the class at some stage; (c) writing a substantial essay. Some of the skills this course develops are notoriously neglected in mathematics courses, and are in great demand in the outside world.

Frankenstein, Marilyn, 'Various uses of history in teaching criticalmathematical literacy', Sergio Nobre (ed), *Proceedings of HPM Meeting, Blumenau*, 1994, 91-98
A prime use of history is for students to examine their personal schooling history. Another is the hidden history, involving peoples' mathematical developments, which can be used to demystify the structure of mathematics and of society.

Freudenthal, Hans, 'Should a mathematics teacher know something about the history of mathematics?', *For the learning of mathematics* 2.1 (1981) 30-33
The teacher's knowledge of history should be integrated knowledge, familiar to the teacher and a cornucopia available for instruction: not hidden in drawers to be opened at pre-established moments. For students and teachers, the history of mathematics should concern the processes rather than the products of mathematical creativity.

Friedelmeyer, Jean-Pierre, 'What history has to say to us about the teaching of analysis', Evelyne Barbin and Régine Douady (eds), *Teaching mathematics: the relationship between knowledge, curriculum and practice*, Topiques éditions 1996, 109-122
Reforms in analysis teaching have attempted to reconcile the apparently irreconcilable needs for rigour and for understanding. Teaching in a historical context enables the meaning and rigour to be interactively constructed along with the student's mathematical insight, by a process which is dynamic and living.

Führer, Lutz, 'Historical stories in the mathematics classroom', *Mathematical gazette* 76, no. 475 (1992) 127-138
The desirability of incorporating history in mathematics teaching is easier to establish than how in practice it may be done. Two stories—Eratosthenes, and ideas of π—illustrate that history is too important to use to bore and perplex pupils: rather, it provides a changed tone for the framework within which mathematics education takes place.

Furinghetti, Fulvia, 'History of mathematics, mathematics education, school practice: case studies in linking different domains', *For the learning of mathematics* 17.1 (1997), 55-61
Experiences of teachers exploring different ways of using history are discussed and taxonomised: informing students' image of mathematics, as a source of problems, as an optional activity, and as a different approach to concepts. 'Integration' is preferable to 'use' of history, to characterise a more methodical development and analysis.

Furinghetti, Fulvia, 'The ancients and the approximated calculation: some examples and suggestions for the classroom' *Mathematical gazette* 76, no. 475 (1992) 139-142
History is a good source of problems for the classroom, particularly in relation to the area of approximated calculation, which is of increasing importance in the practical mathematics curricula of today. These problems are of interest not only from an algorithmic point of view, but also for developing mathematical concepts.

Furinghetti, Fulvia, and Annamaria Somaglia, 'History of mathematics in school across disciplines', *Mathematics in school* 27.4 (1998), 48-51
History of mathematics can help pupils see the genesis of ideas and connections between subjects, with real benefits for their seeing the homogeneity of knowledge as well as mathematical development. Several interdisciplinary projects relate mathematics and philosophy, art, music, &c. Students' mathematical difficulties are addressed by a considered approach drawing upon contexts from the history of mathematics.

Garcia, Paul, 'Dismissis incrutiationibus', *Histoire et épistémologie dans l'éducation mathématique*, IREM de Montpellier (1995), 171-190
Among the reasons why secondary school teachers might consider using history are to show: that today's 'elementary' concepts may not have been obvious to even great past mathematicians; that the personalities of mathematicians have the same problems as everyone else; and that even today there can be disputes about ideas.

Gardiner, Tony, 'Once upon a time' *Mathematical gazette* 76,. 475 (1992) 143-150
History of mathematics has much to offer the teaching of mathematics. Two pitfalls, though, are the temptation to enlist the support of 'history' when trying to change social attitudes, and the uncritical way in which intelligent students respond to pseudo-history.

Gardner, J Helen, '"How fast does the wind travel?": history in the primary mathematics classroom', *For the learning of mathematics* **11.2** (1991) 17-20
Examples of incorporating a historical dimension into multi-ethnic primary education (8 to 10 year-olds).

Gauld, Colin, 'Making more plausible what is hard to believe: historical justifications and illustrations of Newton's third law', *Science and education* 7 (1998): 159-172
Similarity between the notions of young people today and those of pre-Newtonian scientists suggests that a study of attempts to justify Newton's third law from the C17 to the C19 may provide arguments to help students to consider it plausible.

Gerdes, Paulus, 'Examples of incorporation into mathematics education of themes belonging to the history of geometry in Africa', Sergio Nobre (ed), *Proceedings of HPM Meeting, Blumenau, Brazil* 1994, 214-221
Two examples: the living tradition of the originally female geometry of handbags in Mozambique, and the almost disappeared tradition of male geometry of sand drawings from Angola and Zambia.

Graf, Klaus-Dieter and Bernard R. Hodgson, 'Popularizing geometrical concepts: the case of the kaleidoscope', *For the learning of mathematics* **10.3** (1990) 42-50
Historical and pedagogic account of the kaleidoscope (Brewster 1817), a particularly successful example of an instrument which captures the attention of pupils and involves them in mathematics. With further reflections on its transference to computer software.

Grattan-Guinness, Ivor, 'Some neglected niches in the understanding and teaching of numbers and number systems', ZDM **98**/1 (1998), 12-18
Historical examples in the field of number, selected for their possible use in teaching at school or college level, with pedagogic commentary: including fractions and ratios, integers with properties, algorist vs abbacist approaches to calculation, and zero.

Griffiths, H. B., Massimo Galuzzi, Michael Neubrand and Colette Labord, 'The evolution of geometry education since 1900', in C. Mammana and V. Villani (eds) *Perspectives in the teaching of geometry for the 21st century*, Kluwer 1998, 193-234
The roles of geometry in the curriculum over the past century in England, Italy, Germany, and France, compared as a considered exercise in understanding the past better in order to avoid future mistakes of education policy.

Hahn, Alexander J., 'Two historical applications of calculus', *College mathematics journal* **29** (1998), 93-103
L'Hopital's determination of the static geometry of a pulley, and Galileo's experiment with balls rolling down an inclined plane, are two problems pitched at just the right level for students beginning calculus. Through such problems students can both deepen their insights and practise their computational skills.

Hefendehl-Hebeker, Lisa, 'Negative numbers: obstacles in their evolution from intuitive to intellectual constructs', *For the learning of mathematics* **11.1** (1991) 26-32
The intellectual hurdles that blocked the understanding of negative numbers thoughout history may also block the understanding of present-day students. The examples of D'Alembert and Stendhal illustrate the confusions. Among others, Hermann Hankel in 1867 sought to overcome the difficulties by a change of viewpoint.

Heiede, Torkil, 'Why teach history of mathematics?' *Mathematical gazette* **76**, no. 475 (1992) 151-157
Because the history of a subject is part of the subject. If you are not aware that mathematics has a history then you have not been taught mathematics, but have been cheated of an indispensable part of it. You are not a mathematics teacher if you do not teach also the history of mathematics.

Hitchcock, Gavin, 'Dramatizing the birth and adventures of mathematical concepts: two dialogues', in R. Calinger (ed), *Vita mathematica: historical research and integration with teaching*, Washington: MAA 1996, 27-41
The power of dialogue and theatre in reconstructing the historical story of informal mathematics-making is shown in two playlets, about the acceptance in Europe of decimal expansions of irrational numbers (a dialogue between Stifel and Stevin) and of negative roots of equations (Frend, Peacock and De Morgan).

Hitchcock, Gavin, 'Teaching the negatives, 1870-1970: a medley of models', *For the learning of mathematics* **17.1** (1997), 17-25, 42
Six contrasting classroom scenes of good teachers at work: C. Smith (1888), A N Whitehead (1918), E. Landau (1930), T. Apostol (1957), American teacher (1961), English teacher (1966), with prologue (A. De Morgan) and epilogue (F. Klein). With questions and exercises, for teacher-training workshops.

Hitchcock, Gavin, 'The "grand entertainment": dramatising the birth and development of mathematical concepts', *For the learning of mathematics* **12.1** (1992) 21-27
Use of dialogue and theatre is a way to allow the student to share something of the creative tensions and intellectual excitement experienced by human mathematics-makers in their historical problem-situations. An example is given, a synopsis of a six-scene play on the rise of negative numbers. 'Grand entertainment' is Kepler's phrase.

Isaacs, Ian, V Mohan Ram and Ann Richards, 'A historical approach to developing the cultural significance of mathematics amongst first year preservice primary school teachers', Eduardo Veloso (ed), *Proceedings of HEM Meeting, Braga, Portugal* 1996, vol II, 26-33
A course at the Northern Territory University, Australia, set out to modify the belief systems and perceptions of trainee primary teachers about the nature of mathematics and the purpose of school mathematics. Work included geometry from China, India, Egypt and Greece. Results were mixed; many students were unconvinced and more work is needed.

Jones, Charles V., 'Finding order in history learning: defining the history and pedagogy of mathematics', Sergio Nobre (ed), *Proceedings of HPM Meeting, Blumenau, Brazil* 1994, 35-45
Historio-pedagogy will become a discipline when a founding set of assumptions and a research agenda is agreed: for example, seeing the processes of history and of learning as complex systems with emergent order. This view must criticise many assumptions in current pedagogy; teacher and learner might begin to relate as mentoring partners.

Katz, Kaila, 'Historical content in computer science texts: a concern', *Annals of the history of computing* **19.1** (1997), 16-19
Those teaching computer science courses may have little chance or competence to evaluate the historical material found in student textbooks. Yet there are problems with the historical

content of many current texts. The history of the field deserves the same careful treatment in these texts as do other aspects of computer science.

Katz, Victor J., 'Using history in teaching mathematics', *For the learning of mathematics* **6.3** (1986) 13-19
Use of historical materials is profitable both for motivating students and for developing the curriculum, and can give rise to valuable pedagogic ideas. Examples are given from algorithms, combinatorics, logarithms, trigonometry, and mathematical modelling.

Katz, Victor, 'Ethnomathematics in the classroom', *For the learning of mathematics* **14.2** (1994), 26-30
Many important mathematical ideas grew out of the needs of cultures around the world. These are exemplified in examples from combinatorics, arithmetic and geometry. Studying these broadens students' understanding not only of mathematics but also of the world.

Katz, Victor, 'Some ideas on the use of history in the teaching of mathematics', *For the learning of mathematics* **17.1** (1997), 62-63
To discover ways of making learning better for students, teachers need to experiment with various ways of using history and sharing the results. Successful use may require action on a larger scale: setting a series of ideas, or even a whole course, in historical context.

Kleiner, Israel, 'A historically focused course in abstract algebra', *Mathematics magazine* **71** (1998), 105-111
A course in abstract algebra, for an in-service master's programme for mathematics teachers, was based around the theme of showing how abstract algebra originated in, and sheds light on, the solution of concrete problems. The historical material was mainly approached through secondary sources.

Kool, Marjolein, 'Dust clouds from the 16th century', *Mathematical gazette* **76** (1992) 90-96
Working with historical materials in the classroom is a way of motivating pupils. In particular, it can be very useful with students of below average capabilities or with learning difficulties, who are easily distracted in mathematics lessons and have little interest. Here the example is given of working with 16th century Dutch arithmetic texts.

Kool, Marjolein, 'Using historical arithmetic books in teaching mathematics to low attainers', *Histoire et épistémologie dans l'éducation mathématique,* IREM de Montpellier (1995), 215-225
Low-attaining teenagers can be enthused and stimulated by working with carefully selected samples from old arithmetic books and mss. They come to see mathematics as problems done and solved by other people too, with whom they can identify. With several pages of worksheets.

Kubli, Fritz, 'Historical aspects in physics teaching: using Galileo's work in a new Swiss project, *Science and education* **8** (1999), 137-150
A questionnaire about incorporating historical material in their physics programme was sent to students in Swiss high schools, canvassing different types of intervention (eg sporadic recounting, original texts, reconstructed historical experiments). Early results show a difference in the responses of male and female students.

Laubenbacher, Reinhard, and David Pengelley, 'Great problems of mathematics: a course based on original sources', *American mathematical monthly* **99** (1992), 313-317
In this course, aimed at giving students the "big picture", we examine the evolution of selected great problems from five mathematical subjects: area and the definite integral, set theory, solutions of algebraic equations, Fermat's last theorem, and the parallel postulate. The use of original sources allows students to appreciate the progress through time in clarity and sophistication of concepts and techniques.

Laubenbacher, Reinhard, and Michael Siddoway, 'Great problems of mathematics: a summer workshop for high school students', *The college mathematics journal* **25** (1994), 112-114
In a 3-week summer workshop for 22 high school students from across the country we examined, using original sources, the evolution of selected great problems from set theory, number theory, and calculus. The first year used a traditional lecture approach. The second year's teaching style incorporated two pedagogical devices that proved amazingly effective: the 'discovery' method and daily writing.   The discovery method led to far deeper understanding, while the writing was a valuable tool for comprehending and mastering mathematics. The ability of primary source material to engage students' attention and spur their efforts was dramatically evident.

Le Goff, Jean-Pierre, 'Cubic equations at secondary school level: following in Euler's footsteps', Evelyne Barbin and Régine Douady (eds), *Teaching mathematics: the relationship between knowledge, curriculum and practice*, Topiques éditions 1996, 11-34
Whether or not included in the curriculum, cubic equations are important for leading to the emergence of imaginary numbers and to the solution of trigonometric equations. A class of 17-year-olds in Normandy tackled a text of Euler as an investigation, here described in detail. The same text was explored differently in another class.

Lombardi, Olimpia, 'Aristotelian physics in the context of teaching science: a historical-philosophical approach', *Science and education* **8** (1999), 217-239
Aristotelian physics for didactic purposes is sometimes presented in too fragmentary and oversimplified a way. Reading the original texts is a richer intellectual experience and shows the author's thought in action.

Maanen, Jan van, 'L'Hôpital's weight problem', *For the learning of mathematics* **11.2** (1991) 44-47
Classroom use (with 18-year-old pupils in a Dutch gymnasium) of a problem from the first calculus textbook, L'Hôpital's *Analyse des infiniment petits* (1696), with a discussion of the value and purpose of this activity.

Maanen, Jan van, 'New maths may profit from old methods', *For the learning of mathematics* **17.2** (1997), 39-46
Four classroom activities—bisecting an angle, solving a quadratic equation, estimating a logarithm, calculaing the area of a triangle—show how tackling problems from old textbooks can enable school pupils and trainee teachers to gain fresh and invigorating perspectives on what they are learning.

Maanen, Jan Van, 'Old maths never dies', *Mathematics in school* **27** (1998), 52-54

Today's students are intrigued and inspired by C17 textbook problems at a number of levels, from deciphering gothic type (a morale-boosting activity for weaker students) to realising that problems can be solved geometrically as well as algebraically.

Maanen, Jan van, 'Seventeenth century instruments for drawing conic sections', *Mathematical gazette* **76** (1992), 222-230
A consequence of Descartes' new approach to geometry (1637) was an increased interest in instruments for drawing conic sections, taken up particularly by the Dutch mathematician Van Schooten (1615/6-1660).

Maanen, Jan van, 'Teaching geometry to 11 year old "mediaeval lawyers"', *Mathematical gazette* **76,** no. 475 (1992) 37-45
11-year old pupils studying Latin and mathematics studied a 1355 treatise by Bartolus of Saxoferrato on the division of alluvial deposits. Besides integrating the two subjects in the same project, it was a way of encouraging pupils to work together, to see the importance of mathematics in society, and to discover ruler-and-compass constructions.

MacKinnon, Nick, 'Homage to Babylonia' *Mathematical gazette* **76** (1992) 158-178
Some resources on Old Babylonian mathematics that have been used with classes, at various places in the curriculum, in relation to place value, Pythagoras' theorem, and quadratic equations. How the material may be integrated into pupils' general education, and where to see cuneiform mathematics in Britain.

MacKinnon, Nick, 'Newton's teaser', *Mathematical gazette* **76,** no. 475 (1992) 2-27
Leibniz's series for $\pi/4$, and Newton's riposte in his *Epistola posterior* (1676). The latter "makes an excellent peg on which to hang a number of lessons on infinite series, and integration, and in the course of researching this article I found I had touched base with so many A-level topics that my whole teaching at this level has been revolutionised."

McBride, Carl and James H Rollins, 'The effect of history of mathematics on attitudes toward mathematics of college algebra students', *Journal for research in mathematics education* **8** (1977), 57-61
Incorporating ideas from the history of mathematics into a college algebra course produces a significant positive effect on student attitudes towards mathematics.

Menghini, Marta, 'Form in algebra: reflecting, with Peacock, on upper secondary school teaching', *For the learning of mathematics* **14.3** (1994) 9-14
In teaching algebra it is better at a certain level to underline explicitly the transition from arithmetical to symbolic algebra. The work of George Peacock and other C19 English algebraists (Gregory, Babbage, De Morgan and Boole) provides a useful analogy.

Monk, Martin, and Jonathan Osborne, 'Placing the history and philosophy of science on the curriculum: a model for the development of pedagogy', *Science education* **81** (1997), 405-424
Two main issues for those wishing to introduce HPS into science teaching are the justification, and the placement of materials. The justification must point to places where the inclusion of history will directly contribute to the learning of science concepts. Materials must support teachers' main aims, and understanding of science education as epistemological justification, rather than seem bolted on in a context of discovery.

Morley, Arthur, 'Should a mathematics teacher know something about the history of mathematics?', *For the learning of mathematics* **2.3** (1982) 46

Yes, for two reasons: to get student teachers to reflect on the nature of the subject they will teach, and to understand issues of curriculum content.

Mower, Pat, 'Mathematical fiction', *Humanistic mathematics network journal* **19** (1999), 39-46
Students in the history of mathematics class at Washburn University developed their understanding of mathematics and its history through creating imaginative fiction including 'A day in the life of Diophantus' and a newspaper report on the discovery of an ancient document by Diophantus.

Nouet, Monique, 'Using historical texts in the lycée', Evelyne Barbin and Régine Douady (eds), *Teaching mathematics: the relationship between knowledge, curriculum and practice*, Topiques éditions 1996, 125-138
Using primary historical texts has several benefits, enabling students to experience the pleasure of discovery; to see that mathematics has developed and that the same concept can appear in a variety of ways and contexts; and to be reassured, improve their repertoire of approaches and improve their performance. These benefits are seen in the study of texts by Roberval, Pascal, Archimedes, and Arnauld, in the final-year class of a lycée.

Ofir, Ron, 'Historical happenings in the mathematical classroom', *For the learning of mathematics* **11.2** (1991) 21-23
Discussion of activities developed for classroom use (12 to 14 year-olds), in the context of number systems. fractions, and π.

Ofir, Ron, and Abraham Arcavi, 'Word problems and equations: an historical activity for the algebra classroom', *Mathematical gazette* **76,** no. 475 (1992) 69-84
A history of algebra activity for junior high school students (aged 12-14), relating to problems that reduce (in modern terms) to $ax = b$, taking the form of a teacher directed presentation/discussion with accompanying transparencies and worksheet.

Orzech, Morris, 'An activity for teaching about proof and about the role of proof in mathematics', *PRIMUS* **6** (1996), 125-139
A linear algebra class was infused with history and philosophy of mathematics, to help students understand the notion of proof. The method here involved experiencing a historical skit/dialogue about the definition of proof, and looking at some historical proofs to understand the development of the notion.

Perkins, Patricia, 'Using history to enrich mathematics lessons in a girls' school', *For the learning of mathematics* **11.2** (1991) 9-10
Setting mathematics in a historical context, presenting it as part of cultural heritage, has proved a successful strategy for pupils in an independent girls' school, particularly on issues concerning confidence and gender awareness.

Pimm, David, 'Why the history and philosophy of mathematics should not be rated X', *For the learning of mathematics* **3.1** (1982) 12-15
History and philosophy of mathematics can be of use to mathematics education through informing our understanding of mathematics, which is enriched and encouraged by an awareness of its problem sources. It gives a sense of place and meaning from which to learn mathematics, challenging the notion of a static list of accumulated truths.

Ponza, Maria Victoria, 'A role for the history of mathematics in the teaching and learning of mathematics: an Argentinian experience', *Mathematics in school* **27.4** (1998), 10-13
The experience of writing and producing a play about Galois (whose text is reproduced here) had notable effects upon the interest and enthusiasm of pupils for mathematics.

Radford, Luis, 'An historical incursion into the hidden side of the early development of equations', in Joaquim Giménez et al, *Arithmetics and algebra education* (1996), 120-131
A historical case-study, of the rise of the algebraic concept of equation, shows that mathematical reification processes (processes of abstraction and/or generalisation) are socio-culturally related, in this case to the development of writing and of socially elaborated forms of mathematical explanation: equations have always had a *meaning* shaped by the social structures in which they were practised.

Radford, Luis, 'Before the other unknowns were invented: didactic inquiries on the methods and problems of mediaeval Italian algebra', *For the learning of mathematics* **15.3** (1995), 28-38
Didactical-epistemological analysis of problems and methods in Italian algebra from the 12th century onwards helps us understand the meaning of algebraic ideas, and helps draw out information that can be used in teaching: not to follow the same path, but to find new teaching possibilities (e.g. links between algebra and negative numbers).

Radford, Luis, 'On psychology, historical epistemology, and the teaching of mathematics: towards a socio-cultural history of mathematics', *For the learning of mathematics* **17.1** (1997), 26-33
The history of mathematics can be used, in a less naive way than anecdotally or as a source of problems, as an epistemological laboratory to explore the development of mathematical knowledge. This requires critical analysis of how historical and conceptual developments are linked—notably, of the notion of 'epistemological obstacles'—through exploring how knowledge is rooted in its socio-cultural context.

Ransom, Peter, 'A historical approach to maximum and minimum problems', *Mathematical gazette* **76**, no. 475 (1992) 85-89
Finding a minimum, before pupils have met calculus, by studying Fermat's method proves to have several advantages: it encourages library use and practice in algebra as well as following through a mathematical argument and introducing calculus.

Ransom, Peter, 'Navigation and surveying: teaching geometry through the use of old instruments', *Histoire et épistémologie dans l'éducation mathématique,* IREM de Montpellier (1995), 227-239
Report on a workshop showing how to use easily made instruments such as sundials and the cross-staff for teaching trigonometry and geometry, with discussion of the benefits to pupils of becoming involved in practical mathematics in this way.

Rice, Adrian, 'A platonic stimulation: doubling the square or why do I teach maths?', *Mathematics in school* **27.4** (1998), 23-24
Interacting with a mathematics class as, in Plato's *Meno*, Socrates did with Meno's slave-boy is an example of how to stimulate students through introducing problems from history.

Robson, Eleanor, 'Counting in cuneiform', *Mathematics in school* **27.4** (1998), 2-9

Resources for teachers and suggestions for classroom activity involving Babylonian mathematics.

Rogers, Leo, 'History of mathematics: resources for teachers', *For the learning of mathematics* **11.2** (1991) 48-52
Bibliographical survey of resources for teachers interested in the history of mathematics or its use in the classroom.

Rogers, Leo, 'Is the historical reconstruction of mathematical knowledge possible?', *Histoire et épistémologie dans l'éducation mathématique,* IREM de Montpellier (1995), 105-114
By studying the history of mathematics we can examine aspects of the processes and the contexts whereby it was developed. A programme of rational reconstruction of its history is relevant to the communication of mathematics at all levels.

Seltman, Muriel, and P E J Seltman, 'Growth processes and formal logic: comments on history and mathematics regarded as combined educational tools', *Int. J. Math. Educ. Sci. Technol* **9** (1978) 15-29
History of mathematics, seen as permeating through the whole of mathematics, can alleviate some of the teaching problems raised by the formal-logical character of mathematical thinking. Knowledge of the circumstances of mathematical discovery is integral to the access to, appreciation of and performance in mathematics.

Siu, Man-Keung, 'The ABCD of using history of mathematics in the (undergraduate) classroom', *BHKMS* **1** (1997), 143-154
Some teaching experience in using history of mathematics in the undergraduate classroom is shared through selected illustrative examples. These can be roughly categorised into four 'levels' as (1) Anecdotes, (2) Broad outline, (3) Content and (4) Development of mathematical ideas.

Siu, Man-Keung, 'Proof and pedagogy in ancient China: examples from Liu Hui's commentary on *Jiu zhang suan shu*', *Educational studies in mathematics 24* (1993), 345-357
The pedagogical implications of aspects of proof in ancient Chinese mathematics.

Speranza, Francesco and Lucia Grugnetti, 'History and epistemology in didactics of mathematics', Nicolina A Malara, Marta Menghini and Maria Reggiani (eds), *Italian research in mathematics education 1988-1995,* CNR 1996, 126-135
The interaction between mathematical didactics, and its history and epistemology, is rich, and in Italy is institutionalised. In the 1900s the relation was the subject of a rich debate; many writings from that period are still useful. The debate resumed in the 1980s, and now involves many groups across Italy.

Steiner, Hans-Georg, 'Two kinds of "elements" and the dialectic between synthetic-deductive and analytical-genetic approaches in mathematics', *For the learning of mathematics* **8.3** (1988) 7-15
The concept of 'elements', and related words such as 'elementary', in authors such as Euclid, Arnauld, Clairaut and Bourbaki, show how fundamental dualisms between synthesis and analysis, justification and development, representation and operation, &c, have proved a vehicle for epistemological and didactical clarifications consisting in a dialectical synthesis of the original contrasts, based on the elaboration of complementarist views.

Stowasser, Roland, 'A textbook chapter from an idea of Pascal', *For the learning of mathematics* **3.2** (1982) 25-30
Raids on the history of mathematics can contribute to concrete mathematics teaching: exemplified by Pascal's paper relating the divisibility of numbers to the sum of their ciphers .

Stowasser, Roland, and Trygve Breiteig, 'An idea from Jakob Bernoulli for the teaching of algebra: a challenge for the interested pupil', *For the learning of mathematics* **4.3** (1990) 30-38
Jacob Bernoulli's *Ars conjectandi* (1713) has a discussion of sums of powers, arising from 'Pascal's triangle'. A passage from John Wallis's *Arithmetica infinitorum* (1655) can be used in class also, for approximations of power sums which prepare the ground for calculus.

Swetz, Frank, 'Mathematical pedagogy: an historical perspective', Eduardo Veloso (ed), *Proceedings of HEM Meeting, Braga, Portugal* 1996, vol II, 121-127
Analysis of didactical trends in historical texts may explore several aspects, notably organisation of material, use of instructional discourse, use of visual aids and of tactile aids. Examples considered include Babylonian and Chinese texts.

Swetz, Frank, 'To know and to teach: mathematical pedagogy from a historical context', *Educational studies in mathematics* **29** (1995), 73-88
The contents of historical mathematical texts usually embody a pedagogy. Several pedagogical techniques are analysed: instructional discourse, logical sequencing of problems and exercises, employment of visual aids. Many of these have historical antecedents.

Tahta, Dick, 'In Calypso's arms', *For the learning of mathematics* **6.1** (1986) 17-23
Reflections on the role of ancient problems and narrative sensibilities in mathematics teaching. The continuing reflexive generation of the account mathematics gives of its own history is too important to be left to historians, or mathematicians: the challenge for teachers is to recast the historical record knowingly.

Thomaidis, Yannis, 'Historical digressions in Greek geometry lessons', *For the learning of mathematics* **11.2** (1991) 37-43
Two historical digressions (straightedge and compasses constructions, and Ptolemaic trigonometry) in a Greek lyceum (16-17 year-olds), in response to teaching problems, provoked discussion and creative activity. This showed how the distant cultural past of a country can influence its contemporary mathematical education.

Tzanakis, Constantinos, 'Reversing the customary deductive teaching of mathematics by using its history: the case of abstract algebraic concepts', *Proc. of the first European Summer University on history and epistemology in mathematics education*, IREM de Montpellier (1995), 271-273.
The customary deductive approach in mathematics teaching can be reversed by using its history as an essential ingredient, here examined in the case of complex number, rotation group, and morphisms of abstract algebraic structures.

Tzanakis, Constantinos, 'Unfolding interrelations between mathematics and physics, in a presentation motivated by history: two examples', *Int. jour. math. educ. sci. technol.* **30** (1999), 103-118
History plays a prominent role in a genetic approach revealing interrelations between physics and mathematics. The two examples are the derivation of Newton's law of gravitation from Kepler's laws, as an application of differential calculus, and the foundations of special relativity as an example of the use of matrix algebra.

Voolich, Erica Dakin, 'Using biographies in the middle school classroom', Sergio Nobre (ed), *Proceedings of HPM Meeting, Blumenau* 1994, 167-172
Various ways of incorporating biographical material about mathematicians in classroom activities include birthday celebrations, construction of 5-minute biographies in which the name is hidden until the end, and role-playing on mock TV chat shows.

Zaslavsky, Claudia, 'World cultures in the mathematics class', *For the learning of mathematics* **11.2** (1991) 32-36
Introducing multicultural, interdisciplinary perspectives into the mathematics curriculum is of particular benefit for the self-esteem and interest of 'minority' students as well as provoking added appreciation and awareness for all students.

11.6 French

Eliane Cousquer

Barbin, Evelyne, 'Sur les relations entre épistémologie, histoire et didactique', *Repères-IREM* **27** (1997)
Reflections on the links and the oppositions between the different trends of thought in France.

Barbin, Evelyne, '*Les éléments de géométrie* de Clairaut, une géométrie problématisée', *Repères-IREM* **4** (1991)
Teaching geometry by presenting problems is a current theme of thought. From this point of view, Clairaut's book is extremely interesting since his aim was to set up the objects and foundations of elementary geometry in order to solve measurement problems.

Bkouche, Rudolf, 'Enseigner la géométrie, pourquoi ?', *Repères-IREM* **1** (1990)
History enlightens three aspects of geometry in a teaching context: the science referring to solid configurations, geometry in its links with the other areas of knowledge, geometry considered as a language and as a representation.

Bkouche, Rudolf, *Autour du théorème de Thalès,* IREM de Lille, 1994
This booklet investigates different proofs given of the proportional segments theorem (known as *Thales' theorem* by the French), from Greek antiquity to the beginning of the 20th century.

Friedelmeyer, Jean-Pierre, et al, 'Les aires, outil heuristique, outil démonstratif', *Repères-IREM* **31** (1998)
20 activities for secondary education presented from a historical angle.

Friedelmeyer, Jean-Pierre, 'L'indispensable histoire des mathématiques', *Repères-IREM* **5** (1991)
The logarithmic function shows how returning to the past allows the teacher to restore meaning to the words and concepts whose connotations have been lost over the years.

Gaud et Guichard, 'Les nombres relatifs, histoire et enseignement', *Repères-IREM* **2** (1991)

John Fauvel, Jan van Maanen (eds.), *History in mathematics education: the ICMI study,* Dordrecht: Kluwer 2000, pp. 404-405

This thorough article provides teachers with numerous extracts of texts that can be used in class, on the history as well as on the difficulties raised through the centuries in the teaching of signed numbers, and on the models that have been used.

Glaeser, 'Epistémologie des relatifs', *Recherches en didactique des mathématiques* **2** (3), 1981
A detailed study of texts about the rule of signs, from Diophantus to contemporary authors, allows us to localize some of the obstacles which block the comprehension of negative numbers. Educational research should examine whether what troubled Euler or d'Alembert still troubles our young students today.

Groupe Math, IREM de Paris 7, 'Mathématiques, approche par des textes historiques', *Repères-IREM* **3** (1991)
Introducing Pythagorean number triples into middle school with the aid of Diophantus' writings, and natural logarithms into senior classes with the aid of Ozanam's work.

Lefort, Xavier, 'L'histoire de la carte de France de Cassini', *Repères-IREM* **14** (1994)
Interdisciplinary work in the history of mathematics meant for the fourth form and involving librarians, French teachers, history teachers and mathematics teachers.

Métin, Frédérick, 'Legendre approxime π en classe de seconde', *Repères-IREM* **29** (1997)
In-class investigation of a writing by Legendre's estimation of π.

Radford, Luis, 'L'invention d'une idée mathématique: la deuxième inconnue en algèbre', *Repères-IREM* **28**, 1997
The invention of a mathematical idea: the second unknown in algebra, invented by the users of abaci in the Middle Ages and the Renaissance.

Stoll, 'Comment l'histoire des mathématiques peut nous dévoiler une approche possible du calcul intégral', *Repères-IREM* **11** (1993)
How the history of mathematics can reveal a possible approach to the integral calculus: in-class use of historical writings.

11.7 German

Harm Jan Smid

Beutelspacher, A, and Weigand, H.-G., 'Die faszinierende Welt der Zahlen', *Mathematik lehren* **87** (1998), 4-8
The history of numbers, considered from a historic-genetic point of view, can shed light on many learning problems of our pupils today.

Damerow, Peter, 'Vorläufige Bemerkungen über das Verhältnis rechendidaktischer Prinzipien zur Frühgeschichte der Arithmetik', *Mathematica Didactica*, **4** (1981), 131-153

John Fauvel, Jan van Maanen (eds.), *History in mathematics education: the ICMI study*, Dordrecht: Kluwer 2000, pp. 405-411

The starting point for all didactical theories of mathematics today is that mathematical knowledge develops from acting with concrete objects (Piaget), but this starting point is only a rule of thumb when developing learning materials for mathematics. Psychology cannot answer how mathematical techniques developed from material artefacts. The early history of counting provides examples from which we understand better how constructive-additive number systems originated from concrete objects.

Folkerts, M, 'Mathematische Historie und Didaktik der Mathematik', *Praxis der Mathematik* **16** (1974), 322-326
Abstracts of the presentations of 12 participants of the symposium "How can historical elements be incorporated in the teaching of elementary mathematics?", held at the Technical University of Berlin in 1974. In general four points of view were advanced: the possibility of a genetic way of teaching, the history of mathematics as a treasury for all kind of examples, to promote the understanding that mathematics is a human activity and to foster the understanding of relations between mathematics and society.

Freudenthal, Hans, 'Soll der Mathematiklehrer etwas von der Geschichte der Mathematik wissen?', *Zentralblatt für Didaktik der Mathematik* (1978), 75-78
English version ('Should a mathematics teacher know something about the history of mathematics?') referred to in §11.5.1.

Gerstberger, H. 'Irrationalzahlen und Flächenaddition: Wiederentdeckung von Anfang an?', *Mathematik Lehren* **19** (1986), 10-14
By using the Greek method of adding of areas and constructing squares of the same area, the pupils of a class succeeded in proving the theorem of Pythagoras in a more meaningful way than usual.

Glickman, L., 'Warum man historische Notizen in den Stochastik-Unterricht einbauen sollte', *Stochastik in der Schule* **9** (1988), 43-46
The elementary theory of probabilities is known for its difficulties for novices. Historical examples, showing the difficulties the pioneers in this field encountered, like those of the chevalier de Méré, can comfort and help to overcome these difficulties.

Haller, R., 'Zur Geschichte der Stochastik', *Mathematik der Didaktik* **16** (1988), 262-277
History of stochastics can be used to introduce the subject; the original problems, stemming from real life, can be treated; the often interesting lives of founders of stochastics can be told; and the origin of some technical terms and symbols still in use.

Hefendehl-Hebeker, L., 'Die negativen Zahlen zwisschen anschaulicher Deutung und gedanklicher Konstruktion - geistige Hindernisse in ihre Geschichte,' *Mathematik Lehren* **35** (1990), 6-12
Negative numbers can be very problematic for pupils. The history of mathematics can help to understand these difficulties. The transition from the idea of numbers as closely connected with physical quantities to numbers as a system as a logical system of symbols is still a difficult step for pupils today.

Jahnke, H N, 'Mathematik historisch verstehen, oder: Haben die alte Griechen quadratische Gleichungen gelöst?', *Mathematik lehren* **47** (1991), 6-12
Two main problems concerning the use of the history of mathematics in teaching are lack of time, and a lack of expertise by teachers in this field. Toeplitz (1927) argued that by the use of the history of mathematics "the dust of ages would disappear and the mathematical ideas

would seem living creatures again." The debate between Weyl and Unguru on Greek 'geometric algebra' shows that this is a problematic point of view. The history of mathematics can help in the acquiring of mathematical techniques, but this is only useful when teaching historical thinking is taken seriously, and takes account of hermeneutic thinking.

Jahnke, H. N., 'Zahlen und Grössen: Historische und Didaktische Bemerkungen', *Mathematische Semesterberichte* **28** (1981), 202-229
The significance of the history of mathematics for its teaching and learning is still problematic: didactics of mathematics may come to have a more independent place and then history of mathematics might have a more important role. By studying the history of mathematics the role of mathematics for general education can become more clear. In the 19th century mathematics underwent a major change, from a science devoted to objects to a science devoted to functional relations between (formal) objects. This should have consequences for mathematics education.

Jahnke, H.N. 'Al-Khwarizmi und Cantor in der Leherbildung', in R. Biehler et al (eds), *Mathematik allgemein bildend unterrichten*, Köln 1995, 114-136
History of mathematics should play a role in teacher training. Studying historical texts confronts the reader with other points of view and can foster deeper understanding of (school)mathematics. Since history of mathematics, when taken seriously, is difficult, it is not so easy to incorporate history of mathematics in teacher training. Possibilities for doing this were demonstrated in an in-service course for teachers, using primary and secondary sources. The article highlights two examples: solving quadratic equations by the method of Al-Khwarizmi, and the theory of transfinite numbers by Cantor.

Jahnke, H.N., 'Historische Reflexion im Unterricht. Das erste Lehrbuch der Differentialrechnung (Bernoulli 1692) in einer elften Klasse', *Mathematica Didactica* **18** (1995), 30-57
The use of original materials in the classroom offers the opportunity of doing history of mathematics in a hermeneutic way: interpreting these texts respecting the historical context and specific character of the text. It is not important whether or not all students arrive at the same interpretation; the exchange of arguments can promote a better understanding of the mathematical content. A series of five lessons in a German school class used a text of Johann Bernoulli on tangents of a parabola. Later some lessons about a part of Bernoulli's text on points of inflexion were given in the same way.

Jahnke, H.N., 'Mathematikgeschichte für Lehrer: Gründe und Beispiele', *Mathematische Semesterberichte* **43** (1996), 21-46
The idea that history of mathematics should play a role in teaching is not new, nor yet widespread. For a more substantial position, history of mathematics needs to be incorporated in teacher education. Examples from such a course, concerning Newton and Cantor, are discussed. It is essential to have a hermeneutic point of view, ie trying to enter into the understandings of people living in another time and culture.

Kaiser, H. and Nöbauer, W., *Geschichte der Mathematik für den Schulunterricht*, Vienna 1984
This book is the result of in-service courses for teachers, its aim to help teachers to incorporate history of mathematics in their lessons. An overview is given of the history of mathematics, and the historical roots of some topics are presented.

Kronfellner, Manfred, *Historische Aspekte im Mathematikunterricht*, Vienna 1998
The expectations, hopes and limits of introducing history in mathematics teaching are discussed. Then a didactical model of history orientated mathematics teaching is developed: two elements play an important role, the 'genetic principle' and the idea of 'constructive realism'. The third part contains suggestions for lessons with historical content, for instance the development of the differential calculus.

Lehmann, J., '25 historische Mathematikaufgaben', *Math. Lehren* **53** (1992), 6-11
Solving historical problems, and the history of mathematics in general, were liked very much by the students of the author. Here 25 problems are presented, from Babylonia, China, Greece and mediaeval Europe, and from German textbooks.

Lehmann, K, 'Einige Gedanke zur Einbeziehung historische Elemente in dem Mathematikunterricht, dargestelt am Beispiel der Klasse 5', *Mathematik in der Schule*, 26 (1988), 377-384; 452-462; 585-592; 758-769
The curriculum (of the former DDR) indicates that historical elements should be used in teaching. In practice this often doesn't work out, due to lack of time, or lack of expertise by the teacher. The history of mathematics could be used to pursue the following aims: the construction of a scientific world picture, character formation by the examples of historical personalities, making mathematics teaching more interesting by historical examples. Examples of historical material are presented that can be used within the framework of a textbook prescribed in the fifth grade of the former DDR. The articles also give details about the way these examples were used in the classroom.

Malle, G, 'Aus der Geschichte lernen', *Mathematik lehren* **75** (1996), 4-75
The history of the development of the concept of function offers an example of how elements from history can be used: not only for teaching concepts themselves, but also for structuring the way the idea of function is introduced and developed in the curriculum.

Noebauer,V, 'Geschichte der Mathematik im Mathematikunterricht', *Der mathematische und naturwissenschaftliche Unterricht*, **34** (1981), 87-91
Although mathematics is very important for our culture, it is hardly seen as important for general education and is highly isolated from other school topics. Using history of mathematics in teaching could help to improve this situation.

Rieche, A, and J. Maier, 'Mathematikunterricht im historischen Museen: Vorschläge und Bausteine', *Mathematik Lehren* **47** (1991), 14-17
At the museum of Roman excavations in Xanten (North Rhine Westfalia), a mathematics teacher and museum curator have developed games and playful activities around the museum objects, such as inscriptions and abaci, by which children can become acquainted with the Roman numeral system and finger counting.

Riehl, G, 'Quadraturen Eine mathematikhistorisch orientierte Einf,hrung in die Integralrechnung', Mathematik in der Schule, 36 (1998), 347-361; 419-430
A short course on the introduction of integration is presented, based on ideas due to O Toeplitz, about a genetic way of learning new concepts, that is to say taking into account the historical development of a concept.

Rödler, K, 'Die Geschichte der Zahlen und des Rechnens', *Mathematik lehren* **87** (1998), 9-14
During the first years of the primary school, many children have difficulties with the understanding of the decimal system and the place-value system. A project about the

development of various number systems, over four months, had lessons and worksheets on topics as body counting, counting with pebbles and scratches, Egyptian and Roman number systems, the abacus and the introduction of the Hindu-Arabic number system.

Röttel, K, 'Aus der Arbeit der römischen Feldmesser', *Praxis der Mathematik* **23** (1981), 210-215
The methods of the Roman surveyors offer opportunities for geometry teachers. The lay-out of camps, tunnel-surveying and distance between places, and calculations of areas and volumes can motivate pupils of several grades when learning geometry.

Schubring, Gert, 'Historische Begriffsentwicklung und Lernprozess aus der Sicht neuerer mathematikdidaktischer Konzeptionen (Fehler, "Obstacles", Transposition)', *ZDM* (1988), 138-148
The classical justification for using history of mathematics in education, as motivation, is unsatisfactory. There exists a much more fundamental reason. According to modern, subjectivist constructivist views, mathematics has not an unique position: its claim to be a objective fault-free science is not justified. There is a connection between students' mistakes, cognitive obstacles, and problems in the historical development of mathematics. There is not only the problem of transition of mathematics as a scientific object into school mathematics, but also the teaching of mathematics has its influences on the development of mathematics itself.

Schubring, Gert, *Das genetische Prinzip in der Mathematik-Didaktik,* Stuttgart 1978
There are several aspects of the genetic principle, for instance as psychological-genetic or historical-genetic. The nature of scientific knowledge and the social meaning of knowledge play an important role. With extensive case studies from the history of mathematics education.

Scriba, C J, 'Die Rolle der Geschichte der Mathematik in der Ausbildung von Schüler und Lehrer', *Jahresbericht der Deutsche Mathematiker Verein,* **85** (1983), 113- 128
In the first part three axioms are defended: mathematics without history is impossible; mathematics should be taught within a scientific, cultural and social framework; mathematics as a cultural phenomenon cannot be understood without historical considerations. In the second part an outline is given of programmes of history of mathematics in universities and teacher training institutes in several countries. These courses serve as examples how history of mathematics can be integrated in the training of mathematicians and teachers.

Scriba, C J, 'Die Behandlung mathematikgeschichtliche Probleme im Unterricht', *Beiträge zum Mathematikunterricht* (1974), 43-54
Four arguments and ways for using history of mathematics in teaching: to raise interest in mathematics as a form of human activity; to use the historical growth of mathematics for a genetic way of teaching; the use of the history of mathematics as a treasury of examples in teaching; and to explain the interdependence between mathematics and society. Examples, mainly from the theory of series, are given of historical topics that could be used in teaching.

Stowasser, R J K, 'Die Idee der Rekursion und der Isomorphie', *Der Mathematiklehrer* **2** (1983), 2-10
Two classical problems from the history of mathematics for mathematically talented pupils: one by Jacob Steiner concerning the number of regions in which the plane is divided by n

lines; the other by Euler, about the number of ways to put n letters in n addressed envelopes so that all letters are addressed wrongly.

Stowasser, R.J.K., 'Streifzüge durch die Geschichte: Eine Idee von Pascal für das Schulbuch', *Der Mathematiklehrer* 2 (1981), 36-39
 On the basis of an idea from Pascal (taken from his *De numeris multiplicibus ex sola characterum mumericum additione agnoscendi*) children of eleven years old worked on problems concerning remainders with the division of large numbers, introduced by means of a one-handed clock.

Strecker, C. 'Eratosthenes von Kyrene, Columbus von Genua und der Erdumfang: eine fragwürdige Geschichte', *Mathematik in der Schule* 36 (1998), 106-114
History of mathematics can be used to promote critical thinking. For example, the well known story of Eratosthenes measuring the circumference of the earth can give raise to critical doubts over whether this can have happened in the way the story tells us. Columbus's misusing the then known facts about the map of the earth provides an amusing example of how making a mathematical mess can influence world history!

Toeplitz, O., *Die Entwicklung der Infinitesimalrechnung. Eine Einleitung in die Infinitesimalrechnung nach der genetischen Methode*, Berlin 1949
Otto Toeplitz, founding father of the genetic method of teaching, here introduced calculus along these lines: not to present a history of calculus, but to shed light on the origin and genesis of decisive problems and ideas in its development.

Toeplitz.O. , 'Das Problem der Universitätsvorlesungen über Infinitesimalrechnung und ihre Abgrenzung gegenüber der Infinitesimalrechnung an den höheren Schulen', *Jahresbericht der Deutsche Mathematiker Verein* 36 (1927), 88-100
In this classical text the idea of the genetic method is introduced. It is demonstrated in the case of teaching calculus to first year university students. The genetic method, that is going back to the roots of the concepts, can offer a way beyond the dilemma of rigour versus intuition in teaching. It can be applied in a direct way, which implies the use of historic material. It can also be used in an indirect way, which means that historical analysis can help to find didactical diagnosis and therapies for learning difficulties.

Waerden, B.L. van der, 'Die 'genetische Methode' und der Mittelwertsatz der Differentialrechnung', *Praxis der Mathematik* 22 (1980), 52-54
In 1926 Otto Toeplitz advocated the use of the genetic method. Applying this, we see that the mean value theorem did not play an important role in calculus until the middle of the 19th century. Newton, the Bernoullis, Euler etc. could do without it. Using the mean value theorem for proving other theorems that already appear quite obvious without proof could be restricted to the training of future mathematicians. There is no reason to make this theorem a cornerstone in the teaching of calculus for future chemists, physicists, etc.

Windmann, B., 'Methoden des Geschichtsunterrichts im Mahtematikunterricht', *Mathematik Lehren* 19 (1986), 24-31
The connection between history and mathematics has been discussed for more than a century, without much results. Knowledge of history of mathematics can help teacher and pupil to gain insight in the reasons why some topics are taught. It is doubtful if the often heard argument of the 'genetical principle' really is true. More important is that history shows that mathematics is a living subject, created by thinking people.

Zerger, H, 'Historische Aspekte bei der Logarithmus und Exponentalfunktion', *Mathematik Lehren* **19** (1986), 18-23
Including historical elements could enrich teaching of the logarithm function, in two ways: historical side steps, for instance about Bürgi's log table or Bernoulli's problem on the calculation of compound interest, or a complete historical orientated treatment of the logarithmic function.

Zimmerman, B., 'Gudrun auf den Spuren von Gauss und Descartes', *Mathematik lehren* **47** (1991), 30-41
Using history to support the mathematical development of gifted children, the starting point was the counting of squares in a grid, leading to the summation of square numbers. By using an analogy of the well-known summation of the natural numbers by Gauss they tried to solve the problem, which proved to be hard. More historical examples could be used; historical texts, when well chosen, can be made accessible for schoolchildren, motivating them by showing them how mathematics has grown.

11.8 Greek

Yannis Thomaidis and Costas Tzanakis

This bibliography is of papers written in Greek concerning the relation between history of mathematics and mathematics teaching, in chronological order. (Note that the *Euclides* cited here is the Greek journal of that name, published in Athens since 1982, and not the Dutch journal published in Groningen since 1924.)

Lampiris, K., 'Historical remarks in the teaching of mathematics', *The pedagogue* **12-13** (1922), 181-186
The first paper in Greek literature which highlights the positive role of using history in the mathematics teaching process. The author gives many examples that go beyond a mere quotation of dates or biographical information.

Thomaidis, Y.,'The axiomatic method of teaching and the historical reality', *(Greek) Mathematical review* **26** (1984), 2-13
This paper highlights the dichotomy between exposing mathematics axiomatically and discovering mathematics as happened in history. Trigonometry and complex numbers are used as examples to support the argument.

Thomaidis, Y., 'Teaching concepts of the calculus, guided by its historical development', *Euclides* γ **9** (1985), 8-22
This paper explores further the issues raised in (Thomaidis 1984). The teaching of calculus as an axiomatic theory, in Greek upper secondary education, is contrasted with the historical roots of the subject.

Thomaidis, Y., 'Origins and applications of theory in the teaching of mathematics (the case of logarithms)', *Euclides* γ **13** (1986), 1-30

John Fauvel, Jan van Maanen (eds.), *History in mathematics education: the ICMI study*, Dordrecht: Kluwer 2000, pp. 411-414

The theory and applications of logarithms, as they appear in modern Greek textbooks of elementary algebra, are contrasted with the historical development of logarithmic concepts.

Poulos, A., 'The history of mathematics and its importance for secondary school in-service teachers' training', *Contemporary education* **29** (1986), 35-42
The author presents a variety of cognitive, scientific, educational, didactic, cultural and philosophical values of the history of mathematics, which are closely related to the profession of teaching mathematics.

Kastanis, N., 'A case of historical confusion in school geometry textbooks', *Euclides* γ **14** (1987), 71-73
The article calls into question the use of the term 'theorem of Thales' for the theorem concerning the proportional segments formed by parallels on straight lines.

Kastanis, N., 'A frequently encountered mistake in the historiography of mathematics that is incorporated in high school mathematics textbooks', *Euclides* γ **14** (1987), 80-82
The article points out that most textbooks incorrectly use small letters of the Greek alphabet, for representing numbers in ancient Greek mathematical works written before the 3rd century BC.

Thomaidis, Y. & N. Kastanis, 'A historical study of the relation between history and didactics of mathematics', *Euclides* γ **16** (1987), 61-92
This paper examines the development of the relations between history and pedagogy of mathematics from the early 19th century to the present time, both in the international and the Greek educational systems.

Roussopoulos, G., 'History and philosophy of mathematics: their role in teaching mathematics', *Proc. of the 4th Greek Conference on Mathematics Education*, Athens: Greek Mathematical Society (1987), 369-379
The author argues that history and philosophy of mathematics are basic components in the context of a heuristic methodology of teaching mathematics.

Patmanidis, A., 'Revealing the role of the history of mathematics in teaching mathematics', *Diastasi* **3-4** (1988), 102-106
Stemming from pupils' reactions to a historical note on Euclid's 5th postulate in a geometry textbook, this article traces the main steps in the development of non-Euclidean geometries and argues in support of teacher's historical knowledge.

Kastanis, N., 'The concept of space before and after non-Euclidean geometries: an approach for didactic reasons', *Cahiers en didactique des mathématiques* **1** (1988), 15-17
A short account of the author's presentation in the HPM session at ICME-6, Budapest, July 27-August 3, 1988.

Kastanis, N., 'An example of confusion concerning history of mathematics as it appears in the high school textbook', *Euclides* γ **21** (1989), 23-26
The term 'gnomon of Anaximander' is used incorrectly in a Greek geometry textbook.

Thomaidis, Y., N. Kastanis & T. Tokmakidis, 'Relations between history and didactics of mathematics', *Euclides* γ **23** (1990), 11-17

A detailed account of the activities on the theme 'relations between history and didactics of mathematics', at ICME-6, Budapest, July 27-August 3, 1988

Thomaidis, Y., 'Historical digressions in geometry high school course', *Euclides* γ **25** (1990), 27-41
This paper presents an experimental lesson of geometry, motivated by the rich historical background of some, otherwise routine exercises from a Greek geometry textbook. [English version in *For the learning of mathematics* **11** (2) (1991), 37-43]

Tzanakis, C., 'Is it possible to teach abstract algebraic structures in high school? A historical approach', *Euclides* γ **28** (1991), 24-34
On the basis of the historical development of the concepts of a group, ring, field and vector space it is argued that these concepts cannot be understood in their abstract form by high school students. Students would do better by acquaintance with mathematically important specific examples as happened historically.

Tzanakis, C., 'A genetic approach in teaching mathematics and physics', *Proc. of the conference on the didactic use of the history of sciences*, Thessaloniki: Greek Society for the History of Science and Technology 1991, 65-90
A genetic approach is illustrated for mathematics, by describing how concrete examples of algebraic structures may be used at the high school level, to prepare for their subsequent abstract presentation at university level; and for physics, by presenting a teaching sequence for basic concepts of undergraduate quantum mechanics.

Thomaidis, Y., 'Historical problems in mathematics teaching: the case of negative numbers', *Proc. of the conference on the didactic use of the history of sciences*, Thessaloniki: Greek Society for the History of Science and Technology 1991, 127-137
The author argues that the knowledge of historical problems can be beneficial in the planning of didactic situations for introducing mathematical concepts. He offers, as an example, a new interpretation of the history of negative numbers in the early 17th century [English version in *Science & Education* **2** (1993), 69-86].

Christianidis, Y., 'Comments on two historical notes appearing in the high school mathematics textbooks', *Euclides* γ **43** (1995), 1-10
The author, a historian of ancient Greek mathematics, criticises historical notes in two geometry textbooks from the point of view of historiographical accuracy.

Tzanakis, C., 'Relating the teaching of mathematics and physics on the basis of their historical development: a genetic approach', *Proc. of the 1st Greek conference on mathematics in education and society*, T. Exarhakos (ed.), University of Athens (1996), 349-361
Mathematics and physics have always had an intimate connection, which appears in three different ways. This fact should not be ignored in their teaching. Their interconnection in teaching may lead to a deeper understanding of both disciplines.

Thomaidis, Y., 'Is historical parallelism possible in teaching and learning mathematical concepts? The case of the ordering on the number line', *Diastasi (Section on research on the didactics of mathematics)* **2** (1997), 3-38
The findings of an historical study are associated with those of an empirical one with 16 year-old pupils, in order that the controversial relation between the historical evolution of

mathematical concepts and their learning at school be critically discussed. The order-relation and the algebra of inequalities are presented as examples, suggesting a clear distinction between the two domains.

Tsimpourakis, D., 'On the historical notes included in the mathematics textbooks', *Euclides* β **31** (1) (1997), 1-10
The author criticises some historical notes contained in Greek geometry textbooks, from the point of view of historiographic accuracy.

Tzanakis, C.,'Conditions and presuppositions of a constructive role for history of mathematics in understanding and teaching mathematics', *Diastasi* **3** (1998), 58-86
The importance of the history of mathematics in teaching and understanding mathematics is examined, with emphasis on understanding the significance of reasoning by induction and by analogy, on inspiring teaching and on interconnecting the teaching of mathematics and physics. §7.3.2 is based on this article.

11.9 Italian

Fulvia Furinghetti

This bibliography is a supplement to that appearing in a survey article published in 1996, Francesco Spenanza and Lucia Grugnetti, 'History and epistemology in didactics of mathematics', in Nicolina A. Malara, Marta Menghini and Maria Reggiani (eds), *Italian research in mathematics education 1988-1995*, CNR 1996, 126-135. That paper contains a list of papers in the area by Italian authors from 1988 up to 1995.

Bagni, T. G., 'Ma il passaggio non è il risultato. L'introduzione dei numeri immaginari nella scuola superiore', *La matematica e la sua didattica*, **2** (1997), 187-201
The author introduced complex numbers to his high school class through history.

Barozzi, G. C., 'Un esempio di utilizzo del sistema Cabri-Géomètre', *L'insegnamento della matematica e delle scienze integrate,* **17A** (1994), 460-466
Old problems are solved in an alternative way through the software Cabri-géomètre.

Bianchini, S. & Velardi, R., 'Dalla conoscenza dei contenuti alla rielaborazione e sistemazione della matematica: Leonardo Pisano e Maria Gaetana Agnesi', *Scuola e didattica*, **14** (1990)
Authors of the past help to illuminate the passage of mathematics from its birth to its systematisation. The authors specifically notice the work of the Italian woman mathematician Maria Agnesi.

Bottino, R. M., Cutugno, P. & Furinghetti, F.: 1997, 'Progettazione e utilizzo di un sistema ipermediale per la storia della matematica', *L'insegnamento della matematica e delle scienze integrate,* **20A-B** (1997), 839-854

John Fauvel, Jan van Maanen (eds.), *History in mathematics education: the ICMI study*, Dordrecht: Kluwer 2000, pp. 414-416

An activity carried out with university students consisted in projecting and realising in hypermedia the three classical problems.

Brigaglia, A., 'Alcune considerazioni sulle finalità didattiche dell'insegnamento della geometria euclidea', *Archimede,* **48**(1996), 170-184.
Examples taken from Newton, Descartes, Viète show that Euclidean geometry can be an interesting field in which to develop the objectives suggested by official curricula, including links with the use of dynamic and computational software and programming.

De Mattè, A., 'Storia. Pseudostoria. Concezioni', *L'insegnamento della matematica e delle scienze integrate,* **17B** (1994), 269-281
The author investigates beliefs held by secondary students (aged 11-13) about the genesis of mathematical ideas (concepts and processes) and their history. These students have experience in the study of history (events and civilisations) and of mathematics, but not explicit preparation in history of mathematics. This study may serve as a background for studies on the role of history in mathematics teaching.

Dupont, P., 'Storia e didattica della definizione classica della probabilità', *L'educazione matematica,* 7 Suppl. (1986), 1-27
Some important moments of history of probability can provide teachers with hints when faced with the epistemological obstacles for today's students of the classical definition.

Freguglia, P., 'Momenti nella storia dell'algebra', in: *L'algebra tra tradizione e rinnovamento,* Quaderni Ministero Pubblica Istruzione 7 (1994), 131-149.
A brief survey of the theory of algebraic equations developed in sixteenth century before Viète. The links between algorithmic-arithmetic techniques and geometrical questions are studied through the works of the Italian mathematicians Bombelli, Cardano, Ferrari, and Tartaglia. This article is a chapter in a book of a series edited by the Ministry of Education as basic reference in annual training courses for teachers. In each course a different school subject is presented from different points of views and for each subject a chapter is always dedicated to historical issues.

Galuzzi, M. & D. Rovelli: 'Storia della matematica e didattica: qualche osservazione', in: *L'insegnamento della geometria,* Quaderni Ministero Pubbica Istruzione **19/2** (1997), 70-110.
This is the historical chapter in the book for teacher training in geometry. It begins discussing the links of history and mathematics teaching, and afterwards some topics from Euclid, Descartes and Newton.

Grugnetti, L., 'Storia ed epistemologia dell'analisi', in: *Didattica dell'analisi,* Quaderni Ministero Pubblica Istruzione, 24 (1998), 70-105.
Some points of history of mathematics relevant to the history of calculus. The aim is to provide teachers with materials to be discussed in class, in order to confront the epistemological obstacles they encounter in learning and teaching calculus.

Menghini, M., 'Some remarks on the didactic use of the history of mathematics', in: L. Bazzini & H.-G. Steiner (eds.) *Proceedings of the first Italian - German symposium on didactics of mathematics* (1989), 51-58.
The link between art and mathematics is illustrated through the study of the use of conics in Roman Baroque architecture. Other aspects of the connections between mathematics and art are discussed through the work of the Dutch painter M. C. Escher, whose well-known

painting inspired by Poincaré's model of non-Euclidean geometry was used in the classroom to introduce pupils to problems of non-Euclidean geometries and to motivate a discussion on the nature of space as it developed in 19th century.

Palladino, F., 'Planimetri e integrafi', *L'insegnamento della matematica e delle scienze integrate*, **18B** (1995), 51-79

Palladino, F., 'Una rassegna di antichi strumenti di misura per l'insegnamento e le applicazioni della matematica', *L'insegnamento della matematica e delle scienze integrate*, **19A-B** (1996), 594-608.
Some important instruments and models such as integraphs, planimeters, special compasses, were in fashion a century ago. Their use is connected to a particular vision of the teaching and the nature of mathematical knowledge.

Speranza F., 'Perché l'epistemologia e la storia nella formazione degli insegnanti?', *Università e scuola (Periodico Concird)*, **1/R** (1996), 70-72
The author claims the importance of epistemology and history in education of prospective mathematics teachers. This opinion is supported by the conviction that the epistemological reflection intended as a reflection on the construction of knowledge is part of the pedagogical reflection.

11.10 Collections of articles (special issues)

John Fauvel

This section contains bibliographical details of some of the collections which include articles about relations between the history of mathematics and the teaching and learning of mathematics. Such collections have been prepared both as special issues of journals (§11.10.1) and as books (§11.10.2). The annotations in this section do not describe the collections but are confined to listing authors whose papers are in the collections. Some of these papers are annotated above. The listing of names is not always exhaustive: most of the works cited form even richer collections than the short listing of authors implies, often containing further papers about the history of mathematics and the history of mathematics education. Simply looking over these journal and book details gives a strong impression of the remarkable amount of activity in this area over recent years, and it may be hoped and expected that even more activity will take place in the years to come.

11.10.1 Journals (special issues)

L'insegnamento della matematica e delle scienze integrate **14** (1991), no 11/12.
This issue is dedicated to the history of sciences as a help for didactics of sciences. It contains articles of Dupont, Colombo Bozzolo, Balzarini, Sibilla, Saladin, Manara, Brunet

John Fauvel, Jan van Maanen (eds.), *History in mathematics education: the ICMI study*, Dordrecht: Kluwer 2000, pp. 416-418

For the learning of mathematics **11** no. 2 (1991), edited by John Fauvel
Fauvel, Ransom, Perkins, Barbin, Arcavi, Brown, Fowler, Gardner, Ofir, Führer, Zaslavsky, Thomaidis, van Maanen, Rogers

Mathematical gazette **76,** no. 475 (1992), edited by Nick MacKinnon
Crilly, Führer, Furinghetti, Gardiner, Gardiner, Hadley & Singmaster, Heiede, Kool, van Maanen, MacKinnon, MacKinnon, Ofir & Arcavi, Pritchard, Ransom, Smith

Mathematics in school **26** no. 3 (1997), edited by John Bradshaw and Lesley Jones
Sawyer, Joseph, Oliver, French, Downes, Rothman, Taverner

Mathematics in school **27** no. 4 (1998), edited by John Earle
Robson, Ponza, Maher, Barrow-Green, Rice, Ernest, Burn, Weeks, Eagle, Van Brummelen, Fauvel, Furinghetti & Somaglia, van Maanen, Burn

mathematik lehren **19** (December 1986), edited by Lutz Führer, special issue entitled 'Geschichte-Geschichten'
Führer, Kretzschmar, Windmann

mathematik lehren **47** (August 1991), edited by Jürgen Schoenebeck, special issue entitled 'Historische Quellen für den Mathematikunterricht'
Jahnke, Rieche, Rieche & Maier, Zimmermann

mathematik lehren **91** (December 1998), edited by Hans Niels Jahnke, special issue entitled 'Mathematik historisch verstehen'
Biermann, Führer, Gerber, Jahnke, Kaske, van Maanen

11.10.2 Books

Barbin, Evelyne and Régine Douady (eds), *Teaching mathematics: the relationship between knowledge, curriculum and practice,* Topiques éditions 1996
Artigue, Barbin, Chrétien & Gaud, Daniel, Douady, Duperret, Friedelmeyer, Henry, Kuntz, Le Goff, Nouet

Boyé, Anne, François Héaulme, Xavier Lefort (eds), *Contribution à une approche historique de l'enseignment des mathématiques,* Nantes: IREM des Pays de la Loire, 1999
Barbin & Guitard, Bennaceur, Bernard, Bernard, Bernard, Caetano, Collaudin, Cousquer, Delattre, Dorier, Friedelmeyer, Guichard & Gaud, Hauchart, Lakoma, Lamandé, Le Corre, Lefort, Michel-Pajus, Plane, Proust, Provost, Stoll, Stoll, Vassard, Vilain

Calinger, Ronald (ed), *Vita mathematica: historical research and integration with teaching,* Washington: Mathematical Association of America 1996
Aspray & al, Barbin, Bero, Calinger, Cooke, D'Ambrosio, Dadic, Fauvel, Flashman, Grabiner, Heiede, Hensel, Hitchcock, Hughes, Høyrup, Jahnke, Jozeau, Katz, Kidwell, Kleiner, Knorr, Kronfellner, Laubenbacher & Pengelley, Lumpkin, Michalowicz, Rickey, Rowe, Siu, Swetz, Tattersall

Fauvel, John (ed), *History in the mathematics classroom: the IREM papers,* Leicester: The Mathematical Association 1990
Bühler, Friedelmeyer, Hallez, Horain, Jozeau, Lefort, Plane, Plane, Sip

Inter-IREM Commission, *History of mathematics histories of problems*, Paris: Ellipses 1997
Barbin & Itard, Belet & Belet, Bessot & Le Goff, Bkouche & Delattre, Chabert, Chabert, Crubellier & Sip, Daumas & Guillemot, Delattre & Bkouche, Friedelmeyer & Volkert, Friedelmeyer, Grégoire, Guillemot & Daumas, Jaboeuf, Plane

Inter-IREM Commission, *Images, imaginaires, imaginations: une perspective historique pour l'introduction des nombres complexes*, Paris: Ellipses 1998
Boyé, Cléro, Durand-Richard, Friedelmeyer, Friedelmeyer, Friedelmeyer, Hallez & Kouteynikoff, Hamon, Thirion., Verley

IREM de Besançon, *Contribution à une approche historique de l'enseignement des mathématiques*, Besançon: IREM de Franche-Comté, 1996
Bebbouchi, Bernard, Boyé & Borowczyk, Boyé & Lefort, Cousquer, Daumas, Delattre, Dorier, Ferreol, Friedlemeyer, Guichard, Keller, Lanier, Lefebvre, Lefebvre, LeGoff, Martin, Métin, Michel-Pajus, Nicolle, Nordon, Plane, Provost, Verdier, Volkert, Waldegg, Zerner

Jahnke, Hans Niels, Norbert Knoche & Michael Otte, *History of mathematics and education: ideas and experiences*, Göttingen: Vandenhoek & Ruprecht 1996
Aspray, Behr, Bartolini Bussi & Pergola, Chemla, Dauben, Fraser, Jahnke, Menghini, Otte, Panza, Scholz, Sierpinska, Struve, Toepell

Katz, Victor J. (ed), *Using history to teach mathematics: an international perspective,* Washington: Mathematical Association of America 2000
Arcavi & Bruckheimer; Barbin; Barnett; Carvalho e Silva, Duarte & Queiro; D'Ambrosio; Dorier; Furinghetti; Gellert; Giacardi; Grugnetti; Heiede; Heine; Hitchcock; Horng; Isaacs, Ram & Richards; Michel-Pajus; Moreno-Armella & Waldegg; Radford & Guerette; Robson; Siu; Swetz; Tzanakis; Wilson; Winicki

Lalande, Françoise, François Jaboeuf, and Yvon Nouazé (eds), *Histoire et épistémologie dans l'education mathematique*, IREM de Montpellier 1995
Amaro, Bero, Van Brummelen, Garcia, Kool, Ransom, Rogers, Tzanakis, Winicki

Nobre, Sergio (ed), *Proceedings of HPM meeting, Blumenau, Brazil* 1994
Acevedo, Arboleda, Carvalho e Silva, D'Ambrosio, Figoli, Frankenstein, Gerdes, Jones, Kleiner, Morales, Nobre, Pereira da Silva, Sánchez, Saraiva, Sebastiani, Vilela, Visokolskis, Vitti, Voolich

Swetz, Frank, John Fauvel, Otto Bekken, Bengt Johansson and Victor Katz, (eds), *Learn from the masters!*, Washington: Mathematical Association of America 1995
Aiton, Avital, Bekken, Burton & Van Osdol, Fauvel, Gardiner, Helfgott, Jones, Katz, Katz, Kleiner, Lehmann, van Maanen, Mejlbo, Reich, Rickey, Shenitzer, Siu, Siu, Siu, Swetz, Swetz, Swetz

Veloso, Eduardo (ed), *Historía e Educação Matemática,* Braga/Lisbon 1996, 2 vols
Abdounur, Barbin, Barnett, Bebbouchi, Bernard, Bertoni, Bkouche, Brito, Chacko, Charbommeau, Cooper, Cousquer, D'Ambrosio, D'Ambrosio, Dorier, Duarte, Fauvel, Fauvel, Ferreira, Fiorentoni, FitzSimons, Fossa, Furinghetti, Gellert, Gerdes, Giardinetto, Rogers, Ginestier, Gorgan, Guichard, Guichard, Hariki, Heiede, Hitchcock, Horng, Isaacs, Johan, Katz, Katz, Krajcsik, Lefort, van Maanen, Martin, Medes, Miguel, Nobre, Paulo, Radford, Radford, Rajagopal, Robson, Roero, Rosendo, Sheath, Silva, Sirera, Siu, Swetz, Swetz, Testa, Tzanakis, Vilar, Waldegg, Wilson, Zaniratto

NOTES ON CONTRIBUTORS

Abraham Arcavi is Senior Scientist at the Department of Science Teaching, Weizmann Institute of Science, Israel. His Ph.D. was on the design, implementation and evaluation of teacher courses on the historical evolution of several mathematical topics. He has written high school textbooks, and published articles on mathematical education, mathematical cognition, and about ways of integrating history in the teaching and learning of mathematics.

Giorgio T. Bagni has a degree in mathematics from the University of Padova, Italy, and a post-graduate course in numerical analysis (Rende, Italy). He is member of the Nucleo di Ricerca in Didattica della Matematica of Bologna, and professore a contratto of history of mathematics and teacher in the post-graduate course in didactics of mathematics, University of Bologna. He has written and (co-)edited fifteen books, contributed to several international study congresses, and given lectures for university students and teachers in a number of countries. Since 1999 he has been president of Ateneo di Treviso.

Evelyne Barbin is maître des conférences in epistemology and history of science of the IUFM (Institut Universitaire de Formation des Maîtres) of the Academy of Creteil. Her researches concern mainly mathematical proof in history, and history of mathematics in the 17th century. She is director of the inter-IREM National Commission on Epistemology and History of Mathematics, a body which has worked for twenty years on integrating history into the teaching of mathematics.

June Barrow-Green is a research fellow in the history of mathematics at the British Open University. Her main area of research is the history of mathematics in the nineteenth century, and she is the author of *Poincaré and the three body problem* (1997). She is also concerned with the use of databases and the use of the World Wide Web as research tools in the history of mathematics.

Maria G. (Mariolina) Bartolini Bussi (*1948) is professor of elementary mathematics from an advanced standpoint in the University of Modena and Reggio Emilia, Italy. She is director of the laboratory of mathematics at the Science Museum of her University. She has published several research studies on the use of history in the mathematics classroom at all age levels as from primary school, and co-authored a CD-rom with the description of more than 150 mathematical instruments: *<http://museo.unimo.it/theatrum/>*.

Otto B. Bekken is associate professor of mathematics at Agder University College, Kristiansand, Norway. He has authored/co-edited several books for teaching on themes from original historical sources, including *Una historia breve del algebra* (Lima 1983), *Learn from the masters!* (Washington: MAA1995), and *Equacoes de Ahmes até Abel* (Rio de Janeiro 1994).

Paolo Boero is professor of mathematics education at Genoa University. He graduated in mathematics, then he engaged in mathematics education research. Since 1976 he has led a research group (at present, 32 teacher-researchers and 4 university researchers) in mathematics education at Genoa University. His main scientific interests concern historical, epistemological, cognitive and didactical aspects of theoretical knowledge in school mathematics.

Glen Van Brummelen is professor of mathematics at Bennington College, Vermont, and president of the Canadian Society for History and Philosophy of Mathematics. His research interests focus on geometry and astronomy in ancient Greece and medieval Islam. He is the author of a collection of laboratory projects in introductory calculus featuring various historical episodes.

Jaime Carvalho e Silva is associate professor of mathematics in the University of Coimbra, Portugal, researching into partial differential equations; he has strong interest in history of mathematics and mathematics education. He has coordinated mathematics programmes for secondary education in Portugal since 1995, and is the author of the most visited mathematics internet site in Portuguese http://go.to/nonius

Carlos Correia de Sá (*1953) is auxiliary professor in the department of pure mathematics of the University of Oporto, Portugal. His PhD, from the University of Birmingham, UK, was on Poncelet and the creation of complex projective geometry by synthetic methods. His current interests include the history of mathematics in ancient Greece and in seventeenth century Europe, as well as the integration of history and epistemology of mathematics in teacher training.

Eliane Cousquer is lecturer in mathematics at the Science and Technology University of Lille, France. She is the head of LAMIA laboratory for production of multimedia pedagogical tools in the IUFM. Her book *La fabuleuse histoire des nombres* appeared in 1998.

Coralie Daniel is from the University of Otago, New Zealand. Her work and research are concerned with recognising different abilities and aptitudes and with developing strategies to nurture the creative and functional aspects of these in social, educational and workplace situations.

Jean-Luc Dorier is university professor in a teacher training institute (IUFM) in Lyon, France, and head of a research team in didactics of mathematics in Grenoble. His main research subjects deal with the teaching of linear algebra at university level and the connection between history and didactics of mathematics. He edited *L'Enseignement de l'algèbre linéaire en question* (1997), to be published in English (with the title *Teaching and learning linear algebra*) by Kluwer.

Florence Fasanelli is director of the College-University Resource Institute in Washington, DC, USA, through which she establishes mathematical education programmes for teachers and students at tribal colleges. She was the founding chair of the Americas section of HPM in 1984 and was chair of HPM from 1988 to 1992 arranging satellite meetings in Firenze and Toronto. Her research interest is in the relations between the history of art and the history of mathematics.

John Fauvel is senior lecturer in mathematics at the Open University, UK, former president of the British Society for the History of Mathematics, and during 1992-1996 was chair of HPM. He has co-edited several books, including *Darwin to Einstein: historical studies on science and belief* (1980), *The history of mathematics: a reader* (1987), *Let Newton be!* (1988), and *Oxford figures: 800 years of the mathematical sciences* (1999).

Gail FitzSimons has been a teacher of adult and vocational mathematics for almost two decades. She is currently completing a PhD, under the supervision of Alan Bishop, providing a critique of this sector of education in Australia. She is the author and co-author of numerous publications in this field, and the chief organiser for adult education and lifelong learning ICME working groups in 1996 and 2000.

Chun-Ip Fung is lecturer in the department of mathematics at the Hong Kong Institute of Education, Hong Kong, China. He is currently president of the Hong Kong Association for Mathematics Education.

Fulvia Furinghetti is associate professor of elementary mathematics from an advanced standpoint in the University of Genoa, Italy. She is the co-ordinator of a group of mathematics teachers working in the field of mathematics education. Her educational research concerns the integration of history in mathematics teaching, approaches to proof, mathematical beliefs, teacher education. In her historical research she studies the history of elementary mathematics journals.

Hélène Gispert est maitre de conferences en histoire des sciences dans un institut de formation des maitres et a l'universite des sciences d'Orsay, France. Elle travaille sur les mathematiques de la fin du XIXe siecle et leurs cadres institutionnels ainsi que sur l'histoire de l'enseignement primaire, secondaire et superieur des mathematiques.

Lucia Grugnetti is associate professor of foundations of mathematics at the University of Parma, researcher in mathematics education, history of mathematics, history of mathematics in mathematics education, editor of the bilingual journal *L'educazione matematica*, scientific counsellor of a research group in mathematics education in Parma, an international organizer of RMT (Rally Mathematique Transalpin), member of the committee on mathematics education of EMS (European Mathematical Society). She was president of CIEAEM (*Commission internationale pour l' etude et l'amelioration de l'enseignement mathematique*) from 1993 to 1997.

Miguel de Guzmán is professor in the facultad de matemáticas of the Universidad Complutense de Madrid. His mathematical studies were mainly done at the University of Chicago, where he obtained his Ph.D. in mathematics under the guidance of Alberto P. Calderón. He also studied philosophy in Munich. His main mathematical interest has been centered around several areas of harmonic analysis. He has also been involved in several aspects of mathematical education and has acted as president of the ICMI for two periods (1991-1998).

Torkil Heiede (*1931) was educated at Copenhagen University and was from 1963 a senior lecturer in mathematics at the Royal Danish School of Educational Studies. He retired in 1997 but is still active speaking and writing on diverse mathematical subjects and their history, and on the place of history in mathematics education at all levels. He is a co-editor of *Normat*, a member of the Con Amore Problem Group, and secretary of the Danish Society for the History of the Exact Sciences.

Bernard R. Hodgson has been involved for more than 25 years in the mathematical education of primary and secondary teachers at Université Laval, Québec. His mathematical research work is mainly in logic and he has a long-standing interest as an amateur (in the original sense of the word!) in the history of mathematics. He now regularly teaches a history course for secondary school teachers, and in a distance education framework through videoconferencing. He is currently secretary of the International Commission on Mathematical Instruction (ICMI).

Wann-Sheng Horng is professor of mathematics at the National Taiwan Normal University, Taipei, Taiwan. Trained in the City University of New York to be a professional historian of mathematics in the late 1980s, he is now doing research on both the history of Chinese mathematics and how ancient mathematical texts can be used in the classroom.

Abdellah El Idrissi is professeur habilité in the department of mathematics at the Ecole Normale Supérieure of Marrakech, Morocco. His researches are articulated around epistemology and history of mathematics in teaching. He is especially interested by the history of trigonometry and the history of geometrical instruments.

Abdulcarimo Ismael teaches mathematics education in the Pedagogical University, Maputo, Mozambique. He is engaged in introducing ethnomathematics research into teacher training, and his current research is into probability concepts underlying popular games in Mozambique.

Masami Isoda is associate professor of mathematics education at the University of Tsukuba, Japan. He has been researching mathematization in school mathematics from psychological and historical aspects. He has applied history for discussing alternative perspectives in school mathematics, and developed teaching programmes with technology. He has co-edited several books for Japanese teachers, about problem posing and solving and about using technology in mathematics education.

Hans Niels Jahnke is professor of mathematics education at the University of Essen, Germany. He has written, edited and co-edited several books, including *Mathematik in der Humboldtschen Reform* (1990), *History of mathematics and education: ideas and experiences* (1996), *Geschichte der Analysis* (1999).

Lesley Jones is senior lecturer in mathematics education at Goldsmiths University of London. She has research interests in the issues of equity and social justice in education and has published a number of articles on matters concerning gender and the education of ethnic minority groups in the United Kingdom.

Victor Katz is professor of mathematics at the University of the District of Columbia in Washington DC, USA. He is the author of *A history of mathematics: an introduction* (Addison-Wesley, 1998). He has recently directed the Institute in the History of Mathematics and Its Use in Teaching, an NSF-supported project enabling numerous high school and college teachers to study the history of mathematics and how to use it in teaching. This project is currently producing historical/mathematical materials for use in the high school classroom.

Manfred Kronfellner (*1949) is professor at the institute of algebra and computational mathematics of the Vienna University of Technology. His research interests focus on changes of teaching goals, in particular in consideration of applications, computer algebra and history of mathematics. He is the author of *Historical aspects in mathematics education* (in German), co-editor of *The state of computer algebra in mathematics education* and co-author of several textbooks.

Mariza Krysinska is a secondary school teacher at the Collège Saint-Michel in Bruxelles, Belgium. She is a member of the mathematics education group GEM (Groupe d'Enseignement Mathématique) at the Catholic University of Louvain. She has (co-)edited several text books including *De question en question* vols 2, 3, 4 (1994, 1997, 1997) and she collaborated on the teaching project *Approche heuristique d'analyse* (Heuristic approach to calculus, 1999).

Ewa Lakoma is adjunct faculty in the Institute of Mathematics at the Military University of Technology in Warsaw. Her PhD from Warsaw University was on the teaching of probability in relation to the cognitive development of the learner. In her research in probability and statistics education she is especially interested in recognising and understanding the student's natural ways of mathematical thinking. Here the history of mathematics plays an important role. She is author of *Historical development of the probability concept* (in Polish, 1992), addressed to mathematics educators, and co-author of secondary school textbooks.

David Lingard is senior lecturer in mathematics education at Sheffield Hallam University, UK, where he is involved in initial teacher education at both primary and secondary levels. He taught in secondary schools in England for 25 years, including 13 as head teacher of a comprehensive school in South Yorkshire. He is a former honorary secretary of the ATM (Association of Teachers of Mathematics), and currently an active member of the British Society for the History of Mathematics.

Chi-Kai Lit (*1971) teaches mathematics at Cheng Chek Chee Secondary School of Sai Kung & Hang Hau District, New Territories, Hong Kong. He studied for an MPhil thesis at the Chinese University of Hong Kong, whose title is (in English translation) *Using history of mathematics in the junior secondary school classroom: a curriculum perspective*.

Jan van Maanen (*1953) is assistant professor in the mathematics department of the University of Groningen, Netherlands. During his years as a secondary school teacher he had his PhD from the University of Utrecht, on the Dutch audience for the mathematics of René Descartes. His current research is centred around history in relation to mathematics teaching.

Marta Menghini is associate professor in the mathematics department of University 'La Sapienza' in Rome, Italy. Her research centres on the didactics of mathematics and on history of mathematics in relation to mathematics teaching.

Karen Dee Michalowicz is upper school chair, the Langley School, Virginia, USA, and adjunct faculty, George Mason University, and has been involved in education for 37 years. She has undergraduate degrees in mathematics and history, as well as a graduate degree in mathematics education and pedagogy. She is president elect of Women and Mathematics Education. She is treasurer of the Americas Chapter of HPM. Her area of interest is collecting and then using ancient text books in the secondary mathematics classroom. She is a Presidential Awardee in Mathematics Education.

Anne Michel-Pajus is professeur de chaire supérieure in mathematics at Lycée Claude Bernard in Paris, France. She is involved with in-service education at the Université Denis Diderot in Paris, where she works in the IREM (Institut de Recherche sur l'Enseignement des Mathématiques) on the history and epistemology of mathematics and their use in mathematics teaching. She is one of the authors of *A history of algorithms* (Springer 1999, ed J. L. Chabert), and a co-editor of the IREM journal *Mnemosyne*.

Richard Millman is professor of mathematics and president of Knox College, Galesburg, Illinois, USA. He has co-authored three books with George Parker including *Elements of diifferential geometry* (1977) and co-edited two others. In addition to academic appointments, he was Program Director for Geometric Analysis at the National Science Foundation (1984-86). His Ph.D. is from Cornell University. His areas of expertise are mathematics education (especially writing in mathematics) and differential geometry.

Ryosuke Nagaoka is professor of mathematics at the University of the Air, in Chiba, Japan. He is interested in the history of mathematics after 17th century, especially of analysis in the 19th century. His main research concern is the changing paradigm of science after the emergence of modern IT. Among the works he has (co-)edited are *Nyu-ton Shizen Tetsugaku no Keifu* (A genealogy of Newtonian philosophy of nature), 1987; *Suugaku no Rekishi* (A history of mathematics), 1993; *Senkeidaisuu I: Gendaisuugaku heno Houhouron-teki apuro-chi* (Linear algebra: a methodological introduction to modern mathematical thinking), 1999

Mogens Niss is professor of mathematics and mathematics education at Roskilde University, Denmark. He was the secretary of ICMI 1991-98, in which capacity he was a co-editor of the ICMI Study Series. His research interests are in mathematics education at large, with a focus on mathematics education in society and history; applications and modelling of mathematics; assessment; and the characteristics of mathematics education as an academic discipline.

Sergio Nobre (*1957) is assistant professor in the mathematics department of the University of São Paulo, Brazil (Unesp-Brazil) and general secretary of the Brazilian Society of History of Mathematics. His PhD, from the University of Leipzig, Germany, was for research in the history of mathematics. His research interests are the diffusion of mathematics through encyclopaedias, the history of mathematics in Brazil and the relations between history and mathematics education. He has organised international meetings in Brazil on the history of mathematics (in 1997) and on relations between history and pedagogy of mathematics (in 1994), and is the editor of their proceedings.

George Philippou is associate professor of mathematics education in the department of education of the University of Cyprus. He has taught at secondary and higher education institutions and at the University of the Aegean. His research interest includes problem solving, mathematical beliefs and teacher education, examined within a historical perspective.

Joao Pitombeira de Carvalho is associate professor of mathematics at the Catholic University in Rio de Janeiro, Brazil. His main interest in the history of mathematics is the history of mathematics education. He has also participated in several programmes of the Brazilian ministry of education to promote science and mathematics teaching and to evaluate school texts.

María V. Ponza has the Argentinian National Teacher and Professor of Mathematics' degrees. Since 1976 she has taught in govermment secondary schools in Argentina. At present, she coordinates the articulation between primary and secondary schools at Río Ceballos and she is a student of computing and mathematics teaching, in Blaise Pascal University, Córdoba. She gives specialised courses, freely assessed, to teachers of different provinces in the country. She has published several articles in Spain and the UK, and is co-author of the book *Scientific investigation and pluridisciplinary teaching practice.* Her research is centred on the relationship of mathematics with art (especially drama and dance).

Luis Radford is full professor in mathematics education at Laurentian University, Ontario, Canada. His main current research is in the psychology of mathematics, semiotics, epistemology and the history of mathematics in relation to the teaching and learning of mathematics.

Michel Rodriguez has a licentiate in mathematics. He felt the need for further mathematical and historical knowledge when he successively went through the various teaching degrees, from instituteur to professeur agrégé. He taught classes at all ages and level, from kindergarten to higher technical vocational schools. He does research at the IREM of Lille, where he also guides trainee teachers. He is a passionate musician, composer and song-writer.

Leo Rogers is a research supervisor in mathematics education at the University of Surrey, Roehampton, UK. Originally qualified in mathematics and physics, his research interests are in history and philosophy of science and mathematics education, the evolution of human thinking and imagery and visualisation. He has published work in mathematics, history of mathematics and education and is involved in mathematics curriculum development projects in the UK and Europe.

Ernesto Rottoli is an Italian high school teacher. His current research interests centre around using history in relation to the teaching of rational numbers and the development of the multiplicative conceptual field.

Maggy Schneider teaches didactics of mathematics at the University of Namur, Belgium. She also teaches at a secondary school. Her research principally concerns the epistemological aspects of mathematical analysis and of geometry at the secondary school level. This is reflected in the title of her PhD thesis: *Des objets mentaux 'aire' et 'volume' au calcul des primitives* (From the mental objects 'area' and 'volume' to the calculus of primitive functions).

Gert Schubring is a member of the *Institut für Didaktik der Mathematik*, at the University of Bielefeld, Germany. Besides the history of mathematics education, his research interests focus on the history of mathematics and the sciences in the eighteenth and nineteenth centuries and on their broader cultural context. His book publications include *Die Entstehung des Mathematiklehrerberufs im 19. Jahrhundert* (Weinheim 1983/1991), the lecture notes *Analysis of historical textbooks in mathematics* (Rio de Janeiro 1997/1999) and the edited volume *Hermann G. Graßmann (1809-1877) - visionary mathematician, scientist and neohumanist scholar* (Dordrecht 1996).

Anna Sierpinska is professor at the department of mathematics and statistics at the Concordia University in Montréal, Québec, Canada. Her main interest in the years 1984-92 was the notion of epistemological obstacle and its applications in mathematics education, especially with regard to students' understanding of limits and infinity. Her book *Understanding in mathematics* (1994) also appeared in French. Her present interests include the teaching of linear algebra, the use of technology in mathematics teaching and students' difficulties with theoretical thinking in general.

Circe Mary Silva da Silva (*1951) is senior lecturer in the graduate program in education, Federal University of Espirito Santo, Brazil, and vice-president of the Brazilian Society for History of Mathematics. Her masters degree in mathematics is from the Federal University Fluminense, Brazil, and her PhD in education from the University of Bielefeld, Germany. Her main research interests include history of mathematics and mathematical education in the 20th. century, especially in Brazil. She has recently published *Positivist mathematics and its impact in Brazil*.

Man-Keung Siu, who obtained his BSc from the University of Hong Kong and his PhD in mathematics from Columbia University, is a professor of mathematics at his undergraduate alma mater. He has published in the fields of algebra, combinatorics, applied probability, mathematics education and history of mathematics. The Chinese Mathematical Society selected his book *Mathematical proofs* (1990, in Chinese) as one of the seven outstanding books in mathematical exposition in 1991.

Harm Jan Smid is associate professor in mathematics and mathematics education at the Delft University of Technology, Netherlands. His special interest is in the history of mathematics education. His PhD is on the history of mathematics education in the Netherlands during the first half of the 19th century.

Daina Taimina is a docent of the faculty of physics and mathematics of the University of Latvia, where she has taught history of mathematics for more than 20 years. She got her PhD in theoretical computer science. Currently, she is a visiting professor at Cornell University, where she teaches history of mathematics; revises her 1990 textbook (in Latvian) on the history of mathematics; compiles a history of Latvian mathematics from ancient times; and integrates history into the revised second edition of *Experiencing geometry* by her husband David W. Henderson.

Wendy Troy has studied and taught courses in the history and nature of mathematics since 1987 and most recently was a senior lecturer in mathematics education at the University of Greenwich, England. Her work in London included teaching in secondary schools, further education and teacher training, working as an advisory teacher and publishing learning materials. In 1999, she left England for 3 years to begin some research in Bangladesh.

Constantinos Tzanakis (*1956) is associate professor of mathematics at the department of education of the University of Crete, Greece. He studied mathematics at Athens University, Greece; astronomy at Sussex University, UK; and obtained his PhD in theoretical physics from the Université Libre de Bruxelles, Belgium. His area of research is mathematical physics (statistical mechanics, relativity theory and geometrical methods in physics) and mathematics and physics education (the relation between history and epistemology of mathematics and physics and their teaching).

Carlos E. Vasco was born in Medellin, Colombia. He finished a master's degree in theoretical physics, and got his Ph.D. in mathematics in St. Louis, Missouri in 1968. He taught mathematics for 25 years at the Colombian National University in Bogota, and was advisor to the Ministry of Education of Colombia for the improvement of the school mathematics curriculum for 20 years. He is now professor emeritus, and does research in mathematics education at Harvard University in Cambridge, Massachusetts, and in a new doctoral program in mathematics education at the Universidad del Valle in Cali, Colombia.

Chris Weeks is a former secondary school mathematics teacher and lecturer in mathematics education. He has worked with French IREM colleagues in order to make their work on education and history of mathematics more available to an English-speaking readership through translation. The most recent publication is J-L Chabert (ed.) *History of algorithms* (1999). He is currently membership and publicity officer for the British Society for the History of Mathematics.

Dian Zhou Zhang (*1933) is professor in the mathematics department of East China Normal University. His research areas are operator theory, history of mathematics in the 20th century, and mathematics education. Many works published in Chinese, and his paper (in English) 'Mathematical exchanges between the United States and China (1850-1950)' is included in *The history of modern mathematics* vol. iii (1994).

The photograph on page 428 shows the whole team at the ICMI study conference in Luminy (France), April 1998

Index

Printed in the United Kingdom
by Lightning Source UK Ltd.
114695UKS00003B/131